In this major reevaluation of Isaac Newton's intellectual life, Betty Jo Teeter Dobbs shows how his pioneering work in mathematics, physics, and cosmology was intertwined with his study of alchemy. Directing attention to the religious ambience of the alchemical enterprise of early modern Europe, Dobbs argues that Newton understood alchemy – and the divine activity in micromatter to which it spoke – to be a much needed corrective to the overly mechanized system of Descartes. Yet that religious basis was not limited to alchemy, but suffused the rest of his work.

Newton, whose many different studies constitued a unified plan for obtaining Truth, saw value and relevance in all of his pursuits. To him it seemed possible to obtain partial truths from many different approaches to knowledge, be it textual work aimed at the interpretation of prophecy, the study of ancient theology and philosophy, creative mathematics, or experiments with prisms, pendulums, vegetating minerals, light, or electricity. Newton's work was a constant attempt to bring these partial truths together, with the larger goal of restoring true natural philosophy and true religion. Within this broad interpretative strategy, Dobbs traces the evolution of Newton's thought in alchemy, religion, and cosmology, and details his struggles with the interwoven problems of the microscopic spirit of alchemy and the cause of the cosmic principle of gravitation.

A landmark study of the "founder of modern science," *The Janus Faces of Genius* is an important contribution to the history of science.

The Janus faces of genius

The Janus faces of genius

The role of alchemy in Newton's thought

BETTY JO TEETER DOBBS
UNIVERSITY OF CALIFORNIA, DAVIS

PUBLISHED BY THE PRESS SYNDICATE OF THE UNIVERSITY OF CAMBRIDGE
The Pitt Building, Trumpington Street, Cambridge, United Kingdom

CAMBRIDGE UNIVERSITY PRESS
The Edinburgh Building, Cambridge CB2 2RU, UK
40 West 20th Street, New York NY 10011–4211, USA
477 Williamstown Road, Port Melbourne, VIC 3207, Australia
Ruiz de Alarcón 13, 28014 Madrid, Spain
Dock House, The Waterfront, Cape Town 8001, South Africa

http://www.cambridge.org

First published 1991
First paperback edition 2002

A catalogue record for this book is available from the British Library

Library of Congress Cataloguing in Publication data
Dobbs, Betty Jo Teeter, 1930–
The Janus faces of genius: the role of alchemy in Newton's
thought / Betty Jo Teeter Dobbs.
p. cm.
Includes bibliographical references and index.
ISBN 0 521 38084 7
1. Newton, Isaac, Sir, 1642–1727 – Knowledge – Alchemy. 2. Alchemy –
History. I. Title.
QC16.N7D64 1991
540'.1'12–dc20 91-8695

ISBN 0 521 38084 7 hardback
ISBN 0 521 52487 3 paperback

To the memory of my father, Ransom Alexander Teeter, Sr. (1889–1960), Methodist minister, who would not have been surprised at the thesis of this book.

Contents

Illustrations

Acknowledgments

A Danforth Graduate Fellowship for Women and a National Science Foundation Graduate Fellowship provided initial research support for this project some twenty years ago. The North Atlantic Treaty Organization supported a year of postdoctoral work; a fellowship at the National Humanities Center and a Scholar's Award from the National Science Foundation funded further years of study, as did a fellowship from the National Endowment for the Humanities at The Huntington Library. I am grateful to all for their sustained assistance and to The American Council of Learned Societies for two travel grants. For briefer periods of research support I am also indebted to the Northwestern University Research Council, the Center for the Interdisciplinary Study of Science and Technology at Northwestern University, and the Northwestern University College of Arts and Sciences.

For permission to cite, describe, transcribe, or quote from manuscript material in their possession I thank the following: the Syndics of University Library, Cambridge, England; the Provost and Scholars of King's College, Cambridge, England; the Syndics of the Fitzwilliam Museum, Cambridge, England; The Bodleian Library, Oxford, England; The British Library, London, England; The Jewish National and University Library, Jerusalem, Israel; Bibliothèque publique et universitaire de Genève, Geneva, Switzerland; The Dibner Library of the History of Science and Technology, Special Collections Branch, Smithsonian Institution Libraries, Washington, D.C.; Institute Archives and Special Collections Department, Massachusetts Institute of Technology, Cambridge, MA; The Francis A. Countway Library of Medicine, Boston, MA; Medical Historical Library, Cushing/Whitney Medical Library, Yale University, New Haven, CT; Babson College Archives, Babson College, Babson Park, MA; Special Collections, Lehigh University Libraries, Bethlehem, PA; Department of Special Collections, The University of Chicago Library, Chicago, IL; Department of Special Collections, The Stanford University Libraries, Stanford, CA. Special thanks for providing the illustrations for this volume are also due to the following: The Metropolitan Museum of

Art, New York, NY; The Henry E. Huntington Library, San Marino, CA; Rare Books Department, Memorial Library, University of Wisconsin-Madison; The Rare Book and Special Collection Division of The Library of Congress, Washington, D.C.; Zentralbibliothek Zürich, Zürich, Switzerland; The Babson College Archives, Babson Park, MA; The Tate Gallery, London/Art Resource, New York. I am also grateful to Associated University Presses; The University of California Press; Kluwer Academic Publishers; Cambridge University Press; Science History Publications & The Nobel Foundation; The Smithsonian Institution Libraries; Herzog August Bibliothek Wolfenbüttel and Otto Harassowitz; and the editors of *Isis* and *The Southern Journal of Philosophy* for permission to use material previously published under their auspices.

It has been my privilege in recent years to speak on the subject matter of this book to many audiences: as a participant in The History of Science Society's Visiting Historians of Science Program, as an invited guest at other universities and colleges, at research libraries, at some of the many celebrations of the tercentenary of the publication of Newton's *Principia*, and at other academic conferences, to faculty colloquia, and to my own undergraduate and graduate students. I have benefited immeasurably from those encounters, and I thank my auditors for their many perceptive challenges and questions, for their frequent bibliographic advice, and above all for their courteous attention.

Evanston, Illinois
December 1990

1

Isaac Newton, philosopher by fire

Introduction

Isaac Newton studied alchemy from about 1668 until the second or third decade of the eighteenth century. He combed the literature of alchemy, compiling voluminous notes and even transcribing entire treatises in his own hand. Eventually he drafted treatises of his own, filled with references to the older literature. The manuscript legacy of his scholarly endeavor is very large and represents a huge commitment of his time, but to it one must add the record of experimentation. Each brief and often abruptly cryptic laboratory report hides behind itself untold hours with hand-built furnaces of brick, with crucible, with mortar and pestle, with the apparatus of distillation, and with charcoal fires: experimental sequences sometimes ran for weeks, months, or even years. As the seventeenth-century epithet "philosopher by fire" distinguished the serious, philosophical alchemist from the empiric "puffer" or the devious charlatan or the amateur "chymist," so may one use the term to characterize Isaac Newton. Surely this man earned that title if ever any did.

Since my first monograph on this subject appeared in 1975[1] Newton's alchemy has held a prominent position in historiographic debates of some centrality to the history and philosophy of science. Even though Newton's interest in alchemy had often been noted before and had indeed generated a considerable body of scholarly comment,[2] public recognition of it was forced to a new level after 1975 as the full extent of Newton's commitment to alchemical pursuits was made more and more explicit in reviews,[3]

1. B. J. T. Dobbs, *The Foundations of Newton's Alchemy, or "The Hunting of the Greene Lyon"* (Cambridge University Press, 1975).
2. The historiography of Newton's alchemy prior to 1975 is explored in ibid., pp. 6–20.
3. See especially the following reviews of my book: P. M. Rattansi, "Newton as chymist," *Science* 192 (No. 4240, 14 May 1976), 689–90; idem, "Last of the magicians," *Times Higher Education Supplement*, June 1976; Philip

biographical works,[4] articles,[5] and definitive studies of Newton's library.[6]

Yet even so there has remained the possibility of denying the significance and importance of Newton's alchemy for his great achievements

Morrison, *Scientific American* 235 (August 1976), 113–15; Richard S. Westfall, *Journal of the History of Medicine and Allied Sciences* 31 (1976), 473–4; Kathleen Ahonen, *Annals of Science* 33 (1976), 615–17; Henry Guerlac, *Journal of Modern History* 49 (1977), 130–3; Derek T. Whiteside, "From his claw the Greene Lyon," *Isis* 68 (1977), 116–21; A. Rupert Hall, "Newton as alchemist," *Nature* 266 (28 April 1977), 78; Karin Figala, "Newton as alchemist," *History of Science* 15 (1977), 102–37; P. E. Spargo, *Ambix* 24 (1977), 175–6; Marie Boas Hall, *British Journal for the History of Science* 10 (1977), 262–4; Allen G. Debus, *Centaurus* 21 (1977), 315–16; Margaret C. Jacob, *The Eighteenth Century: A Current Bibliography*, n.s. 1 (1975, published 1978), 345–7.

4. Richard S. Westfall, *Never at Rest. A Biography of Isaac Newton* (Cambridge University Press, 1980); Gale E. Christianson, *In the Presence of the Creator. Isaac Newton and His Times* (New York: The Free Press; London: Collier Macmillan, 1984).
5. Richard S. Westfall, "The role of alchemy in Newton's career," in *Reason, Experiment and Mysticism in the Scientific Revolution*, ed. by M. L. Righini Bonelli and William R. Shea (New York: Science History Publications, 1975), pp. 189–232; idem, "Isaac Newton's *Index Chemicus*," *Ambix* 22 (1975), 174–85; idem, "The changing world of the Newtonian industry," *Journal of the History of Ideas* 37 (1976), 175–84; B. J. T. Dobbs, "Newton's copy of *Secrets Reveal'd* and the regimens of the work," *Ambix* 26 (1979), 145–69; Richard S. Westfall, "The influence of alchemy on Newton," in *Science, Pseudo-Science and Society*, ed. by Marsha P. Hanen, Margaret J. Osler, and Robert G. Weyant (Waterloo, Ontario: Wilfrid Laurier University Press, 1980), pp. 145–69; B. J. T. Dobbs, "Newton's alchemy and his theory of matter," *Isis* 73 (1982), 511–28; idem, "Newton's 'Clavis': new evidence on its dating and significance," *Ambix* 29 (1982), 190–202; idem, "Newton's *Commentary* on *The Emerald Tablet* of Hermes Trismegistus: its scientific and theological significance," in *Hermeticism and the Renaissance. Intellectual History and the Occult in Early Modern Europe*, ed. by Ingrid Merkel and Allen G. Debus (Folger Books; Washington, D.C.: The Folger Shakespeare Library; London: Associated University Presses, 1988), pp. 182–91; idem, "Alchemische Kosmogonie und arianische Theologie bei Isaac Newton," tr. by Christoph Meinel, *Wolfenbütteler Forschungern*, 32 (1986), 137–50; idem, "Newton and Stoicism," *The Southern Journal of Philosophy 23 Supplement* (1985), 109–23; idem, "Newton's alchemy and his 'active principle' of gravitation," in *Newton's Scientific and Philosophical Legacy*, ed.

in mathematics, physics, cosmology, and methodology. I argued in 1975 that Newton's alchemy constituted one of the pillars supporting his mature scientific edifice.[7] Nevertheless, since Newton's reputation as one of the founders of modern science rests securely upon achievements in areas of thought still recognized as scientific, and since alchemy has, at least since the eighteenth century, been rejected from the canon of science as hopelessly retrograde, "occult," and false, some scholars have been reluctant to accept the validity of that notion.[8]

Newton's *Philosophiae naturalis principia mathematica*, first published in 1687 and foundational to many later developments in science, has seemed to most readers to be the epitome of austere rationality, and the writer of that remarkable work on "The Mathematical Principles of Natural Philosophy" continues to seem to some of its readers to be a very poor candidate for the epithet "philosopher by fire." Since among Newton scholars I. Bernard Cohen's knowledge of Newton's *Principia* and the manuscript remains associated with it,[9] and since Cohen has been outspoken on that point, one may take his objections as those requiring most serious response. Cohen's detailed examination of the proposition that alchemy made some difference to Newton's science came in a lengthy essay published in 1982.[10] The issue has focused on the origin of Newton's

by P. B. Scheuer and G. Debrock (International Archives of the History of Ideas, 123; Dordrecht: Kluwer Academic Publishers, 1988), pp. 55–80.

6. John Harrison, *The Library of Isaac Newton* (Cambridge University Press, 1978); Richard S. Westfall, "Alchemy in Newton's library," *Ambix* 31 (1984), 97–101.
7. Richard S. Westfall, *Force in Newton's Physics. The Science of Dynamics in the Seventeenth Century* (London: Macdonald; New York: American Elsevier, 1971), esp. pp. 323–423; Dobbs, *Foundations* (1, n. 1), pp. 210–13.
8. Whiteside, "From his claw" (1, n. 3); I. Bernard Cohen, *The Newtonian Revolution. With Illustrations of the Transformation of Scientific Ideas* (Cambridge University Press, 1980), p. 10.
9. Isaac Newton, *Isaac Newton's Philosophiae naturalis principia mathematica. The Third Edition (1726) with Variant Readings. Assembled and Edited by Alexandre Koyré and I. Bernard Cohen with the Assistance of Anne Whitman* (2 vols.; Cambridge, MA: Harvard University Press, 1972); I. Bernard Cohen, *Introduction to Newton's "Principia"* (Cambridge, MA: Harvard University Press; Cambridge University Press, 1971); Cohen, *Newtonian Revolution* (1, n. 8).
10. I. Bernard Cohen, "The *Principia*, universal gravitation, and the 'Newtonian style,' in relation to the Newtonian revolution in science: notes on the occasion of the 250th anniversary of Newton's death," in *Contemporary*

ideas on attractive forces, as Cohen pointed out. For although an attractive force of gravity appeared in the *Principia* and was fundamental to later Newtonian dynamics, ideas of attraction (operating either between small particles of matter or between gross bodies) hardly constituted orthodox mechanical philosophy in 1687. Attractive force smacked of the "occult" to the first generation of mechanical philosophers, writing thirty to forty years before Newton, and they had been careful to substitute for attraction the principles of "impact physics" in which *apparent* attractions (magnetic, electrical, gravitational) were explained by the mechanical encounter of very fine and imperceptible particles of a hypothetical aether with the larger particles of matter. Newton's reintroduction of attraction in the *Principia*, and his dismissal there of an aethereal mechanism as an explanation of gravity, had seemed to Westfall and myself a convincing argument for the influence of alchemy on Newton's thought, for much alchemical literature concerns itself with nonmechanical "active principles" that are conceptually similar to Newton's gravity. Cohen disagreed. Arguing that no documents seemed to exist in which Newton took attractive forces under consideration before 1679–80, when Robert Hooke introduced Newton to a dynamical analysis predicated upon inertia and an attractive central force, Cohen concluded that Newton's subsequent departure from orthodox mechanism derived from his own "style" of mathematical abstraction rather than from the conceptual influence of alchemical "active principles" upon him. Cohen in fact insisted that Newton was able to produce his great work of positive science only by putting aside his alchemical and Hermetic interests temporarily and rising above them.

I have challenged Cohen's argument in my review of the book in which it appeared[11] because his position seemed to be based on the a priori assumption that alchemy could never, by its very nature, make a contribution to science. To accept the premise that alchemy could not do so is to prejudge the historical question of whether it did do so in Newton's case, which is after all the point at issue. Furthermore, Newton's alchemical papers, which were not included in Cohen's analysis, document Newton's interest in alchemical "active principles" for an entire decade before his correspondence with Hooke. But while the presence of "active principles" in Newton's alchemical papers, as well as in the literature of alchemy upon which those papers were based, is hardly to be denied at this stage in the debate, it now seems rather less likely to me that Newton

Newtonian Research, ed. by Zev Bechler (Studies in the History of Modern Science, 9; Dordrecht: D. Reidel, 1982), pp. 21–108.

11. Dobbs, *Isis* 74 (1983), 609–10.

transferred the concept of the "active principle" directly from his alchemical studies to his new formulation of gravity, at least not at first, though he may finally have done so. Rather, as will be argued in detail in the following chapters, all issues of passivity and activity, of mechanical and nonmechanical forces, were enmeshed for Newton in a philosophical/religious complex one can only now begin to grasp. Although Newton's first encounter with attractive and "active" principles may well have been in his alchemical study, his application of such ideas to the force of gravity was almost certainly mediated by several other considerations. And because of Newton's important position in the rise of modern science, and because of the importance of the doctrine of gravity as an "active principle" within his own science, one must strive to understand them all.

My studies since 1975 have yielded hints that Newton was concerned from the first in his alchemical work to find evidence for the existence of a vegetative principle operating in the natural world, a principle that he understood to be the secret, universal, animating spirit of which the alchemists spoke. He saw analogies between the vegetable principle and light, and between the alchemical process and the work of the Deity at the time of creation. It was by the use of this active vegetative principle that God constantly molded the universe to His providential design, producing all manner of generations, resurrections, fermentations, and vegetation. In short, it was the action of the secret animating spirit of alchemy that kept the universe from being the sort of closed mechanical system for which Descartes had argued.[12] These themes will be discussed in detail in the chapters to follow as a way of searching out the relationships of alchemical modes of thought to the general concerns of Newton and his contemporaries.

Nevertheless, the primary goal here is the larger one indicated by the subtitle of this book: the role of alchemy in Newton's thought. Newton stood at the beginning of our modern scientific era and put his stamp upon it irrevocably. He may be seen as a gatekeeper, a Janus figure, for one of his faces still gazes in our direction. But only one of them. Like Janus, who symbolized the beginning of the new year but also the end of the old one, Newton looked forward in time but backward as well. It is the vision seen by the eyes of that second face that I pursue by examining the details of his alchemical labors. It is possible to grasp that vision yet not stop there.

I do not assume the irrelevancy of Newton's pursuit of an ancient, occult wisdom to those great syntheses of his that mark the foundation of modern science. The Janus-like faces of Isaac Newton were after all

12. See my articles cited earlier (1, n. 5).

the production of a single mind, and their very bifurcation may be more of a modern optical illusion than an actuality. Newton's mind was equipped with a certain fundamental assumption, common to his age, from which his various lines of investigation flowed naturally: the assumption of the unity of Truth.[13] True knowledge was all in some sense a knowledge of God; Truth was one, its unity guaranteed by the unity of God. Reason and revelation were not in conflict but were supplementary. God's attributes were recorded in the written Word but were also directly reflected in the nature of nature.[14] Natural philosophy thus had immediate theological meaning for Newton and he deemed it capable of revealing to him those aspects of the divine never recorded in the Bible or the record of which had been corrupted by time and human error. By whatever route one approached Truth, the goal was the same. Experimental discovery and revelation; the productions of reason, speculation, or mathematics; the cryptic, coded messages of the ancients in myth, prophecy, or alchemical tract – all, if correctly interpreted, found their reconciliation in the infinite unity and majesty of the Deity. In Newton's conviction of the unity of Truth and its ultimate source in the divine one may find the fountainhead of all his diverse studies.

Newton's methodology

One cannot agree with Cohen's thesis, then, that an essential part of Newton's methodology was deliberately to create mathematical models

13. For a general statement on Renaissance conceptions of this problem, see Paul Oskar Kristeller, *Renaissance Thought and Its Sources*, ed. by Michael Mooney (New York: Columbia University Press, 1979), pp. 196–210; for views contemporary with Newton, see Arthur Quinn, *The Confidence of British Philosophers. An Essay in Historical Narrative* (Studies in the History of Christian Thought, Vol. 17; ed. by Heiko A. Oberman, in cooperation with Henry Chadwick, Edward A. Dowey, Jaroslav Pelikan, Brian Tierney, and E. David Willis; Leiden: E. J. Brill, 1977), esp. pp. 8–20.
14. Here Newton stood within the mainstream of biblical tradition. See especially Psalm 19, which opens with the statement that "The heavens declare the glory of God; and the firmament sheweth his handywork," and Romans 1:20, where Paul declares, "For the invisible things of him [God] from the creation of the world are clearly seen, being understood by the things that are made, *even* his eternal power and Godhead. . . . " I am indebted to Mr. William Elliott for discussion and references on this point: personal communication, 1 April 1987.

as a first step.[15] Mathematics was only one avenue to Truth, and though mathematics was a powerful tool in his hands, Newton's methodology was much broader than that implied by the creation of mathematical models, and Newton's goal was incomparably more vast than the discovery of the "mathematical principles of natural philosophy." Newton wished to penetrate to the divine principles beyond the veil of nature, and beyond the veils of human record and received revelation as well. His goal was the knowledge of God, and for achieving that goal he marshaled the evidence from every source available to him: mathematics, experiment, observation, reason, revelation, historical record, myth, the tattered remnants of ancient wisdom. With the post-Newtonian diminution of interest in divinity and heightened interest in nature for its own sake, scholars have too often read the Newtonian method narrowly, selecting from the breadth of his studies only mathematics, experiment, observation, and reason as the essential components of his scientific method. For a science of nature, a balanced use of those approaches to knowledge suffices, or so it has come to seem since Newton's death, and one result of the restricted interests of modernity has been to look askance at Newton's biblical, chronological, and alchemical studies: to consider his pursuit of the *prisca sapientia* as irrelevant. None of those was irrelevant to Newton, for his goal was considerably more ambitious than a knowledge of nature. His goal was Truth, and for that he utilized every possible resource.

"Modeling" in the modern sense is incompatible with the pursuit of Truth in any case. To create a speculative system or to devise a mathematical scheme that will "save the phenomena" carries relativistic overtones. A modern scientist may readily admit, with a metaphorical shrug, that while we do not know whether the theoretical superstructure of our science is True, that really does not matter because our science is self-consistent and it works, accounting for known phenomena and predicting new ones. If new phenomena appear that require incorporation, or if a theory predicts falsely, our science will be adjusted, but that is a matter of no great moment. After the adjustment we will have a better science, but we will still not have Truth. As Quinn has convincingly argued, that was not Newton's attitude.[16] Newton may profitably be compared to

15. Cohen, *Newtonian Revolution* (1, n. 8), passim.
16. Arthur Quinn, "On reading Newton apocalyptically," in *Millenarianism and Messianism in English Literature and Thought 1650–1800. Clark Library Lectures 1981–82*, ed. by Richard H. Popkin (Publications from the Clark Library Professorship, UCLA, No. 10; Leiden: E. J. Brill, 1988), pp. 176–92; idem, *Confidence of British Philosophers* (1, n. 13).

such twentieth-century thinkers as G. E. Moore and Bertrand Russell, both intent on Truth, neither skeptical of human capacity to obtain it. In fact both expected to save humanity from skepticism and usher in a millennium. So did Newton.

To save humanity from skepticism was the ambition of many a thinker in the seventeenth century. Quinn reports a serious conversation between Descartes and John Dury in which it was agreed that the emergence of skepticism constituted the profound crisis of their period and that a way needed to be found to counter it with epistemological certainty. Descartes chose mathematics, Dury the interpretation of biblical prophecy, as the most promising response to the crisis.[17] The point, of course, is that no one knew then what would ultimately be established as effective. A modern thinker may be inclined to assume that Descartes chose the better part, but in fact the natural philosophy that he claimed to have established with mathematical certainty was soon overthrown by Newton's. Descartes's mathematico–deductive method was not adequately balanced by experiment, observation, and induction; Newton's was.

Perhaps the most important element in Newton's methodological contribution was that of balance, for no *single* approach to knowledge ever proved to be effective in settling the epistemological crisis of the Renaissance and early modern periods. Newton had perhaps been convinced of the necessity of methodological balance by Henry More, who had worked out such a procedure within the context of the interpretation of prophecy.[18] Since every single approach to knowledge was subject to error, a more certain knowledge was to be obtained by utilizing each approach to correct the other: the senses to be rectified by reason, reason to be rectified by revelation, and so forth. The self-correcting character of Newton's procedure is entirely similar to More's and constitutes the superiority of Newton's method over that of earlier natural philosophers, for others had certainly used the separate elements of reason, mathematics, experiment, and observation before him.

But Newton's method was not limited to the balancing of those approaches to knowledge that still constitute the elements of modern scientific methodology, nor has one any reason to assume that he would deliberately have limited himself to those familiar approaches even if he had been prescient enough to realize that those were all the future would

17. Quinn, "On reading Newton" (1, n. 16), p. 179.
18. Richard H. Popkin, "The third force in seventeenth-century philosophy: Scepticism, science, and biblical prophecy," *Nouvelle République des Lettres* 1 (1983), 35–63.

consider important. Because his goal was a Truth that encompassed not only the "mathematical principles of natural philosophy" but divinity as well, Newton's balancing procedure included also the knowledge he had garnered from theology, revelation, alchemy, history, and the wise ancients. It has been difficult to establish this fact because Newton's papers largely reflect a single-minded pursuit of each and every one of his diverse studies, as if in each one of them lay the only road to knowledge. When he wrote alchemy, he wrote as an alchemist, as Sherwood Taylor long ago observed. [19] But when he wrote chemistry, his concepts conformed to those of contemporary chemists.[20] When he wrote mathematics, no one doubted him to be a pure mathematician.[21] When he adopted the mechanical philosophy, he devised hypothetical aethereal mechanisms with the best.[22] When he undertook to interpret prophecy, his attention to the meaning of the minutest symbol implied that nothing else mattered.[23] In only a few of his papers may one observe his attempt to balance one apparently isolated line of investigation with another.

The characteristic single-mindedness reflected by each set of Newton's papers has led to the modern misunderstandings of Newton's methodology, for study of any one set may lead to a limited view of Newton's interests, goals, and methods, and the papers have all too often been divided up into categories that mesh more or less well with twentieth-

19. Frank Sherwood Taylor, "An alchemical work of Sir Isaac Newton," *Ambix* 5 (1956), 59–84.
20. B. J. T. Dobbs, "Conceptual problems in Newton's early chemistry: a preliminary study," in *Religion, Science, and Worldview. Essays in Honor of Richard S. Westfall*, ed. by Margaret J. Osler and Paul Lawrence Farber (Cambridge University Press, 1985), pp. 3–32.
21. Derek T. Whiteside, "Isaac Newton: birth of a mathematician," *Notes and Records of the Royal Society of London* 19 (1964), 53–62.
22. J. E. McGuire and Martin Tamny, *Certain Philosophical Questions: Newton's Trinity Notebook* (Cambridge University Press, 1983), pp. 362–5, 426–31.
23. See, for example, Newton's notes in his copy of Henry More, *A Plain and Continued Exposition Of the several Prophecies or Divine Visions of the Prophet Daniel, Which have or may concern the People of God, whether Jew or Christian; Whereunto is annexed a Threefold Appendage, Touching Three main Points, the First, Relating to Daniel, the other Two to the Apocalypse* (London: printed by M. F. for Walter Kettilby, at the Bishop's-Head in Saint Paul's Church-Yard, 1681); Harrison, *Library* (1, n. 6), item 1115, p. 196; now BS 1556 M 67 P 5 1681 copy 2, Bancroft Library, University of California, Berkeley.

century academic interests. Only Westfall's recent prize-winning biography of Newton has attempted to deal with all of the papers,[24] and even there no radical reevaluation of Newton's methodology was undertaken. To Westfall, the most important part of Newton's work still seemed to be that directed toward topics that continue to be of central importance to modern science: mathematics, mathematical physics, and scientific methodology. Through the lens of the preconceptions of modern scientific culture, one still sees primarily a Newton who founded modern science.

He did do that, of course, but the historian may ask other questions and construct or reconstruct other lenses. When one sees only the Newton who founded modern science, serious historiographic problems arise, and one is left with the difficulty of explaining, or explaining away, the masses of papers Newton left behind that are focused quite otherwise. My own work on Newton began with one very questionable advantage – questionably advantageous in the opinion of most scholars, that is, because the starting point was Newton's alchemy. Newton's alchemy had almost always been considered the most peripheral of his many studies, the one furthest removed from his important work in mathematics, optics, and celestial dynamics. Most students of Newton's work preferred to ignore the alchemy, or, if not to ignore it, then to explain it away as far as possible. But one may, perhaps pardonably, remain unconvinced that a mind of the caliber of Newton's would have lavished so much attention upon any topic without a serious purpose and without a serious expectation of learning something significant from his study of it. Indeed, working one's way through Newton's alchemical papers, one becomes increasingly aware of the meticulous scholarship and the careful quantitative experimentation Newton had devoted to alchemical questions over a period of many years. Clearly, *he* thought his alchemical work was important. So one is forced to question what it meant to him: if Newton thought alchemy was an important part of his life's work, then what was that life's work? Was it possible that Newton had a unity of purpose, an overarching goal, that encompassed *all* of his various fields of study? The lens that I have attempted to reconstruct, then, is the lens through which Newton viewed himself.

In certain ways Newton's intellectual development is best understood as a product of the late Renaissance, a time when the revival of antiquity had conditioned the thinkers of Western Europe to look backward for Truth. Thanks to the revival of ancient thought, to humanism, to the Reformation, and to developments in medicine/science/natural philosophy prior to or contemporary with his period of most intense study

24. Westfall, *Never at Rest* (1, n. 4).

(1660–84), Newton had access to an unusually large number of systems of thought. Each system had its own set of guiding assumptions, so in that particular historical milieu some comparative judgment between and among competing systems was perhaps inevitable. But such judgments were difficult to make without a culturally conditioned consensus on standards of evaluation, which was precisely what was lacking. The formalized skepticism of Pyrrhonism had been revived along with other aspects of antiquity, but in addition one may trace an increase in a less formal but rather generalized skepticism at least from the beginning of the sixteenth century, as competing systems laid claim to Truth and denied the claims of their rivals. As a consequence, Western Europe underwent something of an epistemological crisis in the sixteenth and seventeenth centuries. Among so many competing systems, how was one to achieve certainty? Could the human being attain Truth?[25]

But Newton was not a skeptic, and in fact his assumption of the unity of Truth constituted one answer to the problem of skepticism. Not only did Newton respect the idea that Truth was accessible to the human mind, but also he was very much inclined to accord to several systems of thought the right to claim access to some aspect of the Truth. For Newton, then, the many competing systems he encountered tended to appear complementary rather than competitive. The mechanical philosophy that has so often been seen as the necessary prelude to the Newtonian revolution probably did not hold a more privileged or dominant position in Newton's mind than did any other system. The mechanical philosophy was one system among many that Newton thought to be capable of yielding at least a partial Truth.

Blinded by the brilliance of the laws of motion, the laws of optics, the calculus, the concept of universal gravitation, the rigorous experimentation, the methodological success, we have seldom wondered whether

25. Richard H. Popkin, *The History of Skepticism from Erasmus to Spinoza* (Berkeley: University of California Press, 1979); Charles G. Nauert, Jr., *Agrippa and the Crisis of Renaissance Thought* (Illinois Studies in the Social Sciences, No. 55; Urbana: University of Illinois Press, 1965); Walter Pagel, *Paracelsus: An Introduction to Philosophical Medicine in the Era of the Renaissance* (Basel: Karger, 1958); idem, *Joan Baptista van Helmont. Reformer of Science and Medicine* (Cambridge University Press, 1982); Kristeller, *Renaissance Thought* (1, n. 13); Ernst Cassirer, *The Platonic Renaissance in England*, tr. by James P. Pettegrove (Austin: University of Texas Press, 1953); John Redwood, *Reason, Ridicule and Religion. The Age of Enlightenment in England 1660–1750* (London: Thames and Hudson, 1976). For this last reference, and for discussions on Arianism (to be considered in later chapters), I am indebted to Mary Louise McIntyre.

the discovery of the laws of nature was all Newton had in mind. We have often missed the religious nature of his quest and taken the stunningly successful by-products for his primary goal. But Newton wished to look through nature to see God, and it was not false modesty when in old age he said he had been only like a boy at the seashore picking up now and again a smoother pebble or a prettier shell than usual while the great ocean of Truth lay all undiscovered before him.[26]

Newton's quest was immeasurably large; it generated questions starkly different from those of modern science. For him, the most important questions were never answered, but in reconstructing them lies our best chance of grasping the focused nature of Newton's work within the comprehensive range of his interests. His questions were not ours. They encompassed fields of knowledge that to us seem to have no relevant points of contact.

But that there was a unity and a consistency in Newton's quest will be a central theme of this book. Evidence for the unity emerges when his alchemical papers are considered in conjunction with his other literary remains, not pushed aside with an a priori assumption of irrelevance. The same may be said for the changes in Newton's explanatory "mechanisms" over the long decades of his search. Although those changes have often appeared erratic and inconsistent to later scholars, from the viewpoint of Newton's primary goal the consistency appears. The consistency lies in his overwhelming religious concern to establish the relationship between Creator and creation. The pattern of change results from the slow fusion and selective disentanglement of essentially antithetical systems: Neoplatonism, Cartesian mechanical philosophy, Stoicism; chemistry, alchemy, atomism; biblical, patristic, and pagan religions. I shall argue that it was precisely where his different lines of investigation met, where he tried to synthesize their discrepancies into a more fundamental unity, when he attempted to fit partial Truth to partial Truth, that he achieved his greatest insights.

Not only was Newton's goal a unified system of God and nature, it was also his conviction that God *acted* in the world. Though Newton avoided most hints of pantheism and though his Deity remained wholly "other" and transcendental, Newton had no doubt that the world was created by divine fiat and that the Creator retained a perpetual involvement with and control over His creation. The remote and distant God of the deists, a Deity that never interacted with the world but left it to

26. Cf. David Brewster, *Memoirs of the Life, Writings, and Discoveries of Sir Isaac Newton* (2 vols.; Edinburgh: Thomas Constable and Co.; Boston: Little, Brown, and Co., 1855), vol. II, 407–8.

operate without divine guidance, was antithetical to Newton . Newton's God acted in time and with time, and since He was so transcendent, He required for His interaction with the created world at least one intermediary agent to put His will into effect. Just such an agent was the alchemical spirit, charged with animating and shaping the passive matter of the universe.

But was the alchemical spirit God's only agent? Newton's conviction that God acted in creating and maintaining the world never wavered, but his explanations of God's manner of acting underwent drastic and multiple revisions as he examined certain questions. Is the agent of vegetation the same as the agent of cosmogony? Is that agent light, or very similar to light? Is there a different, distinctive spirit for prophetic inspiration? How much activity does God manage by "mechanical principles" and how much requires "active principles"? Is gravity mechanical or active, and will it play a role in the final consummation of all things promised by Scripture? May not one distinguish two sorts of chemistry, the one mechanical and the other vegetative and demonstrative of God's nonmechanical powers? Will the rediscovery of the pure, potent fire that is the ultimate secret of the active alchemical principle lead to the restoration of true religion and the ushering in of the millennium?

Alchemical documentation

That alchemy played a role in Newton's thought is no longer to be denied; one can now trace its contributions with some exactitude. Any hope, however, of reconciling Newton's alchemical work with his recognized achievements in mathematics, optics, and celestial dynamics must first be launched from a firm textual framework of his own alchemical manuscripts, within which framework one may readily see the overriding importance of the alchemical spirit to him.

Newton's alchemical papers have been more widely scattered than any other set of his manuscripts.[27] Major collections of them exist at King's

27. A complete listing of chemical/alchemical papers that were retained by the family until 1936, when they were sold at auction, can be found in *Catalogue of the Newton Papers Sold by Order of the Viscount Lymington to Whom They Have descended from Catherine Conduitt, Viscountess Lymington, Great-niece of Sir Isaac Newton* (London: Sotheby and Co., 1936), pp. 1–19, and in Dobbs, *Foundations* (1, n. 1), pp. 235–48. These papers, as well as others of a chemical/alchemical nature, primarily experimental, are also listed in *A Catalogue of the Portsmouth Collection of Books and Papers*

College, Cambridge, at Babson College, Babson Park, Massachusetts, at the Smithsonian Institution in Washington, D.C., at the Jewish National and University Library in Jerusalem, and at University Library, Cambridge. Other items have been acquired by a variety of British and American institutions: the British Library, Clifton College, St. Andrews University, Trinity College (Cambridge), the Royal Society, and the Bodleian Library; Stanford University, Yale University, Harvard University, the University of Texas, the University of Wisconsin, the Massachusetts Institute of Technology, Columbia University, and the University of Chicago. The Sotheby sale of 1936 that dispersed Newton's alchemical papers left some in the hands of private collectors; others have disappeared.

Of necessity one must be selective here in the alchemical documentation for this study, for a more extensive use of Newton's alchemical papers could swamp the book in esoteric detail without accomplishing its primary goal of relating alchemical themes to Newton's developing thought in other areas. What is needed is a series of carefully dated anchor points within the alchemical corpus from which one may survey the larger issues. For that purpose the following manuscripts will serve well.

From the decade of the 1660s I have chosen two radically different items that in themselves mark Newton's transition from exoteric to esoteric chemistry. The first,[28] a practical chemical dictionary of about 7,000 words that seems to have been completed in about 1667, offers specific definitions that sometimes tell what Newton thought about a substance or a process before he began to read alchemy. The second,[29] a brief series of alchemical propositions, was written about 1669. It represents one of Newton's earliest attempts to order the chaotic alchemical material he was encountering, and it adumbrates themes of continuing interest. Extensive selections from it appear below. The principal discovery from these two documents is of Newton's full acceptance of a living, "vegetable" chemistry, identifiable as alchemy, distinct from the more mechanical sort of chemistry. He was to make the distinction quite explicit early in the 1670s.

In the decade of the 1670s Newton wrote a treatise of about 4,500 words, "Of Natures obvious laws & processes in vegatation."[30] Probably

written by or belonging to Sir Isaac Newton, the scientific portion of which has been presented by the Earl of Portsmouth to the University of Cambridge. Drawn up by the Syndicate appointed the 6th November, 1872 (Cambridge University Press, 1888), pp. 11–24.

28. Sotheby lot no. 16; Bodleian Library, Oxford, MS Don. b. 15; Dobbs, "Conceptual problems" (1, n. 20).
29. Sotheby lot. no 1; King's College, Cambridge, Keynes MS 12A.
30. Sotheby lot no. 113; Dibner Library of the History of Science and Tech-

completed before 1672 or during that year, this document is an invaluable conceptual source for the entire decade and beyond, and the entire English section of the treatise is transcribed in Appendix A. It tells more about vegetation, including mineral vegetation, and it leads into the closely related topics of putrefaction and fermentation. One must refer to it again and again. For the second half of the 1670s I point once more to a manuscript of about 1,200 words, the "Clavis" or "Key,"[31] in which Newton originally seemed to me to be reporting some experimental alchemical success. This document has already been published in full, with some hesitation, as Newton's own,[32] but a non-Newtonian source for it has recently been located among the papers of Robert Boyle.[33] Nevertheless, even though one now knows for certain that Newton did not compose the document, the basic alchemical process it describes continues to appear in Newton's later papers, and so the "Key" continues to provide a useful point of reference. The evidence suggests that Newton accepted, and continued to accept, the validity of the experimental report in this document, and that he used it quite literally as a "key" to unlock the secrets of other alchemical processes.[34]

A commentary by Newton on the *Emerald Tablet* of Hermes Trismegistus[35] brings one into the decade of the 1680s. About 1,000 words in length and probably written between 1680 and 1684, this manuscript shows filiation with other papers written before the *Principia*

nology of the Smithsonian Institution Libraries, Washington, D.C., Dibner Collection MSS 1031 B (formerly Burndy MS 16). Cf. *Manuscripts of the Dibner Collection in the Dibner Library of the History of Science and Technology of the Smithsonian Institution Libraries* (Smithsonian Institution Libraries, Research Guide No. 5; Washington, D.C.: Smithsonian Institution Libraries, 1985), p. 7 (No. 80) and Plate VII (facing p. 46); B. J. T. Dobbs, "Newton manuscripts at the Smithsonian Institution," *Isis 68* (1977), 105–7. The entire manuscript is reproduced in facsimile as an appendix in B. J. T. Dobbs, *Alchemical Death & Resurrection: the Significance of Alchemy in the Age of Newton. A lecture sponsored by the Smithsonian Institution Libraries in conjunction with the Washington Collegium for the Humanities Lecture Series: Death and the Afterlife in Art and Literature. Presented at the Smithsonian Institution, February 16, 1988* (Washington, D.C.: Smithsonian Institution Libraries, 1990).

31. Sotheby lot no. 11; King's College, Cambridge, Keynes MS 18.
32. Dobbs, *Foundations* (1, n. 1), pp. 251–5.
33. William Newman, "Newton's 'Clavis' as Starkey's 'Key,' " *Isis 78* (1987), 564–74.
34. Cf. Dobbs, "Newton's 'Clavis' " (1, n. 5).
35. Sotheby lot no. 31; King's College, Cambridge, Keynes MS 28; Dobbs, "Newton's *Commentary*" (1, n. 5).

of 1687. It may be found in Appendix B and it will give some curious insights into the Newton of the early 1680s regarding alchemy and cosmogony. Themes from the pre-*Principia* period continue, however, into the latter part of the decade, as the next selection shows: "Out of La Lumiere sortant des Tenebres."[36] "Out of La Lumiere," to be found in Appendix C, consists of abstracts, translated by Newton, from a book published in 1687.[37] Since the material is abstracted, one may assume that it comprises what Newton found important in the book and in places one does find here a remarkably clear exposition of late seventeenth-century alchemical theory, especially as it delineates the themes of vegetation and divine cosmogonic action.

For the decade of the 1690s I have chosen first of all a set of papers less theoretical in orientation: "Experiments & Observations Dec. 1692 & Jan 169⅔"[38] When Newton's scientific papers were presented to the University of Cambridge in the nineteenth century, his chemical/alchemical notebook and a number of loose sheets were included. The present manuscript was among the loose sheets and had been precisely dated in

36. Sotheby lot no. 40 and lot no. 41; Jewish National and University Library, Jerusalem, Yahuda MS Var. 1, Newton MS 30, and The Sir Isaac Newton Collection, Babson College Archives, Babson Park, MA, Babson MS 414 B.

37. *La lumière sortant par soy même des tenebres ou veritable theorie de la Pierre des Philosophes écrite en vers Italiens, & amplifiée en Latin par un Auteur Anonyme, en forme de Commentaire; le tout traduit en François par B. D. L.* (Paris: Chez Laurent D'Houry, ruë S. Jacques, devant la Fonteine S. Severin, au S. Esprit, 1687); Harrison, *Library* (1, n. 6), item 1003, p. 184; now NQ. 16. 117 in that portion of Newton's own books now returned to Trinity College, Cambridge. The book shows many signs of dog-earing, presumably by Newton himself.

38. University Library, Cambridge, Portsmouth Collection, MS Add. 3973.8. Seventeenth-century English dating practices frequently evoke confusion for two principal reasons: (1) England adhered to the Julian calendar until 1752, whereas the Gregorian calendar had been established on the continent since 1582; (2) the English new year legally began on 25 March rather than 1 January. The first factor meant that English dates were consistently ten days behind continental dates throughout the century. The second factor entailed various ways of indicating the year for dates between 1 January and 25 March. Many Englishmen, including Newton, resolved the latter problem by utilizing a number combining the dates for both the old and new years for dates between 1 January and 25 March, as in this instance when Newton wrote "169⅔" for a date that would be January 1693 in modern notation. For further discussion, see Westfall, *Never at Rest* (1, n. 4), p. xvi.

the title by Newton himself. In fact, later entries toward the end of the manuscript, also dated by Newton, carry it forward to June 1693. The complete manuscript is reproduced in Appendix D. In it one will find the clearest sort of emphasis upon mineral fermentations, showing Newton's continued interest in that topic even after a quarter of a century. One will also find in it an experimental record of his search for the "fermental virtue" in metallic reactions that is related to the "Key" of the late 1670s and also to the experimental procedures he recommends in the next manuscript, the "Praxis."[39] The "Praxis" is a formal treatise, reproduced in Appendix E (omitting earlier drafts of the treatise that also survive with this set of papers). It was probably written about the middle of the 1690s, which would place it at or near the end of Newton's serious experimental alchemical work. Although he continued to study and to rework his papers, there is no evidence that he performed alchemical experiments after he left Cambridge for London and the Royal Mint in 1696. The final, climactic synthesis of the "Praxis," however, is thoroughly demonstrative of the continuity in Newton's alchemical concerns, even ten years after the writing of the *Principia*.

These eight manuscripts then – two from each of the four decades in which Newton was most intensely involved with alchemy – provide the framework for the present study. Nevertheless, it will be necessary to draw upon other papers from Newton's alchemical hoard occasionally, either to corroborate major points or to mention specific items not contained in those selected. Such additional material will be cited as required. The interested reader is also reminded that previous publications, which inform the present work, were based on other sets of Newton's alchemical papers, and that much particular and specific information is available there: on all the pre-1680 alchemical manuscripts in the Keynes Collection, King's College, Cambridge;[40] on a series of papers and annotations on the regimens of the great work of alchemy that spans almost all of Newton's alchemical career.[41]

Prospectus

This book is predicated upon the conviction that to Newton himself all his diverse studies constituted a unified plan for obtaining Truth, and

39. Sotheby lot no. 74; The Sir Isaac Newton Collection, Babson College Archives, Babson Park, MA, Babson MS 420 (part).
40. Dobbs, *Foundations* (1, n. 1); idem, "Newton's 'Clavis' " (1, n. 5).
41. Dobbs, "Newton's copy of *Secrets Reveal'd*" (1, n. 5).

it is organized around a religious interpretation of Newton's alchemy, but more than that, a religious interpretation of all his work.[42] It is in Newton's belief in the unity of Truth, guaranteed by the unity and majesty of God, that one may find a way to reunite his many brilliant facets, which, however well polished, now remain incomplete fragments. It is my hope thus to make him again into one whole and historical human being.

For this attempt at recreating the historical Newton, the book falls naturally into two related approaches, one conceptual and the other chronological. While it has not always been possible, or even desirable, to keep the two approaches entirely separate, Chapters 2–3 are primarily conceptual and Chapters 4–7 are primarily chronological.

In the conceptual chapters one must first of all inquire about the meaning Newton attached to the alchemical enterprise. There one will first find it related to divine providence, then to divine activity at the time of creation, then finally to the Judeo–Christian vision of providential history. In the chronological chapters, on the other hand, with this religious interpretation of alchemy in hand, it will be appropriate to see all of Newton's work as a study of the modes of divine activity in the world, and it will be possible to trace the evolution of his thought in several areas, paying special attention to his changing interpretations of active and passive principles. At the end, one will see that his alchemy was indeed there all along, but never quite in the way one had previously supposed.

42. I owe an apology to Mary S. Churchill, who first suggested a religious interpretation of Newton's alchemy in her "*The Seven Chapters*, with explanatory notes," *Chymia 12* (1967), 29–57. I discounted her suggestion for many years, as in Dobbs, *Foundations* (1, n. 1), pp. 14–15. However, I develop ideas similar to hers in Chapter 5 of this book.

2

Vegetability and providence

Introduction

Isaac Newton has seldom been viewed as a thinker concerned with "the problem of life," a problem that had attracted and perplexed the best minds since antiquity and continued to challenge his contemporaries.[1] The definition of life, the distinction of living forms from the nonliving may not at first have concerned him, but eventually he came to believe that "all matter duly formed is attended with signes of life."[2]

Newton probably developed an interest in the subject only indirectly, through matter theory and the significant theological questions associated with the definition of matter and motion in the seventeenth century, especially issues of cohesion and organization in matter. It is possible to follow Newton's developing thought on questions of cohesion and life within the general context of his matter theory.

Sometime during his student years Newton became an eclectic corpuscularian, a second-generation mechanical philosopher, choosing elements of matter theory from Descartes, Gassendi (via Charleton), Boyle, Hobbes, Digby, and More, and leaving a record of his thoughts in the *Questiones quaedem philosophicae* of his student notebook. "First matter," he noted, was clearly not composed of mathematical points and

1. C. U. M. Smith, *The Problem of Life. An Essay in the Origins of Biological Thought* (A Halsted Press Book; New York: John Wiley & Sons, 1976); Hans Regnéll, *Ancient Views on the Nature of Life. Three Studies in the Philosophies of the Atomists, Plato and Aristotle* (Library of Theoria, No. 10; Lund: CWK Gleerup, 1967).
2. Isaac Newton, Cambridge University Library, Portsmouth Collection MS Add. 3970, f. 619r, quoted in J. E. McGuire and P. M. Rattansi, "Newton and the 'Pipes of Pan,' " *Notes and Records of the Royal Society of London* 21 (1966), 108–43, on p. 118; a variant of this quotation is also quoted in Henry Guerlac, "Theological voluntarism and biological analogies in Newton's physical thought," *Journal of the History of Ideas* 44 (1983), 219–29, on p. 229.

parts, nor of a "simple entity before division indistinct," nor was it infinitely divisible. He argued that the "first matter must be homogeneous," but subsequently cancelled that passage; nevertheless he concluded that there must be "least parts of matter," or atoms, that "were either created so or divided by means of a vacuum."[3]

At first, Newton, like other mechanical philosophers of his time, placed considerable faith in the existence of an all-pervasive material medium which served as an agent of change in the natural world. By postulating a subtle aether, a medium imperceptible to the senses but capable of transmitting effects by pressure and impact, mechanical philosophers had devised a convention that rid natural philosophy of incomprehensible occult influences acting at a distance (e.g., magnetic attraction and lunar effects). For Newton just such a mechanical aether, pervading and filling the whole world, became an unquestioned assumption. By it he explained gravity and, to a certain extent, the cohesion of particles of matter.[4] But because of the general passivity of matter in the mechanical philosophy certain problems arose for many contemporary philosophers regarding cohesion and life, and eventually, for Newton, regarding gravity also.

The problem of cohesion and life

The question of cohesion had always plagued theories of discrete particles, atomism having been criticized even in antiquity on this point. The cohesion of living forms seems intuitively to be qualitatively different from anything that the random, mechanical motion of small particles of matter might produce. Nor does atomism explain even mechanical cohesion in inert materials very well, for it requires the elaboration of ad hoc,

3. A. Rupert Hall, "Sir Isaac Newton's note-book, 1661–65," *Cambridge Historical Journal* 9 (1948), 239–50; Richard S. Westfall, "The foundations of Newton's philosophy of nature," *British Journal for the History of Science* 1 (1962/63), 171–82; idem, *Force in Newton's Physics* (1, n. 7), pp. 324–6; idem, *Never at Rest* (1, n. 4), pp. 89–97; McGuire and Tamny, *Certain Philosophical Questions* (1, n. 22), pp. 336–45, 420–5. In these and subsequent quotations from Newton's student notebook I have employed the modernized English version provided by McGuire and Tamny. The lengthy commentary, "Infinity, indivisibilism, and the void," provided by McGuire and Tamny (ibid., pp. 26–126) seriously overemphasizes Newton's early commitment to the void, as can be seen later in Chapters 4 and 5.
4. Ibid., pp. 362–5, 426–7 (on gravity), 348–51 (on "Conjunction of bodies"). See also the essay by McGuire and Tamny, "Gravitation, attraction, and cohesion," ibid., pp. 275–95.

unverifiable hypotheses about the geometric configurations of the atoms or else speculation about their quiescence under certain circumstances. In the various forms in which corpuscularianism was revived in the seventeenth century, the problems remained and variants of ancient answers were redeployed. Descartes, for example, held that an external pressure from surrounding subtle matter just balanced the internal pressure of the coarser particles that constituted the cohesive body. Thus no special explanation for cohesion was required: the parts cohered simply because they were at rest close to each other in an equilibrated system. Gassendi's atoms, on the other hand, stuck together through the interlacing of antlers or hooks and claws, much as the atoms of Lucretius had before them. Charleton found not only hooks and claws but also the pressure of neighboring atoms and the absence of disturbing atoms necessary to account for cohesion.[5] Francis Bacon introduced certain spirits or "pneumaticals" into his speculations. In a system reminiscent of that of the Stoics, those ancient critics of atomism, Bacon concluded that gross matter must be associated with active, shaping, material spirits, the spirits being responsible for the forms and qualities of tangible bodies, producing organized shapes, effecting digestion, assimilation, and so forth.[6]

For Newton during his student years, with his mechanical aether ready at hand, a pressure mechanism seemed sufficient to explain cohesion; he rejected quiescence but affirmed that "the close crowding of all the matter

5. Lancelot Law Whyte, *Essay on Atomism: From Democritus to 1960* (reprint of the 1961 ed., Harper Torchbooks; New York: Harper & Row, 1963); E. C. Millington, "Theories of cohesion in the seventeenth century," *Annals of Science* 5 (1941–7), 253–69.
6. Bacon's early atomism gave way in his mature years to a position thought to have been influenced by his chemical/alchemical readings, a position much closer to Stoicism than has generally been recognized: Robert Hugh Kargon, *Atomism in England from Hariot to Newton* (Oxford: Clarendon Press, 1966), pp. 43–54; J. C. Gregory, "Chemistry and alchemy in the natural philosophy of Sir Francis Bacon, 1561–1626," *Ambix* 2 (1938), 93–111; Charles W. Lemni, *The Classic Deities in Bacon: A Study in Mythological Symbolism* (reprint of the 1933 ed.; New York: Octagon Books, 1971), esp. pp. 74–109; A. A. Long, *Hellenistic Philosophy: Stoics, Epicureans, Sceptics* (London: Duckworth, 1974), pp. 152–6; Graham Rees, "Francis Bacon's semi-Paracelsian cosmology," *Ambix* 22 (1975), 81–101; idem, "Francis Bacon's semi-Paracelsian cosmology and the Great Instauration," ibid., 22 (1975), 161–73; idem, "The fate of Bacon's cosmology in the seventeenth century," ibid., 24 (1977), 27–38; idem, "Matter theory: a unifying factor in Bacon's natural philosophy?" ibid., 24 (1977), 110–25; idem, "Francis Bacon on verticity and the bowels of the earth," ibid., 26 (1979), 202–11.

in the world" might account for it. He noted the occasional geometric approach of Descartes, but did not himself develop it: "Whether hard bodies stick together by branchy particles folded together."[7]

It was to be a long, circuitous, even tortuous journey that carried Newton away from those reflections on matter in his student *Questiones*. Within a very short time he had begun to modify his mechanical philosophy with a chemical and then an alchemical one.

Newton had been concerned with the unity of matter during his brief years of strict mechanism, had called the "first matter" homogeneous, but then had cancelled that passage, possibly through the influence of chemical thought. His early chemical dictionary, probably completed about 1667, provides much practical detail on chemical operations and substances: there he seemed to treat each substance sui generis and not at all as variant forms of a common matter.[8] Investigators have often tacitly assumed that Newton accepted the unity of matter from contemporary mechanical philosophy. Certainly talk among the mechanical philosophers of the particles of one catholic and universal matter did nothing to undermine ancient doctrines of prime matter, even though among most pre-Socratics, Aristotelians, and alchemists *materia prima* had not been considered particulate. Yet neither the catholic matter of the mechanical philosophers nor any hint that matter is particulate is in evidence in Newton's chemical dictionary, even though he cites the great corpuscularian Robert Boyle as the source for some of his chemical information.[9]

On the other hand, one does find Newton's lucid expression for the unity of matter in an early alchemical context.[10] About 1669 he prepared

7. McGuire and Tamny, *Certain Philosophical Questions* (1, n. 22), pp. 348–51 ("Conjunction of bodies") and 360–1 ("Of softness, hardness, flexibility, ductility, and tractility").
8. MS Don. b. 15 (1, n. 28). For example, the entry for "Mercury," f. 4r: "Mercury being put upon Lead or Tin when almost cold after fusion becoms by their steame coagulated into a malliable & fusible body. In Scicely its oare lys in veines hard as stone but more weighty of a liver colour or y^{t} of Crocus Metallorum, As much as can be is washed out of y^{e} ore wch they call Virgin mercury. The rest is distilld. Virgin mercury Amalgamd wth Gold & evaporated carrys away the ☉ wch common ☿ will not doe. Somtimes tis found runing. Somtimes in round clods wth gold in specks." ☉ = gold; ☿ = mercury."
9. Dobbs, *Foundations* (1, n. 1), p. 46; idem, "Conceptual problems" (1, n. 20).
10. On Newton's turn to alchemy in 1668, cf. Dobbs, *Foundations* (1, n. 1), pp. 121–5.

a short paper containing a series of alchemical propositions. Gold, silver, iron, copper, tin, lead, mercury, and "magnesia" are all the species of the art, he said, and all of them are from one root.[11] Whether he had first absorbed the notion in a mechanical or other philosophical context or not, his early alchemical work evidently secured in Newton a conviction from which he never subsequently wavered regarding the unity of matter.

The doctrine of the unity of matter and its transmutability became a part of the published record of his views. The first edition of the *Principia* (1687) carried the most explicit statement: "Any body can be transformed into another, of whatever kind, and all the intermediate degrees of qualities can be induced in it."[12] There is a passage in all editions of the *Principia* in which Newton speculated that matter falling to earth from the tails of comets might be condensed into all types of earthly substances.[13] In his later years Newton stated the doctrine a number of times: in his small tract *On the Nature of Acids*,[14] in the *Opticks*,[15] and

11. Keynes MS 12A (1, n. 29), f. 1r: "☉ ☽ ♂ ♀ ♃ ♄ ☿ & magnesia sunt omnes species artis.... 2. Omnes species sunt ex una radice...."
12. Isaac Newton, *Philosophiae naturalis principia mathematica* (Londini: Jussu Societatis Regiae ac Typis Josephi Streater. Prostat apud plures Bibliopoles, 1687), p. 402, Hypothesis III. Subsequent changes are given in Newton, *Principia* (Koyré–Cohen) (1, n. 9), vol. II, 550–3. See also Alexandre Koyré, "Newton's 'Regulae Philosophandi,'" in Alexandre Koyré, *Newtonian Studies* (Cambridge, MA: Harvard University Press, 1965), pp. 261–72.
13. Isaac Newton, *Sir Isaac Newton's Mathematical Principles of Natural Philosophy and His System of the World* (1729, tr. by Andrew Motte), ed. by Florian Cajori (reprint of the 1934 ed.; 2 vols.; Berkeley: University of California Press, 1962), vol. II, 542; Newton, *Principia* (Koyré–Cohen) (1, n. 9), vol. II, 758.
14. The published variants of the tract are: (1) John Harris, *Lexicon Technicum or an Universal English Dictionary of Arts and Sciences* (facsimile of 1704–10 ed.; 2 vols.; New York: Johnson Reprint Corp., 1966), vol. II, sig. b3v-b4v; (2) Isaac Newton, *Isaaci Newtoni Opera quae exstant omnia. Commentariis illustrabat Samuel Horsley, LL. D. R. S. S. Reverendo admodum in Christo Patri Roberto Episcopo Londinensi a Sacris* (5 vols.; Londini: Excudebat Joannes Nichols, 1779–85), vol. IV, 395–400; (3) idem, *The Correspondence of Isaac Newton*, ed. by H. W. Turnbull, J. P. Scott, A. R. Hall, and Laura Tilling (7 vols.; Cambridge: published for the Royal Society at the University Press, 1959–77), vol. III, 205–14.
15. Isaac Newton, *Opticks, or A Treatise of the Reflections, Refractions, Inflections & Colours of Light*, foreword by Albert Einstein, intro. by Sir

to David Gregory, who duly recorded it among his memoranda.[16] Although that most uncompromising statement from the *Principia* of 1687 disappeared in subsequent editions, and although the later *Opticks* passages demonstrate some possible ambiguities, the consensus of recent studies is that Newton maintained to the end his belief in the inertial homogeneity and transformability of matter.[17] His insistence on this point in fact intensified the theological issue he had already recognized and which one must consider below. If all the ultimate passive particles of matter are alike, how is it possible for them to become organized by mechanical impact into the immense variety of living forms?

Vegetation

Although alchemy and mechanism do share the doctrine of the ultimate unity of matter, it seems impossible to find a mechanical counterpart for the active, vitalistic alchemical agent Newton introduced into his "Propositions" about 1669. There he called the agent by its code name *magnesia*, a term that evoked for the alchemists all the mysterious properties of the magnet and expressed their understanding that certain substances had the capacity to draw into themselves the active vivifying celestial principle necessary for life. Newton aligned "magnesia" with the metals in being from "one root," but he added that magnesia is the only species that revivifies.[18]

Newton had become preoccupied with a process of disorganization and reorganization by which developed species of matter might be radically reduced, revivified, and led to generate new forms. The alchemical agent responsible for these changes is vitalistic and universal in its actions;

Edmund Whittaker, preface by I. Bernard Cohen, analytical table of contents by Duane H. D. Roller (based on the 4th London ed. of 1730; New York: Dover, 1952), pp. 266–9, 394.

16. W. G. Hiscock, Ed., *David Gregory, Isaac Newton and Their Circle: Extracts from David Gregory's Memoranda 1677–1708* (Oxford: printed for the editor, 1937), pp. 30–1.
17. Dobbs, *Foundations* (1, n. 1), pp. 199–204, 231–2; J. E. McGuire, "Transmutation and immutability: Newton's doctrine of physical qualities," *Ambix* 14 (1967), 69–95; Arnold Thackray, *Atoms and Powers: An Essay on Newtonian Matter-Theory and the Development of Chemistry* (Cambridge, MA: Harvard University Press, 1970), esp. pp. 8–42.
18. Keynes MS 12A (1, n. 29), f. 1r: "Sola est magnesia quae species revivificat. . . ." Cf. Dobbs, *Foundations* (1, n. 1), pp. 159–60.

it is a "fermental virtue"[19] or "vegetable spirit"[20] and is eventually to become the force of fermentation of the *Opticks*.[21] In the "Propositions" it is the agent that confounds into chaos and then aggregates anew the particles of matter.

> This and only this is the vital agent diffused through all things that exist in the world.
>
> And it is the mercurial spirit, most subtle and wholly volatile, dispersed through all places.
>
> This agent has the same general method of operating in all things, namely, excited to action by a moderate heat, it is put to flight by a great one, and once an aggregate has been formed, the agent's first action is to putrefy the aggregate and confound it into chaos. Then it proceeds to generation.
>
> And the particularities of its method are many, according to the nature of the subject in which it operates. For it accommodates itself to every nature. From metallic semen it generates gold, from human semen men, etc.
>
> And it puts on various forms according to the nature of the subject. In metals it is not distinguished from the metallic substance, in men, not from the human substance, etc.
>
> In the metallic form it is found most plentifully in Magnesia.
>
> And from this one root came all species of metals.
>
> And that in this order: mercury, lead, tin, silver, copper, iron, gold.[22]

19. Keynes MS 18, in Dobbs, *Foundations* (1, n. 1), pp. 251–5. The term *fermental virtue* in the last paragraph of the manuscript (p. 255) is not present in the non-Newtonian original: cf. Newman, "Newton's 'Clavis' " (1, n. 33).
20. Dibner Collection MSS 1031 B (1, n. 30), f. 6r. In this and subsequent quotations from this manuscript, I have modernized Newton's English; for the original English, see Appendix A.
21. Newton, *Opticks* (2, n. 15), p. 401.
22. Keynes MS 12A (1, n. 29), ff. 1v–2r:

    Idem et unicus est agens vitalis per omnia quae in mundo sunt diffuses

    Estque spiritus mercurialis subtilissimus et summè volatilis per omnia loca dispersus.

    Hujus agentis eadem est methodus generalis operandi in omnibus, nempe modico calore excitatus ad agendum, fugatur magno, et proposito substantiarum aggregato primus ejus actio est putrefacere & in chaos confundere, deinde ad generationem procedit.

    Ejus autem particulares methodi sunt plures pro naturâ subjecti in quo operatur. Nam omni naturae se accommodat. Ex semine metallico generat aurum, ex humano hominem &c

    Et pro naturâ subjecti varias etiam formes induit. In metallis non distin-

From what sources has Newton derived his ideas on the universal vital agent that he is here busily attaching to seventeenth-century mechanism? Quite possibly from alchemy only at this early stage in his development, though his vitalistic ideas were soon reinforced by other sources.

Vitalism seems to belong to the very origins of alchemy. In the early Christian centuries, when alchemical ideas were taking shape, metals had not been well characterized as distinct species. They were thought to have variable properties, like modern alloys, or, more frequently, they were thought to be like a mix of dough, into which the introduction of a leaven might produce desired changes by a process of fermentation, or even as similar to a material matrix of unformed matter, into which the injection of an active male sperm or seed might lead to a process of generation. By analogy alchemists referred to this critical phase of the alchemical process as fermentation or generation, and the search for the vital ferment or seed became a fundamental part of their quest.[23] Similar ideas occur in Aristotle and were commonplace in Newton's time.[24]

Inspired by his interest in a vital agent, Newton had begun to grope his way toward mending the deficiencies of ancient atomism and contemporary corpuscularianism. He had concerned himself with life and cohesion. He now sought the source of all the apparently spontaneous processes of fermentation, putrefaction, generation, and vegetation – that is, everything associated with normal life and growth, such as digestion and assimilation, "vegetation" being originally from the Latin *vegetare*, to animate, enliven. These processes produced the endless variety of living forms and could not be relegated to the mechanical actions of gross

guitur a substantia metallica, in homine non ab humana substantia. &c
In maxima copia sub formâ metallicâ reperitur in Magnesia
Et hac unica radice sunt omnes metallorum species
Idque in hoc ordine ☿ ♄ ♃ ☽ ♀ ♂ ☉

I have expanded Newton's abbreviations for *que* in this quotation but otherwise have followed the manuscript as closely as possible.

23. Marcellin Pierre Eugene Berthelot, *Les origines de l'alchimie* (reprint of the 1885 ed.; Paris: Librairie des Sciences et des Arts, 1938), pp. 240–1.

24. See, e.g., Aristotle, *De generatione animalium*, 1.1, 715b26; 1.2, 716a17; 1.16, 721a8; 2.3, 736b30; 2.4, 738b23, in Aristotle, *The Works of Aristotle Translated into English under the Editorship of J. A. Smith and W. D. (Sir David) Ross* (12 vols.; Oxford: University Press and Clarendon Press; London: Humphrey Milford, 1908–60), Vol. 5; Rosalie L. Colie, *Light and Enlightenment. A Study of the Cambridge Platonists and the Dutch Arminians* (Cambridge University Press, 1957), esp. pp. 117–44. Other contemporary examples appear in later notes.

corpuscles, a point he emphasized in the 1670s and to which one must return later. Mechanical action could never account for the process of assimilation, in which food stuffs were turned into the bodies of animals, vegetables, and minerals. Nor could it account for the sheer variety of forms in this world, all of which had somehow sprung from the common matter.

The most comprehensive answer to such problems in antiquity had been given by the Stoics. The Stoics postulated a continuous material medium, the tension and activity of which molded the cosmos into a living whole and the various parts of the cosmic animals into coherent bodies as well. Compounded of air and a creative fire, this medium was the Stoic *pneuma* and was related to the concept of the "breath of life" that escapes from a living body at the time of death and allows the formerly coherent body in which it had resided to disintegrate into its disparate parts. Although always material, the *pneuma* becomes finer and more active as one ascends the scale of being, and the (more corporeal) air decreases as the (less corporeal) fire increases. The Stoic Deity, literally omnipresent in the universe, is the hottest, most tense and creative form of the cosmic *pneuma* or aether, pure fire or nearly so. The cosmos permeated and shaped by the *pneuma* is not only living, it is rational and orderly and under the benevolent, providential care of the Deity. Though the Stoics were determinists, their Deity was immanent and active in the cosmos, and one of their most telling arguments against the atomists was that the order, beauty, symmetry, and purpose to be seen in the world could never have come from random, mechanical action. Only a providential God could produce and maintain such lovely, meaningful forms. The universe, as a living body, was born when the creative fire generated the four elements; it lived out its lifespan, permeated by vital heat and breath, cycling back to final conflagration in the divine active principle, and always regenerated itself in a perpetual circle of life and death.[25]

The original writings of the Stoics were mostly lost, but not before

25. Shmuel Sambursky, *Physics of the Stoics* (reprint of the 1959 ed.; London: Hutchinson, 1971); Stephen Toulmin and June Goodfield, *The Architecture of Matter* (reprint of the 1962 ed.; Harper Torchbooks; New York: Harper & Row, 1966), esp. pp. 92–108; Mary B. Hesse, *Forces and Fields: The Concept of Action at a Distance in the History of Physics* (reprint of the 1962 ed.; Westport, CT: Greenwood Press, 1970), esp. pp. 74–9; John M. Rist, ed., *The Stoics* (Berkeley: University of California Press, 1978), esp. Robert M. Todd, "Monism and immanence: the foundations of Stoic physics," pp. 137–60, and Michael Lapidge, "Stoic cosmology," pp. 161–85.

ideas of *pneuma* and *spiritus* came to pervade medical doctrine, alchemical theory, and indeed the general culture with form-giving spirits, souls, and vital principles. Spiritualized forms of the *pneuma* entered early Christian theology in discussions of the immanence and transcendence of God and of the Holy Ghost, just as the Stoic arguments that order and beauty demonstrate the existence of God and of providence entered Christianity as the "argument from design." The creative emanations of Stoic fire melded with the creative emanations of light in Neoplatonism. In addition to this broad spectrum of at least vaguely Stoic ideas, excellent, though not always sympathetic, summaries of philosophical Stoicism were available in Cicero, Seneca, Plutarch, Diogenes Laertius, Sextus Empiricus, and others. By the seventeenth century ideas compatible with Stoicism were very widely diffused, and latter-day Stoics, Pythagoreans, Platonists, medical men, chemists, alchemists, and even Peripatetics vied with each other in celebrating the occult virtues of a cosmic aether that was the vehicle of a pure, hidden, creative fire.[26]

Nonetheless, such a vital aether was to be found in its most developed form in philosophical Stoicism. It is possible, as Newton's concern for the processes of life and cohesion grew apace in the early 1670s, that he amplified his mechanical philosophy further by a close reading of the available literature on the Stoics. Virtually all of the scanty fragments of ancient Stoicism known today had already been recovered by Western Europe during the Renaissance, and Newton had most of them. He owned two different editions of Cicero's *Opera omnia*, Diogenes Laertius in Greek and Latin, Seneca's collected works as well as two separate editions of his tragedies, and the *Opera quae extant* of Sextus Empiricus.[27] He had four volumes of Epictetus, works by the sixteenth-century Stoic Justus

26. Mary Anne Atwood, *Hermetic Philosophy and Alchemy: A Suggestive Inquiry into "The Hermetic Mystery" with a Dissertation on the More Celebrated of the Alchemical Philosophers*, intro. by Walter Leslie Wilmhurst (rev. ed.; New York: The Julian Press, 1960), p. 78; D. P. Walker, *Spiritual and Demonic Magic from Ficino to Campanella* (reprint of the 1969 ed.; Notre Dame: University of Notre Dame Press, 1975); idem, *The Ancient Theology: Studies in Christian Platonism from the Fifteenth to the Eighteenth Century* (London: Duckworth, 1972); idem, "Medical spirits: four lectures," Boston Colloquium for the Philosophy of Science, 27 October–19 November 1981; Rosaleen Love, "Some sources of Herman Boerhaave's concept of fire," *Ambix* 19 (1972), 157–74; idem, "Herman Boerhaave and the element-instrument concept of fire," *Annals of Science* 31 (1974), 547–59.
27. Harrison, *Library* (1, n. 6), items 381–82, 519, 1486–90, 1503, pp. 119, 133, 236–7.

Lipsius, and two sixteenth-century editions of Plutarch that together comprised a total of fifteen volumes.[28] Even some passages preserved by Christian apologists were in Newton's possession, as in the works of Clement of Alexandria and Eusebius.[29] A full search of Newton's library would perhaps reveal other material of Stoic provenance, but enough has been said to establish the point that Newton surely could have reconstructed for himself a reasonably sophisticated and comprehensive knowledge of Stoic thought.

Such reading would have affected Newton's alchemy only in reinforcing certain critical ideas, for most of his early alchemical sources were distinctly Neoplatonic in tone, and in them the universal spirit or soul of the world already permeated the cosmos with its fermental virtue.[30] But Stoic ideas would have affected his views on the mechanical aether of his student years. It seems one may conclude that if Newton had not read the Stoics, then he must independently have reached answers similar to theirs when confronted with similar problems, for by about 1672 the original mechanical aether of his *Questiones* had assumed a strongly Stoic cast.

The new vital aether is described in the alchemical treatise, "Of Natures obvious laws & processes in vegetation." The earth is "a great animal," Newton said, "or rather an inanimate vegetable [that] draws in aethereal breath for its daily refreshment and vital ferment and transpires again with gross exhalations." He described this aethereal breath as a "subtle spirit," "nature's universal agent, her secret fire," and the "material soul of all matter." The similarity between this particular Newtonian aether and the Stoic *pneuma* is unmistakable: they are both material and both somehow inspire the forms of bodies and give to bodies the continuity and coherence of form that is associated with life. Newton expanded upon that theme, saying that the earth, "according to the condition of all other things living, ought to have its times of beginning, youth, old age, and perishing," and that the subtle aethereal agent is the "only ferment and principle of all vegetation."[31] One may trace the vivid imagery of the earth–animal back through Stoic and Neoplatonic commentators on Plato,[32] as one may trace the "perishing" of the earth in

28. Ibid., items 561–4, 959–60, 1330–1, pp. 138, 180, 219.
29. Ibid., items 398, 588–92, pp. 120, 140.
30. Dobbs, *Foundations* (1, n. 1), *passim.*
31. Dibner Collection MSS 1031 B (1, n. 30), f. 3v. Cf. Appendix A.
32. Plato, *Plato's Cosmology: The Timaeus of Plato*, ed. and tr. by Francis MacDonald Cornford (reprint of the 1937 ed.; Indianapolis: Bobbs Merrill, n.d.), p. 332.

Newton's comment backward and forward to his own convictions regarding a final cosmic conflagration.[33]

In his treatise "Of Natures laws" Newton made a sharp distinction between vegetation and mechanism. "Nature's actions," he said, "are either vegetable or purely mechanical," and as mechanical he listed, among other things, "vulgar chemistry." Vulgar chemistry may readily be identified with the kind of chemistry he discussed in his chemical dictionary and in the *Opticks*, the operations of which take place only among the "grosser" particles of matter. As he said in 1672, "all the operations in vulgar chemistry (many of which to sense are as strange transmutations as those of nature) are but mechanical coalitions or separations of particles... and that without any vegetation." Newton admitted that to many it may seem that all the "changes made by nature" may be done the same way, "that is by the sleighty transpositions of the grosser corpuscles, for, upon their disposition only, sensible qualities depend." But he argued that such is far from being the case. There is a "vast and fundamental" difference between vulgar chemistry and vegetation, which requires that we have recourse to some further cause.[34]

He continued with a development of themes from his earlier "Propositions" on the subtle nature of the vital agent, on its sensitivity to heat, and on its universality – emphasizing all the while the great differences between vulgar and vegetable chemistry.

> 6 There is, therefore, besides the sensible changes wrought in the textures of grosser matter, a more subtle, secret, and noble way of working in all vegetables which makes its products distinct from all others; and the immediate seat of these operations is not the whole bulk of the matter, but rather an exceeding subtle and unimaginably small portion of matter diffused through the mass, which if it were separated, there would remain but a dead and inactive earth. And this appears in that vegetables are deprived of their vegetable virtue by any small excess of heat, the tender spirit being either put to flight or at least corrupted thereby (as may appear in an egg), whereas those operations which depend upon the texture of the grosser matter (as all those in common chemistry do) receive no damage by heats far greater....

33. David Charles Kubrin, "Newton and the cyclical cosmos: providence and the mechanical philosophy," *Journal of the History of Ideas* 28 (1967), 325–46; idem, "Providence and the Mechanical Philosophy: The Creation and Dissolution of the World in Newtonian Thought. A Study of the Relations of Science and Religion in Seventeenth Century England" (Cornell University: Ph. D. dissertation, 1968); Dobbs, "Newton and Stoicism" (1, n. 5).
34. Dibner Collection MSS 1031 B (1, n. 30), f. 5r, v. Cf. Appendix A.

7 It is the office therefore of those grosser substances to be medium or vehicle in which rather than upon which those vegetable substances perform their actions.

8 Yet those grosser substances are very apt to put on various external appearances according to the present state of the invisible inhabitant, as to appear bones, flesh, wood, fruit, etc. Namely, they consisting of differing particles, watery, earthy, saline, airy, oily, spiritous, etc., those parts may be variously moved one among another according to the acting of the latent vegetable substances and be variously associated and concatenated together by their influence.[35]

In these distinctions between mechanical and vegetable chemistry Newton was working toward his famous hierarchical system of parts and pores arranged in three-dimensional netlike patterns. One may take his statement about the "grosser" corpuscles from 1672: "upon their disposition only, sensible qualities depend." It is identical in meaning with one from the *Opticks* in which Newton described the building up of the hierarchies of matter "until the Progression end in the biggest Particles on which the Operations in Chymistry, and the Colours of natural Bodies depend, and which by cohering compose Bodies of a sensible Magnitude."[36]

Or one may take another statement from 1672: grosser substances "consisting of differing particles, watery, earthy, saline, airy, oily, spiritous, etc., these parts may be . . . variously associated and concatenated together. . . . " These smaller units of watery, saline, and the like particles form subunits of the largest ones, and their characteristics are drawn from contemporary chemical systems of (usually) five or six chemical elements or principles derived from sixteenth- and seventeenth-century combinations of Aristotelean matter theory (four elements: earth, air, fire, water) with Paracelsian (three principles: salt, sulfur, mercury). Similar intermediate particles appear in both the *Opticks* and in *On the Nature of Acids*.[37]

Furthermore, when Newton discussed putrefaction in 1672, he not only expanded upon the view expressed in the "Propositions" a few years earlier but also adumbrated the final section of *On the Nature of Acids*.

Nothing can be changed from what it is without putrefaction . . .

No putrefaction can be without alienating the thing putrefied from what it was.

35. Ibid., ff. 5v–6r.
36. Newton, *Opticks* (2, n. 15), p. 394.
37. See Dobbs, *Foundations* (1, n. 1), pp. 219–21, for a discussion of Newton's attempt to achieve a significant union between chemistry and mechanical philosophy with his concept of "chymical" subunits.

> Nothing can be generated or nourished (but of putrefied matter).
>
> ....
>
> Her [Nature's] first action is to blend and confound mixtures into a putrefied chaos.
>
> Then they are fitted for new generation or nourishment.[38]

It is putrefaction that reduces matter to its ultimate state of disorganization, where the particles of matter are all alike and hence can be remodeled in any form whatsoever by the vegetable spirit. Although the later notion of pores is missing in the 1672 tract on vegetation, in other respects this version of Newton's matter theory closely resembles that in *On the Nature of Acids*. There, if a menstruum could adequately penetrate the pores of gold, or if gold could "ferment," it could be reduced to its most primordial particles. Then

> it could be transformed into any other substance. And so of tin, or any other bodies, as common nourishment is turned into the bodies of animals and vegetables.[39]

Newton's early alchemical papers thus contain both the doctrine of the ultimate unity of matter and the rudiments of a hierarchical system of subvisible parts in matter, but they do not exhibit all aspects of his final theory of matter. Missing are the pores devoid of gross matter which later become an essential feature of his structured hierarchies. In fact, in 1672 Newton's matter theory could hardly accommodate empty pores. His universe was still filled to capacity with a circulating aether, not unlike that of the *Questiones* in some ways but in other ways most unlike, bearing as it did a "secret fire," a "vegetable spirit," an agent that is the "only ferment and principle of all vegetation" which supplies the cohesiveness of bodies as their "material soul." Albeit that Newton's early aethers seem to have been particulate, in contrast to the tense continuum of the Stoics, they nevertheless permeated the cosmos and filled all interstices. He had actually discussed pores in regard to the interactions of light and matter in some of his earliest optical papers, where the pores were filled with a fluid medium of subtle matter with which the light particles interacted,[40] but even those pores filled with an "optical aether" failed to appear explicitly in Newton's early chemical–alchemical papers. Missing also is any mention of forces, although the addition of forces to

38. Dibner Collection MSS 1031 B (1, n. 30), f. 5r (Newton's parentheses, my brackets). Cf. Appendix A.
39. Newton, *Correspondence* (2, n. 14), vol. III, pp. 207, 211.
40. McGuire and Tamny, "The origin of Newton's optical thought and its connection with physiology," in *Certain Philosophical Questions* (1, n. 22), pp. 241–74.

matter theory proved to be Newton's most significant modification of corpuscularianism, especially in regard to cohesion.

But before pursuing these later developments in Newton's thought, one must attempt to clarify, first, the relationship he saw between vegetation and providence, and, second, the implications of the processes of illumination and fermentation, which he was convinced were essential features of vegetability.

Providence

From the 1660s Newton was troubled by a theological problem. He was, as were his older contemporaries Isaac Barrow, Henry More, and Ralph Cudworth, alarmed at the atheistic potentialities of the revived corpuscularianism of their century, particularly of Cartesianism.[41] Although the ancient atomists had not really been atheists in any precise modern sense, they had frequently been so labeled because their atoms in random mechanical motion received no guidance from the gods. Descartes, Gassendi, and Charleton had been at pains to allay the fear that the revived corpuscular philosophy would carry the stigma of atheism adhering to ancient atomism. They had solved the problem, they thought, by having God endow the particles of matter with motion at the moment of creation. All that resulted then was due not to random corpuscular action but to the initial intention of the Deity.[42]

Later writers, going further, had carefully instated a Christian Providence among the atoms (where the ancients of course had never had it). Only Providence could account for the obviously designed concatenations of the particles, and so, via Christianity, a fundamental Stoic critique actually came to be incorporated into seventeenth-century atomism. This development was all to the good in the eyes of most Christian philosophers: atomism now supported religion, because without the providential

41. Dobbs, *Foundations* (1, n. 1), esp. pp. 100–5; Guerlac, "Theological voluntarism" (2, n. 2); idem, "Newton et Epicure," in Henry Guerlac, *Essays and Papers in the History of Modern Science* (Baltimore: The Johns Hopkins University Press, 1977), pp. 82–106: originally published as *Newton et Epicure. Conférence donnée au Palais de la Découverte, Université de Paris, le 2 Mars 1963, Histoire des Scicnce* (Paris: Sur les presses de l'imprimerie Alençonnaise, 1963).
42. Kargon, *Atomism* (2, n. 6), pp. 64, 67–8, 87–9; Margaret J. Osler, "Descartes and Charleton on nature and God," *Journal of the History of Ideas* 40 (1979), 445–56.

action of God the atoms could never have assumed the lovely forms of plants and animals so perfectly fitted to their habitats. Though traceable ultimately to Plato, used by Aristotle, amplified by the Stoics and Philo, and present in Christianity from a very early period, this argument from design assumed unparalleled importance in the seventeenth century, and if the new astronomy had raised doubts about the focus of Providence upon such an obscure corner of the cosmos, the new atomism seemed to relieve them.[43]

The difficulty came when one began to wonder *how* Providence operated in the law-bound universe emerging from the new science, and that difficulty was especially severe in the Cartesian system, where only matter and motion were acceptable explanations. Even though Descartes had argued that God constantly and actively supported the universe with His will, in fact it seemed to Henry More and others that Descartes's God was in danger of becoming an absentee landlord, one who had set matter in motion in the beginning but who then had no way of exercising His providential care.

Newton faced this theological difficulty squarely and directly. The mechanical action of matter in motion was not enough. Granted that such mechanical action existed among the particles and could account for large classes of phenomena, yet it could not account for all. It could not account for the processes of life, where cohesive and guiding principles were clearly operative. It could not account for the manifold riches of the phenomenal world. All forms of matter, never mind how various, could be reduced back to a common primordial matter, but how had they been produced in the first place? The production of variety from unity seemed to posit an effect greater than its cause. Newton's problem was similar to but broader than the general problem of the origin of forms and qualities for the corpuscularian. From the particles of a catholic matter with only primary mathematical properties, there seemed no "suf-

43. Robert H. Hurlbutt III, *Hume, Newton, and the Design Argument* (rev. ed.; Lincoln: University of Nebraska Press, 1985), pp. 3–132; Jacob Viner, *The Role of Providence in the Social Order: An Essay in Intellectual History. Jayne Lectures for 1966*, foreword by Joseph R. Strayer (Memoirs of the American Philosophical Society Held at Philadelphia For Promoting Useful Knowledge, Vol. 90; Philadelphia: American Philosophical Society, Independence Square, 1972), pp. 8–9. For a more extended discussion of the supernatural ontology explicit in some seventeenth-century mechanical philosophies, see also Keith Hutchinson, "Supernaturalism and the mechanical philosophy," *History of Science* 21 (1983), 297–333.

ficient reason" for forms and qualities to emerge at all.[44] But emerge they did, and in such incredible and well-crafted plenitude that causal explanations based on mechanical interactions seemed totally insufficient. As Newton was finally to say in the General Scholium to the *Principia*, "Blind metaphysical necessity [i.e., mechanical action], which is certainly the same always and everywhere, could produce no variety of things."[45] Variety requires some further cause: it is produced by vegetation. As he said in 1672:

> So far therefore as the same changes may be wrought by the slight mutation of the tinctures of bodies in common chemistry and such like experiments, many may judge that such changes made by nature are done the same way, that is by the sleighty transpositions of the grosser corpuscles, for, upon their disposition only, sensible qualities depend. But, so fast as by vegetation such changes are wrought as cannot be done without it, we must have recourse to some further cause. And this difference is vast and fundamental because nothing could ever yet be made without vegetation which nature useth to produce by it.[46]

Newton's distinction between mechanical and vegetable chemistry thus emerges as crucial to his solution of the theological problem posed by his Cartesian inheritance. Mechanical chemistry may be accounted for simply by matter and motion, where only large corpuscles are rearranged. Changes in that realm may *seem* to be "strange transmutations" but they are only "mechanical coalitions or separations of particles" and require no further explanation.[47] But for all that great class of beings that nature produces by vegetation – "all that diversity of natural things which we find suited to different times and places"[48] – we must have recourse to some further cause. Ultimately, the cause is God, God who in His wisdom and with His dominion, His providential care, and the final purposes known only to Himself, produces all the variety: natural diversity "could arise from nothing but the ideas and will of a Being necessarily existing."[49] Within the realm of vegetable chemistry, where God does these things,

44. Leroy E. Loemker, *Struggle for Synthesis. The Seventeenth Century Background of Leibniz's Synthesis of Order and Freedom* (Cambridge, MA: Harvard University Press, 1972), pp. 219–21.
45. Newton, *Principia* (Motte–Cajori) (2, n. 13), vol. II, 546; Newton, *Principia* (Koyré–Cohen) (1, n. 9), vol. II, 763 (my brackets).
46. Dibner Collection MSS 1031 B (1, n. 30), f. 5v. Cf. Appendix A.
47. Ibid.
48. Newton, *Principia* (Motte-Cajori) (2, n. 13), vol. II, 546; Newton, *Principia* (Koyré-Cohen) (1, n. 9), vol. II, 763.
49. Ibid.

lies one arena of His continuing guidance of the world, an area of providential care. Newton's God, for all His transcendence, was not an absentee landlord.

If God was the ultimate cause, however, what Newton wanted to find in the natural world was the more proximate cause of the phenomena of life that was God's agent in these matters. Newton's theological papers, in fact, reveal a God so wholly "other" from His creation that some such agent is to be expected whenever the Deity acts in the world:

> That God the ffather is an infinite, eternal, omniscient, immortal, & invisible spirit whom no eye hath seen nor can see . . . & God does nothing by himself wch he can do by another.[50]

It seemed good to Newton that God should use a "creature" to work His will, for that enhanced the divine power.

> If any think it possible that God may produce some intellectual creature so perfect that he could, by divine accord, in turn produce creatures of a lower order, this so far from detracting from the divine power enhances it; for that power which can bring forth creatures not only directly but through the mediation of other creatures is exceedingly, not to say infinitely, greater.[51]

It has been argued that Newton emphatically rejected intermediary agents between God and the world,[52] but both the alchemical and the theological papers seem to invalidate that conclusion. In the theological papers Newton wrestled for several decades with the precise definition

50. Isaac Newton, Jewish National and University Library, Jerusalem, Yahuda MS Var. 1, Newton MS 15.4, f. 67r. I have omitted all indications of Newton's deletions and interlineations in this quotation. Cf. also Westfall, *Never at Rest* (1, n. 4), pp. 315–18.
51. Isaac Newton, Cambridge University Library, Portsmouth Collection MS Add. 4003, in Isaac Newton, *Unpublished Scientific Papers of Isaac Newton: A Selection from the Portsmouth Collection in the University Library, Cambridge. Chosen, edited, and translated by A. Rupert Hall and Marie Boas Hall* (Cambridge University Press, 1962), pp. 108, 142.
52. J. E. McGuire, "Neoplatonism and active principles: Newton and the *Corpus Hermeticum*," in Robert S. Westman and J. E. McGuire, *Hermeticism and the Scientific Revolution: Papers Read at a Clark Library Seminar, March 9, 1974* (Los Angeles: Clark Memorial Library, University of California, 1977), pp. 93–142. To support his point of view McGuire cites (p. 107) a portion of the preceding quotation from Portsmouth Collection MS Add. 4003 but omits Newton's clear statement that an intermediary "so far from detracting from the divine power enhances it. . . . " Cf. also Guerlac, "Theological voluntarism" (2, n. 2), where the earlier literature on this continuing debate is reviewed.

of one such agent,[53] though he did reject some types of intermediaries and quite explicitly refused to view the Deity through Neoplatonic lenses as the soul of the world.[54] But in the alchemical papers the alchemical spirit, the universal vital activator, was in some sense acting for God in the world in Newton's estimation. It was an intermediate agent in God's continuing governance of the world.

Newton had no theological qualms regarding an active universal agent that acted for God in the world to produce and guide the processes of vegetation; on the contrary he welcomed the existence of this spirit as certifiable evidence of Providence. That was what alchemy was all about. In the words of an anonymous alchemist Newton studied with care early in the eighteenth century, the alchemical spirit was God's "Vicegerent," and it was through alchemy that man could "*pierce through the external shell of things, to the internal working Spirit... and so become an Opener and Manifester of the Wonders of God in Nature.*"[55] The bond between vegetable chemistry and divine activity was unquestionable, and Newton understood alchemy to be one of the most, if not the most, important of his many studies. If all went well, he could demonstrate God's action in the world in an absolutely irrefutable fashion by demonstrating the operations of the nonmechanical vegetable spirit, and thus lay the specter of atheism to rest forever more.

Illumination

In his treatise "Of Natures obvious laws & processes in vegetation," Newton used the term "vegetation" in quite a general sense, applying it to all three kingdoms of nature – mineral and animal as well as vegetable. He said, furthermore, that his "vegetable spirit is radically the same in

53. This issue will be discussed in more detail in Chapter 3.
54. Newton, *Principia* (Motte–Cajori) (2, n. 13), vol. II, 544: "This Being governs all things, not as the soul of the world, but as Lord over all...."; Newton, *Principia* (Koyré–Cohen) (1, n. 9), vol. II, 760.
55. Cleidophorus Mystagogus, *Mercury's Caducean Rod: Or, The great and wonderful Office of the Universal Mercury, or God's Vicegerent, Displayed. Wherein is Shewn His Nativity, Life, Death, Renovation and Exaltation to an Immutable State; Being A true Description of the Mysterious Medicine of the Ancient Philosophers* (London: printed by W. Pearson, and sold by T. Northcott, in George-Ally in Lombard-street, 1702), sig. A3v; Harrison, *Library* (1, n. 6) item 1138, p. 198, now Trinity College, Cambridge NQ. 16. 131[1].

all things,"[56] and he was greatly interested in working out the similarities and dissimilarities of its actions in the three kingdoms. Metals are the only part of the mineral kingdom that vegetate in his opinion, other mineral substances being formed mechanically.[57] Vegetation in metals was thus the simplest case for study, the vegetation of the animals and vegetables in the other kingdoms being obviously more complex. So in the vegetation in metals lay the most accessible key to the problem of nonmechanical action.

At first Newton suggested that the vegetable spirit was none other than the aether that pervades all things, and that congealed aether, interwoven with the grosser texture of sensible matter, constituted the "material soul of all matter, which, ... if incited by a gentle heat, actuates and enlivens it...."[58] This is the same material aether, so similar to that of the Stoics, that he had said was daily inspired by the earth–animal and which also accounted in this treatise for the action of gravity.[59] But then Newton decided that the aether was more probably only the vehicle for "some more active spirit" entangled in it, and suddenly he saw what that more active spirit might be: perhaps it was "the body of light." In a rapture of insight he set out parallels and relationships between the vegetable spirit and light.

> This spirit perhaps is the body of light because both have a prodigious active principle, both are perpetual workers. 2 Because all things may be made to emit light by heat. 3 The same cause (heat) banishes also the vital principle. 4 It is suitable with infinite wisdom not to multiply causes without necessity. 5 No heat is so pleasant and bright as the sun's. 6 Light and heat have a mutual dependence on each other, and no generation without heat. Heat is a necessary condition to light and vegetation. [Heat excites light and light excites heat; heat excites the vegetable principle and that increases heat.] No substance so indifferently, subtly, and swiftly pervades all things as light, and no spirit searches bodies so subtly, peircingly, and quickly as the vegetable spirit.[60]

Newton never fully abandoned the moment of insight he captured in

56. Dibner Collection MSS 1031 B (1, n. 30), f. 6r. Cf. Appendix A. See also Keynes MS 12A, f. 1v (2, n. 22), for Newton's earlier expression of the same idea.
57. Dibner Collection MSS 1031 B (1, n. 30), f. 3r. Cf. Appendix A.
58. Ibid., f. 3v.
59. This interconnection between the aethers for gravitation and vegetation will be considered in detail in Chapter 4.
60. Dibner Collection MSS 1031 B (1, n. 30), f. 4r (the parentheses and brackets are Newton's). Cf. Appendix A.

that remarkable passage. He continued to think that light at least contributed to the activity of matter, that it could enter into the composition of bodies and in turn be emitted by bodies. "Are not gross Bodies and Light convertible into one another," he asked in the *Opticks*, "and may not Bodies receive much of their Activity from the Particles of Light which enter their Composition?"[61] How much richer the meaning of that familiar passage now that one begins to grasp the alchemical context in which it was first formulated. But in the *Opticks* Newton said "much of their Activity," not all of it, whereas in the treatise on vegetation he had been quite clear that, if the vegetable spirit were separated from the mass of the body, "there would remain but a dead and inactive earth." Perhaps, he concluded finally, something more was involved in vegetation.[62]

That the vegetable spirit might be identified with light is implied by a suggestive term used by the alchemists – illlumination. The anonymous Eirenaeus Philalethes, for example, one of Newton's early favorites, reported that illumination was crucial at two places in the work. The first illumination produced an "acuated" mercury to begin the work, the second completed the whole process: "at the last, by the will of God, a light shall be sent upon thy Matter, which thou canst not imagine; then expect a sudden end. . . ."[63] Illumination was a process of activation of "acuation," as Philalethes called it. Newton only explored the connotations of the word itself when he considered that the activating principle might be "the body of light." But it would be an oversimplification to suppose that the alchemists had ever intended such a narrow interpretation of their concept, and Newton himself did not hold to his limited identification indefinitely.

The alchemical concept of illumination was really quite broad. Activation might be effected by light, and illumination was sometimes associated with the light of Genesis. The light was God's creature, and to

61. Newton, *Opticks* (2, n. 15), p. 374; Ernan McMullin, *Newton on Matter and Activity* (Notre Dame: University of Notre Dame Press, 1978), pp. 84–94.
62. These issues will be considered in more detail in Chapter 7.
63. Eirenaeus Philalethes, *Secrets Reveal'd: or, An Open Entrance to the Shut-Palace of the King: Containing The greatest Treasure in Chymistry. Never yet so plainly Discovered. Composed By a most famous English-man, Styling himself Anonymous, or Eyraeneus Philaletha Cosmopolita: Who, by Inspiration and Reading, attained to the Philosophers Stone at his Age of Twenty three Years, Anno Domini, 1645. Published for the Benefit of all English-men, by* W. C. *Esq.; a true Lover of Art and Nature* (London: printed by W. Godbid for William Cooper in Little St. Bartholomews, near Little-Britain, 1669), pp. 73, 108.

many alchemists it seemed an obvious candidate for the active agent God used in the work of creation. Tract after tract on the alchemical process used illumination to explain God's actions with respect to matter at the beginning of time. But the active principle could also be, and was, identified with the spirit of God that moved upon the face of the waters in Genesis. Activation might be effected by an agent totally imperceptible to the senses. There were perhaps hundreds of alchemical names for it, for it was one of the great secrets of alchemy, disguised willy-nilly in the texts to keep the knowledge of it from the noninitiate. In Newton's very first paper on the regimens of the work, written before 1678, he opened with the statement that for the work in common gold the mercury must first be illuminated.[64] In later regimen papers there is always a process of activation though Newton did not always call it illumination.[65] By the time he compiled his last treatise on the subject he was able to discuss nearly thirty names for the activating agent and devoted a full chapter to it, "De agento primo." It is called, he said, Mercury's caducean rod, a saline spirit, the salt of nature, the waterbearer, the winged dragon, a water, a moist fire, our Cupid – to give only a few examples.[66] But however it is named, it is the first agent, that without which nothing is done, for it activates or illuminates the matter.

Illumination, then, may be said to have had a symbolic, methaphorical significance in alchemy. Light represented the power of God to activate or reactivate lifeless matter. The alchemical usage was closely akin to the inconographic tradition of representing divine power by light rays. In many Annunciations, for example, the Christ–child entered Mary's womb on a beam of light (Plate 1).[67] In symbolism made especially appropriate by the beauty of sunlight streaming through the brilliant

64. Isaac Newton, "The regimens described wth ye times & signes," Sotheby lot no. 87 (part); The Francis A. Countway Library of Medicine, Boston Medical Library/Harvard Medical Library, Boston, MA, Newton MS, Part A, f. 1r.
65. Cf. Dobbs, "Newton's copy" (1, n. 5).
66. Babson MS 420 (1, n. 39), pp. 10–12. Cf. Appendix E.
67. Annunciation, central panel from Triptych of the Annunciation by the Flemish Robert Campin (active by 1406 – d. 1444), New York, Metropolitan Museum of Art, The Cloisters Collection. See also James Hall, *Dictionary of Subjects and Symbols in Art*, intro. by Kenneth Clark (rev. ed.; Icon Editions; New York: Harper & Row, 1979), s.v. "Annunciation," pp. 18–20; *The International Style. The Arts in Europe around 1400. October 23–December 2, 1962. The Walters Art Gallery, Baltimore* (Baltimore: The Walters Art Gallery, 1962), The Annunciation by the Florentine Bicci di Lorenzo (1373–1452), pp. 2–4 and Plate XXIX.

Plate 1. The Annunciation, central panel from Triptych of the Annunciation by the Flemish Robert Campin (active by 1406–d. 1444), Metropolitan Museum of Art, The Cloisters Collection, 1956. (56.70) Reproduced by permission of the Metropolitan Museum of Art, New York, NY.

colors of the glazed windows of medieval churches – windows penetrated but not injured by the solar rays – the Virgin came to be regarded as the undamaged window through which the spirit of God passed to earth. Earlier symbols for God's creative power – dew, rain, vegetable florescence – generated by a more strictly agricultural culture – were partly replaced in the Middle Ages by the image of light through glass, a symbol dependent upon technology yet bearing still the full force of archaic belief.[68] Similarly, divine grace poured into the prophet, bearing the in-

68. Millard Meiss, "Light as form and symbol in some fifteenth-century

spired words he was to utter (Plate 2).[69] The same iconographic tradition holds in an alchemical emblem Newton studied about 1689, where God's power enters the alchemical flask, itself surmounted by symbols of resurrection, by means of the rays of the sun and the moon (Plate 3).[70] The "sublimed stone... receives efficacy," Newton wrote, "from ye Sun & Moon wch are its father & mother whence it inherits speedily ye first crown of perfection."[71]

As it was in alchemy with regard to the activation of mineral matter by God's power, so it was also in questions of generation in the other kingdoms of nature. Spontaneous generation had been an accepted fact for time out-of-mind, sanctioned by Aristotle and vouched for by a host of empirical observations, not only in alchemy but in the world at large.[72]

paintings," in *Renaissance Art*, ed. Creighton Gilbert (reprint of 1970 ed.; New York: Harper & Row, 1974), pp. 43–68. On earlier symbolism see also Edmund Reiss, *The Art of the Middle English Lyric. Essays in Criticism* (Athens, GA: University of Georgia Press, 1972), pp. 158–64, where God's power fell upon Mary "As dew in Aprylle/That fallyt on the gras." I am grateful to Hugh Ormsby-Lennon for these references.

69. Frontispiece for *Tractatus primi, Sectionis II, Portio I: De mentis humanae scientia, hoc est, de vaticinio, & Prophetis seu hominibus spiritu divino afflatis, in libros tres distributa*, in Robert Fludd, *Utriusque Cosmi Maioris scilicet et Minoris Metaphysica, Physica atqve Technica Historia In duo Volumina secundum Cosmi differentiam diuisa. Avthore Roberto Flud aliàs de Fluctibus, Armigero, & in Medicina Doctore Oxoniensi* (2 vols. in 3; Oppenhemii: AEre Johan-Theodori de Bry Typis Hieronymi Galleri, 1617–21), vol. III, 3 (of the separately paginated *Tomi secvndi, Tractatus primi, Sectio secunda, De technica Microcosmi historia, in Portiones VII. divisa*). In Fludd's representation the divine presence is triply assured by the beam of light, the anointing with holy oil, and the descent of the spirit-bearing dove that recalls Jesus's baptism (Matthew 3:16).
70. Frontispiece, [A. T. Limojon (Sieur de St. Didier)], *Le triomphe hermetique, Ou La Pierre Philosophale victorieuse. Traité Plus complet & plus intelligible, qu'il y en ait eu jusques ici, touchant le Magistère hermetique* (Amsterdam: Chez Henry Wetstein, 1699). Newton's copy of the 1689 edition of this book still exists: Harrison, *Library* (1, n. 6), item 1642, p. 252, now Trinity College, Cambridge, NQ. 16. 123. The frontispiece is, however, missing from that copy, and I am indebted to the Rare Books Department, Memorial Library, University of Wisconsin-Madison for this illustration. See also, Dobbs, *Alchemical Death* (1, n. 30), where this alchemical emblem is analyzed in greater detail.
71. Isaac Newton, "Commentary on Didier's 'Six Keys,' " Sotheby lot no. 17; King's College, Cambridge, Keynes MS 21, f. 6r.
72. John Farley, *The Spontaneous Generation Controversy from Descartes to*

Plate 2. Frontispiece for Tractatus primi, Sectionis II, Portio I of Robert Fludd, *Utriusque Cosmi Maioris scilicet et Minoris* (2 vols. in 3; Openhemii: AEre Johan-Theodori de Bry Typis Hieronymi Galleri, 1617–21). Reproduced by permission of The Henry E. Huntington Library, San Marino, CA.

Yet even though it was an almost unquestioned natural event, spontaneous generation was still thought to need a precipitating trigger – the warmth of the sun or an act of God. In addition to Aristotle and observation, there was the authority of Genesis. God had said, "Let the earth bring forth," and it was so.[73] What God had done in the beginning, He could still do. The most orthodox interpretation of the generation of eels from mud or of maggots from putrefying flesh was

Oparin (Baltimore: The Johns Hopkins University Press, n.d. [1974]), pp. 1–15; Laurinda S. Dixon, *Alchemical Imagery in Bosch's Garden of Delights* (Studies in the Fine Arts: Iconography, 2; Linda Seidel, Series Ed.; Ann Arbor, MI: UMI Research Press, 1981), pp. 21–3, 38.

73. Genesis 1: 24.

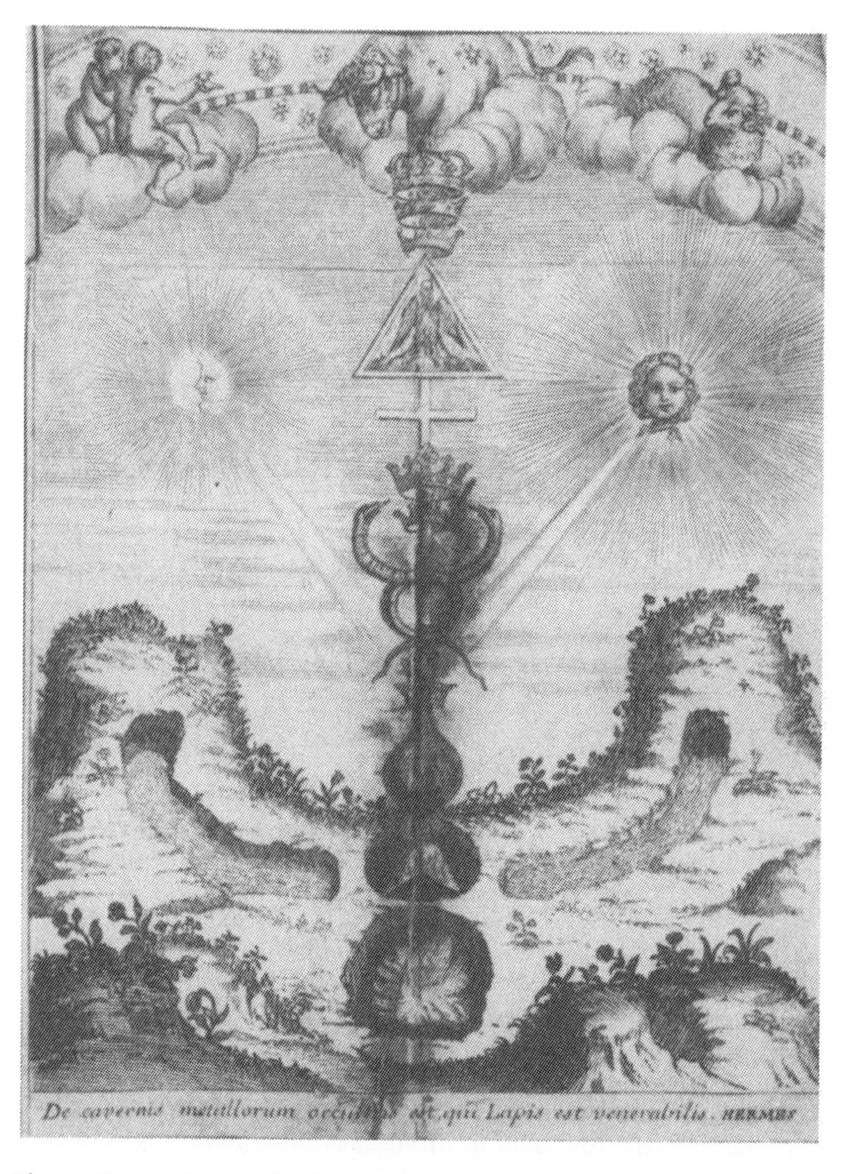

Plate 3. Frontispiece, [A.T. Limojon (Sieur de St. Didier)], *Le triomphe hermetique* (Amsterdam: Chez Henry Wetstein, 1699). Reproduced by permission of the Rare Books Department, Memorial Library, University of Wisconsin-Madison.

that God's creative power brought them forth, that the spirit of God had activated lifeless matter. In sexual generation the authority of William Harvey stood foursquare behind the same concept. It is not the cock's semen per se that conveys fertility to the hen but rather "the spirit and virtue of a divine agent" with which the semen is imbued, Harvey said.[74] The spirit and virtue of a divine agent, without which the work of generation cannot proceed, without which there will be no new life – how similar Harvey's statement is to much of what one finds in alchemy on the activating, vegetative principle without which the life of metals is lost. As Newton had observed, the vital agent was the same in all three kingdoms.

One cannot stress the vital nature of the process of activation too strongly, and one must also insist that the activating agent in illumination was broadly conceived and was related to visible light rather more metaphorically than literally. As the vegetative spirit, as the agent of fertility, however, it was associated especially with the light received by the earth during the season of spring. The presence of the alchemical spirit was sometimes even symbolized by the spring zodiacal signs, as indeed it is in Plate 3. Upon their presence in the emblem he was studying, Newton commented:

> And above all is a part of ye Zodiaque wth ye 3 signes of Aries taurus & Gemini, . . . to denote ye spring time of ye year wherein ye Philosophical nuptials are celebrated. W^{ch} thing is also signified in Cosmopolites AEnigma by rams & bulls kept in a pasture by two young men.[75]

One may explain Newton's remarks by quoting from Virgil's *Georgics*, as William Harvey did, a passage that would have been familiar to every educated person in the seventeenth century.

> Earth teems in Spring, and craves the genial seed.
> The almighty father, AEther, then descends,
> In fertilizing showers, into the lap
> Of his rejoicing spouse, and mingling there
> In wide embrace sustains the progeny
> Innumerous that springs. The pathless woods

74. William Harvey, *Anatomical Exercises on the Generation of Animals*, in *Great Books of the Western World*, Robert Maynard Hutchins, Editor-in-Chief (54 vols.; Chicago: William Benton for Encyclopedia Britannica, 1952), Vol. XXVIII, 404.
75. Keynes MS 21 (2, n. 71), f. 6r. I have omitted all indications of Newton's deletions and interlineations in this quotation.

Then ring with the wild bird's song, and flocks and herds
Disport and spend the livelong day in love.[76]

Fermentation

Once the matter was illuminated, it began to ferment. If illumination was the process of activation, then fermentation was the activity produced. As inert matter began to move and change its form, that is, to ferment, the activity itself became a sign of life, a signal that the matter had been activated and enlivened by the mysterious undetectable vital agent. "There is plainly a life and a ferment in that composition," Nicolas Fatio de Duillier wrote to Newton in 1693, describing the activity of a mineral preparation.[77] And in 1672 Newton explained how either an excess or defect of heat might cause the vital agent "to cease acting forever . . . unless it receive new life from a fresh ferment."[78] Life and fermentation were indissolubly linked in alchemy.

The term fermentation of course included much more than its alchemical meaning. When the leavening of bread, the brewing of beer, and the fermenting of wine were everyday experiences for most people, the term fermentation covered any type of process in which a substance changed its properties due to an internal "working." In the seventeenth century the word was frequently used in context with putrefaction, generation, and vegetation, as well as to refer to processes started by the addition of yeast, barm, or leaven. Originally from the Latin *fervere*, to boil, fermentation could also apply to any effervescent process.[79] In 1667 Newton had given it a general definition as the "working of liquors, wherby they are further digested & seperated from their faeces &c."[80] And in 1672 he said air could be generated "by any means where the parts of a body are set aworking among themselves." One means of so

76. Virgil, *Georgics*, Book II, 323–9, as quoted in Harvey, *Anatomical Exercises on the Generation of Animals* (2, n. 74), Vol. XXVIII, 346.
77. Newton, *Correspondence* (2, n. 14), vol. III, 265–7 (Fatio to Newton, 4 May 1693).
78. Dibner Collection MSS 1031 B (1, n. 30), f. 3v. Cf. Appendix A.
79. Joseph S. Fruton, "From ferments to enzymes," in Joseph S. Fruton, *Molecules and Life, Historical Essays on the Interplay of Chemistry and Biology* (New York: Wiley-Interscience, 1972), pp. 22–86, esp. pp. 23–42; *The Compact Edition of the Oxford English Dictionary. Complete Text Reproduced Micrographically* (2 vols.; Oxford University Press, 1971), s.v. "fermentation."
80. MS Don. b. 15 (1, n. 28), f. 7r.

doing was through fermentation: "Hence ebullition, flying of bottle beer, etc., swelling after a stroke."[81]

Though one cannot always demonstrate an intentional vitalism in the general usage of the word, there can be little doubt that the aura of life hung over it. In a tradition that perhaps went back to Anaximander, life had come from a ferment in moist soil warmed by the sun;[82] in Aristotle the heart – the seat of life itself – was brisk with fermentation, alternately swelling and collapsing in its pulsation like a boiling or fermenting liquid.[83] Van Helmont had given the concept pride of place in his reformed physiology where no fewer than six different fermentations were necessary to maintain life; some of his vitalistic ideas were adapted to circulation physiology by followers of Harvey in England.[84]

Yet fermentation had early been adopted as a central concept also by first-generation mechanical philosophers, and therein lies considerable ambiguity. Dropping out any hint of celestial activation or guidance for the process, Digby made it operative nonetheless in the generation of new life in plants. When the moisture of the earth presses upon the seed, Digby said, the hot parts of the seed, imprisoned in cold and dry ones, are stirred up, and mingling with the moisture they ferment and distend.[85]

81. Dibner Collection MSS 1031 B (1, n. 30), f. 3v. Cf. Appendix A.
82. Charles H. Kahn, *Anaximander and the Origins of Greek Cosmology* (New York: Columbia University Press, 1960), pp. 109–13. See also the comparable discussions of motion or fermentation, heat, and moisture for other ancient thinkers in Regnéll, *Ancient Views* (2, n. 1), pp. 33–6, 76–7, and in W. K. C. Guthrie, *In the Beginning. Some Greek Views on the Origins of Life and the Early State of Man* (Ithaca, NY: Cornell University Press, 1957), pp. 29–45.
83. Aristotle, *De partibus animalium*, III. 4. 666a10–667b14, in Aristotle, *Works* (2, n. 24), vol. 5; idem, *De respiratione*, in *Parva naturalia*, Chap. XXVI. 479b17–480a16, in Aristotle, *Works* (2, n. 24), vol. 3.
84. Pagel, *Van Helmont* (1, n. 25); Audrey B. Davis, *Circulation Physiology and Medical Chemistry in England 1650–1680* (Lawrence, KA: Coronado Press, 1973); Robert G. Frank, Jr., *Harvey and the Oxford Physiologists. Scientific Ideas and Social Interaction* (Berkeley: University of California Press, 1980).
85. Kenelm Digby, *Of Bodies, and of Mans Soul. To Discover the Immortality of Reasonable Sovls. With two Discourses Of the Powder of Sympathy, and Of the Vegetation of Plants* (London: printed by S. G. and B. G. for John Williams, and are to be sold in Little Britain over against St. Buttolphs-Church, 1669), pp. 290–1 of the separately paginated *First Treatise: Declaring the Nature and Operation of Bodies*. Newton owned a copy of Digby's *Two Treatises* (those on bodies and on souls) in the London ed. of 1658: Harrison, *Library* (1, n. 6), item 516, p. 133, now missing.

Like Digby, Descartes wished to discuss the processes of life in physical rather than psychic terms, that is, to eliminate souls, spirits, and divine agents. In Descartes's opinion life could be explained by the corporeal structure of the organism and the motion of its corporeal particles, the particles having an internalized principle of motion – which was, not surprisingly, fermentation. In animal generation the male and female liquors served each as a ferment to the other. The fermental action excited heat; the heat caused particles to gather differentially and begin to form the heart, etc. Descartes's analogies for these fermental actions are the rising of dough, the brewing of beer, the effervescing of wines, and the composting of moist hay.[86]

At the very center of Descartes's mechanical system of generation there is thus the ambiguous concept of fermentation. Fermentation is the necessary internalized principle of motion for the particles, and the process of fermentation produces the heat that serves Descartes as an agent of embryonic development. For Digby, heat had been prior, acting with moisture to produce fermentation. That inversion, however, is less significant than the fact that for both men the principles of action in the generative process were the same: heat, moisture, and fermentation. But had those not always been the principles of generation? The concept of fermentation carried with it so many nonmechanical associations that one may reasonably question whether either man was being as fully mechanistic as he thought.[87] The early mechanical philosophers often restated common cultural assumptions in terms of corpuscularian mechanisms that disguised but by no means eliminated their vitalistic components.[88]

86. Thomas S. Hall, *Ideas of Life and Matter, Studies in the History of General Physiology 600* B.C.–*1900* A.D. (2 vols.; Chicago: The University of Chicago Press, 1969), vol. I, 250–63; idem, "Descartes' physiological method: position, principles, examples," *Journal of the History of Biology 3* (1970), 53–79.
87. This issue has previously been raised for Descartes: Ann Wilbur Mackenzie, "A word about Descartes' mechanistic conception of life," *Journal of the History of Biology 8* (1975), 1–13.
88. Richard S. Westfall, *The Construction of Modern Science. Mechanisms and Mechanics* (reprint of the 1971 ed.; Cambridge University Press, 1977), pp. 82–104; idem, "Newton and the Hermetic tradition," in *Science, Medicine, and Society in the Renaissance. A Festschrift in Honor of Walter Pagel*, ed. by Allen G. Debus (2 vols.; New York: Neale Watson Academic Publications, 1972), vol. II, 183–98. For an excellent, wide-ranging discussion of contemporary cultural context that includes poets, alchemists, and several early Fellows of the Royal Society, all of whom were still

In van Helmont the vitalism was of course not disguised at all. Van Helmont's reform of physiology appeared just as the first generation of mechanists was struggling to reduce all to matter and motion, but for their program van Helmont could scarcely have cared less. Ferments and fermentation occupy central positions in his natural philosophy, but in this he is conceptually affiliated with Paracelsus, the alchemists, and Renaissance Neoplatonists, rather than with the early mechanists. He had nonetheless one problem in common with the mechanical philosophers, the problem of the origin of forms and qualities from a common matter. For although van Helmont eschewed the one catholic matter of the corpuscularians, a matter with quantitative properties only, he did believe in a common material base for all objects. Taking his cue from Genesis, where water was the first matter to exist, he became convinced that water was the general undifferentiated substratum for everything. Water itself is "empty" but "all bodies are the fruit of water." Ferments dispose the "empty" matter to receive specific forms of individuated bodies. As the spirit of the Lord hovered on the waters in Genesis, so the ferments carry to the water an "odor" and image of God's plan of life for every specific creature. In van Helmont's natural philosophy, fermentation thus serves as the principle of differentiation for all objects that come from the common matter. The ferment originates in a divine idea and, as it operates upon the "empty" matter, the ferment itself is internalized and becomes the archeus, the internal governing principle of the created being that insures the working out of God's plan for its existence.[89]

It is possible that Newton was influenced by these Helmontian concepts, for his use of fermentation as a principle of specification bears a certain resemblance to van Helmont's, as one may see below. Newton owned van Helmont's *Ortus medicinae* in an edition that included almost all his works,[90] and one of Newton's manuscripts contains notes on van Helmont's work.[91]

drawing upon very ancient ideas, see also Elizabeth Mackenzie, "The growth of plants. A seventeenth-century metaphor," in *English Renaissance Studies Presented to Dame Helen Gardner in Honour of her Seventieth Birthday* (Oxford: Clarendon Press, 1980), pp. 194–211.

89. Pagel, *Van Helmont* (1, n. 25), esp. pp. 60–81.
90. Jan Baptista van Helmont, *Ortus medicinae, id est initia physicae inaudita ... edente F. M. van Helmont.... Ed.* 4ª (Lugduni, 1667); Harrison, *Library* (1, n. 6), item 751, p. 158, now missing. Cf. Pagel, *Van Helmont* (1, n. 25), pp. 209–14.
91. Sotheby lot no. 9; King's College, Cambridge, Keynes MS 16. According

It is also possible that Newton was influenced by the restatement of Helmontian ideas on fermentation by Willis and Leibniz. Newton did not own Willis's *De fermentatione* in which fermentation was given its classical chemico-corpuscular explication: "an intestine motion of Particles or, the Principles of every Body, either tending to the Perfection of the same body or because of its change into another."[92] But Willis's definition quickly became the standard one in dictionaries and encyclopedias and Newton would almost certainly have encountered it somewhere.[93] On the other hand, Newton did have a copy of Leibniz's *Hypothesis physica nova* of 1671 in which a divine aether penetrates and activates matter through fermentative processes.[94]

Discussion of fermentation as the paradigmatic natural active process was so widespread in the seventeenth century, however, that the sources of Newton's ideas can probably not now be isolated. He might have encountered fermentation in what was at least superficially mechanical thought; he might have encountered it in Helmontian vitalism. He might

to a note from P. M. Rattansi preserved with the manuscript, Keynes MS 16 consists of notes on van Helmont's *Causae et initia naturalium*, from the 1667 edition of *Ortus Medicinae* (2, n. 90).

92. Thomas Willis, *Diatribae Duae Medico-Philosophicae* (London, 1659) and subsequent editions. Of Willis's works, Newton had only the *Pathologiae cerebri* in the 1668 Amsterdam ed.: Harrison, *Library* (1, n. 6), item 1741, p. 262, now missing. The definition is quoted from Thomas Willis, *Dr Willis's Practice of Physick, Being all the Medical Works of that Renowned and Famous Physician: Containing These Ten several Treatises, viz. I. Of Fermentation. II. Of Feavours. III. Of Urines. IV. Of the Accension of the Bloud. V. Of Musculary Motion. VI. Of the Anatomy of the Brain. VII. Of the Description and Use of the Nerves. VIII. Of Convulsive Diseases. IX. Pharmaceutice Rationalis the 1st and 2^d Part. X. Of the Scurvy. Wherein most of the Diseases belonging to the Body of Man are treated of, with excellent methods and Receipts for the Cure of the same. Fitted to the meanest Capacity by an Index for the Explaining of all the hard and unusual Words and Terms of Art, derived from the Greek, Latine, or other Languages, for the benefit of the English Reader, with a large Alphabetical Table to the whole. With Thirty two Copper Plates. Done into English by S. P. Esq*; (London: printed for T. Dring, C. Harper, and J. Leigh, and are to be sold at the Corner of Chancery-lane, and the *Flower-de-Luce* over against St Dunstans Church in Fleet-street, 1681), p. 9.

93. Davis, *Circulation Physiology* (2, n. 84), pp. 82–4 et passim; Pagel, *Van Helmont* (1, n. 25), pp. 83–5; Hansruedi Isler, *Thomas Willis 1621–1675. Doctor and Scientist* (New York: Hafner, 1968), esp. pp. 45–68.

94. Pagel, *Van Helmont* (1, n. 25), pp. 85–6; [G. W. von Leibniz], *Hypothesis physica nova . . . Autore G. G. L. L.* (Londini, 1671): Harrison, *Library* (1, n. 6), item 826, p. 166, now missing.

have encountered it also in the same places van Helmont had found it – in the literature of alchemy. In John Webster's *Metallographia* of 1671 he almost certainly encountered it in a scholarly discussion on the vegetability of metals, where Webster allowed Edward Jorden to summarize the views of the alchemists.

> 'There is a seminary spirit of all Minerals in the bowels of the Earth, which meeting with convenient matter, and adjuvant causes, is not idle, but doth proceed to produce Minerals, according to the nature of it, and the matter which it meets withal: which matter it works upon like a ferment, and by its motion procures an actual heat, as an instrument to further its work, which actual heat is increased by the fermentation of the matter. The like we see in making of Malt, where the grains of Barley being moistned with water, the generative spirit in them is dilated, and put in action. . . .'[95]

Newton's copy of Webster's book[96] is heavily dog-eared, and this work may well have been the immediate stimulus behind "Of Natures obvious laws & processes in vegetation," where Newton's language often seems to echo Webster's. Regarding the concept of fermentation in general, however, Newton clearly might have been stimulated from every side.

What Newton took from the surrounding milieu he bent to his own ends. As a corpuscularian he found Willis's "intestine motion of Particles or Principles" attractive, and his subunits of particles "watery, earthy, saline, airy, oily, spirituous, etc.," are quite similar to Willis's chemical principles. Yet for Newton those particulate "grosser substances" move only at the urging of "the invisible inhabitant" to become bones or flesh or wood or fruit. In Newton's hierarchical system of matter the ultimate

95. John Webster, *Metallographia: or, An History of Metals, Wherein is declared the signs of Ores and Minerals both before and after digging, their kinds, sorts, and differences; with the description of sundry new Metals, or Semi Metals, and many other things pertaining to Mineral knowledge. As also, The Handling and shewing of their Vegetability, and the discussion of the most difficult Questions belonging to Mystical Chymistry, as of the Philosophers Gold, their Mercury, the Liquor Alkahest, Aurum potabile, and such like. Gathered forth of the most approved Authors that have written in Greek, Latine, or High-Dutch; With some Observations and Discoveries of the Author himself* (London: printed by A. C. for Walter Kettilby at the Bishopshead in St. Pauls Churchyard, 1671), p. 66. On Jorden, see Allen G. Debus, "Edward Jorden and the fermentation of the metals: an iatrochemical study of terrestrial phenomena," in *Toward a History of Geology*, ed. by C. E. Schneer (Cambridge, MA: Harvard University Press, 1969), pp. 100–21.
96. Harrison, *Library* (1, n. 6), item 1718, p. 260, now Trinity College, Cambridge NQ. 16. 150.

particles are structured into the chemical subunits, but the chemical subunits themselves are still common to all substances. It is only as "the latent vegetable substances" guide the penultimate particles into various associations and concatenations that true differentiation occurs and "specificateness" is achieved. The vegetable principle for Newton is thus not unlike the odor of the ferment in van Helmont's thought, for it is the principle that produces variety from a common matter. Though Newton allowed for change to occur in his system by the mechanical interaction of corpuscles, mechanical change was limited by its very nature and could not serve to organize specific creatures that followed a plan of growth until they reached maturity. Only vegetable changes would do that.

> All vegetables have a disposition to act upon other adventitious substances and alter them to their own temper and nature. And this is to grow in bulk, as the alternation of the nourishment may be called growth in virtue and maturity or specificateness.[97]

The spirit that guided the intestine motion of the particles in such a fermenting, vegetating mass revealed nature acting in her nonmechanical mode.

So Newton thought in the early 1670s on the problems associated with vegetability and providence. In the 1680s his distinction between vegetable and mechanical action in chemistry was amplified and transmuted into a broader distinction between active and passive forces in general, the active–passive distinction serving exactly the same theological function as that between vegetable and mechanical chemistry.

97. Dibner Collection MSS 1031 B (1, n. 30), f. 5v. Cf. Appendix A.

3

Cosmogony and history

Cosmogony as alchemy

Any identification of the vegetative principle with light, such as that which Newton made in his 1672 treatise on vegetation, had in turn some rather obvious implications for any cosmogonist reared on the creation account in Genesis. Furthermore, Newton must have become aware of potential alchemical interpretations of Genesis quite early in his alchemical reading, for Eirenaeus Philalethes, whom Newton began to study in the late 1660s, explicitly compared the alchemical work to creation, quoting the critical passages from Genesis.

> In the beginning God created the heavens and the earth. And the earth was waste and void; and darkness was upon the face of the deep: and the Spirit of God moved upon the face of the waters. And God said, Let there be light: and there was light.[1]

Philalethes said the "Sophi" agree unanimously on the significance of alchemy for an understanding of creation. He was probably correct in his assessment, or nearly so, for recent scholarship has defined a widespread chemical philosophy in the sixteenth and seventeenth centuries in which creation was viewed as a chemical or alchemical separation.[2] Newton came to hold a quite similar view in the early 1680s if not before, and ended by comparing the illumination of matter in the alchemical process with God's use of light at the beginning of the world.

Newton's work on that aspect of alchemy was perhaps stimulated by Thomas Burnet late in 1680. Burnet had completed a manuscript of his *Telluris theoria sacra* and had sent it to Newton for comment.[3] There

1. Eirenaeus Philalethes, *Secrets Reveal'd* (2, n. 63), p. 9; Genesis 1: 1–3.
2. Allen G. Debus, *The Chemical Philosophy. Paracelsian Science and Medicine in the Sixteenth and Seventeenth Centuries* (2 vols.; New York: Science History Publications, 1977); Dixon, *Alchemical Imagery* (2, n. 72), pp. 65–8 and Fig. 167.
3. Burnet published the first part of this work the next year as Thomas Burnet,

followed an exchange of letters, in one of which Newton offered Burnet several suggestions concerning the "generation of hills" and other irregularities out of the originally uniform chaos of creation.[4] The problem was to describe what had happened in a plausible and "Philosophical" way without conflicting with the Mosaic description. Newton decided that Moses had simply left out some steps so as not to make his account tedious and laughable to the vulgar. Moses had not bothered to describe the separation of the chaos into the several parcels of the sun and planets, for example, but Newton thought that to have been the work of the first day. Moses had dealt with the important "divisions" and "separations" within the earthly chaos, on the other hand, and as Newton reviewed the stages in the generation of our globe, his terminology acquired the overtones of contemporary chemical philosophy. The analogies he offered for the irregular formation of hills bespeak the chemical laboratory: the coagulation of dissolved saltpeter into irregular bars, the congelation of melted tin into nonuniform lumps, and the curdling of milk.

Within a very few years, probably by 1684, Newton carried his application of chemical analogy forward into the deeper waters of alchemical analogy in a *Commentarium* on the *Emerald Tablet* of Hermes Trismegistus. The search by the chemical philosophers for an alchemical explication of the creation of the world derived ultimately from this *Emerald Tablet* of Hermes. Because of its supposed antiquity, and because of the quasi divinity attributed to Hermes himself, the mysterious text had acquired a lustrous patina of authority. In it Hermes concluded his description of the alchemical process with "*Sic mundus creatus est*" – "Thus was the world created."[5] It is accordingly only natural that one finds Newton's amplification of the parallels between creation and alchemy in his *Commentarium* on the *Emerald Tablet.*

At the root of Newton's new venture into analogical reasoning was one of the most fundamental presuppositions of alchemy, that of the undifferentiated first matter from which all things could be derived, the "philosophical chaos." One has already seen Newton's early use of the term "chaos" in Keynes MS 12A, the "Propositions" paper, in which he

Telluris Theoria Sacra: Orbis Nostri Originem & Mutationes Generales, quas Aut jam subiit, aut olim subiturus est, complectens. Libri duo priores De Diluvio & Paradiso (Londini: Typis R. N. Impensis Gualt. Kettilby, ad Insigne Capitis Episcopi in Coemetrio Paulino, 1681).

4. Newton, *Correspondence* (2, n. 14), vol. II, 329–34 (Newton to Burnet, January 1680/1).
5. Cf. Appendix B; see also Jack Lindsay, *The Origins of Alchemy in Graeco-Roman Egypt* (New York: Barnes & Noble, 1970), pp. 157–93, esp. 185–6.

said that the mode of acting of the alchemical agent was first to putrefy and confound into chaos, then to proceed to generation. The chaos is the essential analogical element linking the alchemical work to cosmogony on the one hand and to spontaneous and sexual generation on the other. One may look a little more closely at the concept of the chaos that seemed in the seventeenth century to make all these processes fundamentally similar.

In spontaneous generation the analogy with the "philosophical chaos" of alchemy and with the initially unformed matter of creation is quite clear: decaying flesh and stinking mud are in a state of putrefaction, in a process of degradation from organized forms to an inchoate mass, from which new, unrelated forms may spring. In the 1670s Newton casually assumed that putrefaction precedes spontaneous generation.

> Or if a carcass be put in a glass and kept warm in Balneum Mariae that it may putrefy and breed insects, are not those insects as natural as others bred in a ditch without any artifice?[6]

The analogue of an unformed primal matter in sexual generation is not now quite so easy to grasp, for one now conceives organized life to pass in an unbroken chain from parent to offspring with no intervention of death and decay. But in the seventeenth century there was on the contrary just such a sense of discontinuity as is now lacking. Although there were competing theories as to whether male or female emissions were more important, it was rather generally agreed that after coition there was a mingling of male and female "sperms" (or perhaps male semen and female menstrual blood). This confused mass underwent decay and resolution into a "primordium" – simple and undifferentiated – from which the new individual arose. Harvey was at pains to deny the ancient theories because his anatomical researches revealed none of the required masses.[7] His conclusion – that the vector of life was a divine spirit – nevertheless emphasized a fundamental discontinuity between the generations.

Apparently never having read Harvey, Newton rested in unperturbed assurance with antiquity on the matter of sexual generation. His views, vaguely Galenic, come from his paper on the vegetation of metals, where he compared and contrasted metallic generation with that of animals.

> ...[S]o is an infant generated from the mixture of male and female seed.... Also there ought to be most of the female seed.[8]

6. Dibner Collection MSS 1031 B (1, n. 30), f. 5r. Cf. Appendix A.
7. Harvey's arguments against the ancients are well summarized in Elizabeth B. Gasking, *Investigations into Generation 1651–1828* (Baltimore: The Johns Hopkins University Press, 1967), pp. 19–24.
8. Dibner Collection MSS 1031 B (1, n. 30), f. 1v. Cf. Appendix A.

Furthermore, he was certain that putrefaction was essential to all generation. His statements on putrefaction are so simple and general that they take on the character of laws of nature. Indeed, he said they are "only observable universals."[9]

> Nothing can be changed from what it is without putrefaction. . . .
> No putrefaction can be without alienating the thing putrefied from what it was.
> Nothing can be generated or nourished (but of putrefied matter).
> All putrefied matter is capable of having something generated out of it and in motion toward it.
> All nature's operations are between things of different dispositions. The most powerful agent acts not upon itself.
> Her first action is to blend and confound mixtures into a putrefied chaos.
> Then they are fitted for new generation or nourishment.
> All things are corruptible.
> All things are generable.
> . . . .
> Putrefaction is the reduction of a thing from that maturity and specificateness it had attained by generation.[10]

In its generality, Newton's conviction that putrefaction was an essential step that necessarily came before new generation could begin applied also to life in the vegetable kingdom. For plants, the accepted paradigm was somewhat different, but the belief that decay preceded new life in the fields was widespread, and indeed had biblical justification: "Except a corn of wheat fall into the ground and die, it abideth alone: but if it die, it bringeth forth much fruit."[11] Many alchemists quoted that passage to explicate the necessity for death and putrefaction in the mineral kingdom.[12] In common contemporary understanding of generation in the vegetable kingdom, "Mother Earth" herself played the female role as matrix, the seed the male role of impregnation, and the fructifying breath of springtime served as the divine activator for the new life to come.[13]

In summary, one may clearly see the overall thrust of Newton's dis-

9. Ibid., f. 1r.
10. Ibid., f. 5r, v (Newton's parentheses).
11. John 12: 24.
12. Dobbs, *Foundations* (1, n. 1), pp. 30–1 and the literature cited there. See also Appendix C.
13. See John Farley, *Gametes & Spores. Ideas about Sexual Reproduction 1750–1914* (Baltimore: The Johns Hopkins University Press, 1982), p. 22, for a discussion of that commonly accepted paradigm in a compendium of sexual lore that remained popular into the eighteenth century, *Aristotle's Compleat and Experienc'd Midwife*.

cussion on the three kingdoms of nature in his treatise on vegetation. There are differences as one moves from one kingdom to another, yet the differences are not great and the characteristic process of generation in each kingdom is similar to that process in the others. Putrefaction is essential at an early stage, for it produces the chaos from which new substances are generated or old ones nourished under the guidance of the vegetable spirit.

With respect to cosmogony, the concept of the prerequisite chaos provided a natural focus for analogy to the time when the earth was "without form." In alchemy, metallic putrefaction produced a chaotic, dark, unformed matter – a chaos such as existed "in the beginning." Upon it the spirit of God moved, and, as light had come upon the first creation, the alchemical matter was "illuminated" and endowed with the potentiality of life and growth. In similar fashion inchoate seminal matters were also activated for generation in the vegetable and animal kingdoms. Newton reiterated these analogous changes as he understood them in his *Commentary* on the *Emerald Tablet*, as one may soon see.

The hexaemeral tradition

Newton's speculations in the letter to Burnet on the received Judeo–Christian account of creation were produced in the context of a venerable tradition that was being modulated but reinvigorated by his contemporaries.[14] Creation myths have arisen in every culture, and they have been elaborated or modified or displaced as that culture evolved. One can

14. Frank Egleston Robbins, *The Hexaemeral Literature. A Study of the Greek and Latin Commentaries on Genesis. A Dissertation Submitted to the Faculty of the Graduate School of Arts and Literature in Candidacy for the Degree of Doctor of Philosophy (Department of Greek)* (Chicago: The University of Chicago Press, 1912); Katherine Brownell Collier, *Cosmogonies of Our Fathers. Some Theories of the Seventeenth and Eighteenth Centuries* (Studies in History, Economics and Public Law, ed. by the Faculty of Political Science of Columbia University, No. 402; New York: Columbia University Press; London: P. S. King & Son, 1934; New York: Octagon Books, 1968); Arnold Williams, *The Common Expositor. An Account of the Commentaries on Genesis 1527–1633* (Chapel Hill: The University of North Carolina Press, 1948); Nicholas H. Steneck, *Science and Creation in the Middle Ages. Henry of Langenstein (d. 1397) on Genesis* (Notre Dame: University of Notre Dame Press, 1976); Redwood, *Reason, Ridicule and Religion* (1, n. 25), pp. 116–32.

discover much about the individual's perception of the universe from a study of the cosmogony his or her group accepts. But recent analysts of creation myths argue that there is even more to be learned from a study of the dialectic process of mythopoeia in speculations on cosmic origins; that one has in the development of cosmogonic views a record of the development of intelligence in mankind. Creating a story about his or her origins, the human being tests that story against the world and changes it as required by a developing perception and ability to manipulate, in what may perhaps best be seen as a cultural counterpart to the development of intelligence in a child.[15]

In Christendom, the traditional format for such discussions had always been the so-called hexaemeral literature, commentaries on the six days of creation as described in Genesis. However, the complexity of the Christian tradition had grown to be simply immense. In the fourteenth century Henry of Langenstein cited by name in his *Lecturae super Genesim* a total of sixty-four authorities. Although most of his sources were Christian predecessors, cosmogonic writers from Arabic, Jewish, Roman, and Greek cultures were well represented. Henry's touchstone was still Genesis, of course, but the diverse sources from which he drew comprise a valid indicator of the distance of late medieval worldviews from those of the primitive. In fact by Henry's time the commentary on the six days of creation had become a way of doing science, a vehicle for reflection upon the rational adequacy of received explanations of the natural order of the created world.[16] In Newton's time Genesis was still the touchstone, but the problems of reconciling the Mosaic account with philosophic ones had been enormously aggravated during the three centuries since Henry delivered his *Lecturae*.

15. Charles Doria and Harris Lenowitz, Eds., *Origins. Creation Texts from the Ancient Mediterranean. A Chrestomathy*, tr. with introduction and notes by Charles Doria and Harris Lenowitz, and with a preface by Jerome Rothenberg (Anchor Books; Garden City, NY: Anchor Press/Doubleday, 1976), "Introduction," pp. xxii–xxiii.
16. Steneck, *Science and Creation* (3, n. 14), pp. 24–5, 55 et passim. The number of Henry's authorities were: Latin, fourteenth century, 3; Latin, twelfth and thirteenth centuries, 15; Latin, early medieval, 4; Arabic, 10; Jewish, 4; Roman, 9; early Christian, 8; Greek, 11. By contrast one may note that the well-read Robert Grosseteste, writing probably in the 1230s, cited only thirty-six separate authors in his *Hexaëmeron* (although utilizing a total of ninety-eight titles): see Robert Grosseteste, *Hexaëmeron*, ed. by Richard C. Dales and Servus Gieben (Auctores Britannici Medii Aevi, VI; London: published for The British Academy by The Oxford University Press, 1982), "Introduction," esp. pp. xix–xxv.

In part the problems had been enlarged by the recovery of yet more material from antiquity, especially the cosmogonic stories in the Hermetic corpus. In part the aggravation arose from an increasingly detailed knowledge of the natural world. There is nothing in Genesis, for example, on the beginnings of diurnal rotation or annual revolution, and even Newton confessed himself to be at a loss to see where they should fit into the six-day span of creation, saying to Burnet,

> Where natural causes are at hand God uses them as instruments in his works, but I doe not think them alone sufficient for ye creation & therefore may be allowed to suppose that amongst other things God gave the earth it's motion by such degrees & at such times as was most suitable to ye creatures.[17]

Problems of the latter sort could be resolved by the application of one variety of exegetical "accommodation," however. The term "accommodation" has broader meanings when applied to ethical issues or to Christian missionary activities,[18] but here one may restrict the discussion to its significance in reconciling physical descriptions in, or absent from, the Mosaic account of creation with philosophical ones, or of reconciling any biblical passage containing physical description with the emerging new science.

Accommodationist techniques of exegesis on physical questions had been in use at least since the time of Augustine, who had struggled in his *De Genesi ad litteram* to reconcile the literal meaning of the texts with his own rather sophisticated knowledge of ancient natural philosophy, saying, for example, in one place that one may presume a certain passage (Genesis 1:2)

17. Newton, *Correspondence* (2, n. 14), vol. II, 334 (Newton to Burnet, January 1680/1).
18. F. L. Cross and E. A. Livingston, Eds., *The Oxford Dictionary of the Christian Church* (2nd ed.; Oxford: Oxford University Press, 1983), s.v. "Accommodation," p. 10; *Encyclopedic Dictionary of Religion*, ed. by Paul Kevin Meagher, OP, S.T.M.; Thomas C. O'Brien; and Sister Consuelo Maria Akerne, SSJ (3 vols.; Washington, D.C.: Corpus Publications, 1979), s.v. "Accommodation," vol. I, 26; *The New Schaff-Herzog Encyclopedia of Religious Knowledge, Embracing Biblical, Doctrinal, and Practical Theology, and Biblical, Theological, and Ecclesiastical Biography from the Earliest Times to the Present Day. Based on the Third Edition of the Realencyklopädie Founded by J. J. Herzog, and Edited by Albert Hauck*, Samuel Macauley Jackson, Editor-in-Chief (12 vols.; New York: Funk and Wagnalls Co., 1908–1912), s.v. "Accommodation," vol. I, 22–4. I am indebted to John D. Woodbridge for discussions and suggestions on this issue.

> refers to the visible creation but implies its unformed state in terms that are adapted to the unlearned. For these two elements, earth and water, are more pliable than the others in the hands of an artisan, and so with those two words it was quite fitting to indicate the unformed matter of things.[19]

"Accommodation," in the restricted sense appropriate to this discussion then, means that the form in which the divinely inspired Word had been cast had been determined by the need to accommodate it to the understanding of ignorant ordinary people. Physical descriptions in the Bible thus conformed to commonsense perceptions and might appear to be far removed from the more refined perceptions generated by sophisticated philosophers. The application of accommodation theory to the exegesis of physical questions left the moral lessons and soteriological message of the Bible untouched. Galileo summed up the position of most orthodox accommodationists by quoting a contemporary: "That the intention of the Holy Ghost is to teach us how one goes to heaven, not how heaven goes."[20] Or, as Augustine had commented in the fifth century regarding the failure of Holy Writ to provide a definitive description of the shape of heaven, "the sacred writers knew the truth, but... the Spirit of God, who spoke through them, did not wish to teach men these facts that would be of no avail for their salvation."[21]

Of special interest in the early modern application of exegetical accommodation to physical questions were Augustine's comments regarding the divine creation of the sun and the moon: "And God made two great lights; the greater light to rule the day, and the lesser light to rule the night: he made the stars also."[22] Augustine was well aware that many Greek and Roman writers had speculated that the stars might be as large as, or even larger than, the sun, even though appearing smaller because

19. Augustine, *The Literal Meaning of Genesis* I, 15. 30. In this and subsequent quotations I have utilized the English translation from *Ancient Christian Writers. The Works of the Fathers in Translation*, vols. 41 and 42: *St. Augustine. The Literal Meaning of Genesis*, tr. and annotated by John Hammond Taylor, S.J. (2 vols.; New York: Newman Press, 1982).
20. Galileo Galilei, "Letter to Madame Christina of Lorraine, Grand Duchess of Tuscany, Concerning the Use of Biblical Quotations in Matters of Science [1615]," in *Discoveries and Opinions of Galileo, Including The Starry Messenger (1610), Letter to the Grand Duchess Christina (1615), And Excerpts from Letters on Sunspots (1613), The Assayer (1623)*, tr. with intro. and notes by Stillman Drake (Doubleday Anchor Books; Garden City, NY: Doubleday & Co., 1957), p. 186.
21. Augustine, *The Literal Meaning of Genesis* (3, n. 19), II, 9.20.
22. Genesis 1: 16.

of their greater distance.[23] The literal sense of the scriptural passage, however, suggests that the sun and moon are larger than the stars. To Augustine the most important point was the religious one that God was the creator of all the heavenly bodies "whatever may be the true account of all this [their relative sizes]," but he concluded with the argument that the biblical text was *phenomenologically* accurate, that is, that the passage spoke the truth from the human, observational point of view.

> They [non-Christian philosophers] will certainly grant this at least to our eyes, that these two lights obviously shine more brightly upon earth, that day is illumined only by the light of the sun, and that night with all its stars does not shine as bright without the moon as when lighted by her rays.[24]

It was Augustine's phenomenological argument that was to be elaborated by a succession of sixteenth- and seventeenth-century theologians and natural philosophers in a concerted effort to reconcile new scientific discoveries with biblical language.[25] Kepler, especially, argued that biblical authors accommodated their descriptions to the human sense of sight; following Kepler's lead in England John Wilkins and Isaac Newton did likewise.[26]

23. Augustine, *The Literal Meaning of Genesis.* (3, n. 19), II, 16.33.
24. Ibid., II, 16.33–16.34.
25. R. Hooykaas, *Religion and the Rise of Modern Science* (Edinburgh: Scottish Academic Press, 1973), pp. 114–35; "History of the interpretation of the bible," in *The Interpreter's Bible*, ed. by George A. Buttrick, et al. (12 vols.; Nashville, TN: Abington Cokesbury Press, 1951–7), vol. I, 125 (on Calvin); David Bass, " 'The Errors of our Conceits': Accommodation in the Wilkins-Ross Debate" (Trinity Evangelical Divinity School: Th. M. thesis, 1989); Robert S. Westman, "The Copernicans and the Churches," in *God and Nature. Historical Essays on the Encounter between Christianity and Science*, ed. by David C. Lindberg and Ronald L. Numbers (Berkeley: University of California Press, 1986), pp. 76–113, esp. 89–103, where Westman distinguishes between "absolute accommodationism" (very similar to the Augustinian variety) and "partial accommodationism" in which the interpreter gave a heliostatic or geomotive construal of passages conventionally read as geostatic.
26. Bass, " 'The Errors of our Conceits' " (3, n. 25) suggests a connection between the invention of the telescope, as well as early modern optical and perspective studies, and the increased emphasis on varieties of visual perception. For a general overview of the place of accommodation in Newton's thought, see Richard Stoddard Brooks, "The Relationships between Natural Philosophy, Natural Theology, and Revealed Religion in the Thought of Newton and Their Historiographic Relevance" (Northwestern University: Ph. D. dissertation, 1976), esp. pp. 116–20.

The Counter-Reformation Church rejected Galileo's efforts to move it into the modern era with a fresh application of accommodationist techniques, but Protestant commentators who wished to take that approach encountered no such organized resistance. Newton used accommodation freely, and he certainly found it preferable to Burnet's unfortunate tendency to doubt Moses' veracity, giving Burnet a pithy lecture on his error.

> Consider therefore whether any one who understood the process of ye creation & designed to accommodate to ye vulgar not an Ideal or poetical but a true description of it as succintly & theologically as Moses has done, without omitting any thing material wch ye vulgar have a notion of or describing any thing further then the vulgar have a notion of, could mend that description wch Moses has given us.[27]

Accommodationist exegesis was effective. With it, the Mosaic account could be pulled and stretched until it was made to fit the new astronomical theories or the new facts constantly being presented to the theologians by the natural philosophers. Even the growing sense that a time span longer than six ordinary days was necessary could be handled by it. Newton offered Burnet as much time as he (Burnet) felt he needed for God to do all that work, by first withholding and then applying "an eaven force" to the terrestrial globe to bring the rate of its diurnal revolution from zero up to 365 spins per year.

> [Y]ou may make ye first day as long as you please, & ye second day too if there was no diurnal motion till there was a terraqueous globe, that is till towards ye end of that days work. And then if you will suppose ye earth put in motion by an eaven force applied to it, & that ye first revolution was done in one of our years, in the time of another year there would be three revolutions, of a third five, of a fourth seaven, &c & of the 173d year 365 revolutions, that is as many as there are days in one year. . . .[28]

Such numerical speculations could be used to lengthen the time of creation as much as required. Copernicanism could not be negated by the simple but true phrases Moses had composed for the common people. In short, Mosaic and philosophic accounts could readily be reconciled by accommodationist techniques, and since that was the case, one must ask what then produced the radical divergence of the scientific and traditional accounts of cosmic origins so characteristic of the post-Newtonian period. The hexaemeral format had allowed the scientific and the traditional to coexist, but the containing bound-

27. Newton, *Correspondence* (2, n. 14), vol. II, 333 (Newton to Burnet, January 168$^{0}/_{1}$).
28. Ibid., vol. II, 333–34.

aries of the hexaemeron began to dissolve under the impact of two developments.

One was the flood tide of higher biblical criticism that eventually led to serious questions about the direct delivery by God to humanity of knowledge about the natural world. The waters of higher criticism began to rise in Newton's lifetime, and he himself owned works by Richard Simon,[29] but the full impact of this development came somewhat later. The second development that so eroded the old ways of reconciling Moses and philosophy, on the other hand, had reached maturity by the second half of the seventeenth century. Descartes, Mersenne, Burnet, Newton, and many another were engaged in it.

Briefly, it was that the simple divine fiat no longer seemed adequate. Although of course all depended on the fiat, a technique, a mechanism, a detailed procedure describing God's action was wanted, which entailed the elaboration of secondary causes. Increasingly, as the centuries of hexaemeral discussion lengthened, and as more and more authorities required incorporation, the authors of this class of literature became more eclectic, but, significantly, they also came to treat the six days as stages in a concrete physical process that resulted from the working out of continuous natural causes.[30] Once God said fiat, that causal Word entered into nature and continued to operate in building up the final result. William of Conches, for example, gave primal heat a role in the creation of living beings, and he countered the criticism that this detracted from God's power with the argument that it rather exalted God's power to say that he was able to give to things such a nature that they could then work for him.[31] Henry of Langenstein struggled to reconcile the fiat separating the waters from the dry land with the operation of the Aris-

29. Harrison, *Library* (1, n. 6), items 945, 1514–17, pp. 178 and 239. John D. Woodbridge, "German responses to the biblical critic Richard Simon: from Leibniz to J. S. Semler," *Wolfenbütteler Forschungen* 41 (1988), pp. 65–87, demonstrates the early knowledge of Simon's work in Germany and its use by English Deists. On the pre-Adamite theorist, who earlier raised problems regarding the Mosaic authorship of all of the Pentateuch, and who probably influenced Simon, see Richard H. Popkin, *Isaac La Peyrère (1596–1676). His Life, Work and Influence* (Brill's Studies in Intellectual History, Vol. 1, General Ed.: A. J. Vanderjagt; Leiden: E. J. Brill, 1987). On Newton's own studies on the authorship of the Bible and on the corruption of biblical texts, see Richard H. Popkin, "Newton's biblical theology and his theological physics," in *Newton's Legacy* (1, n. 5), pp. 81–97.

30. Robbins, *Hexaemeral Literature* (3, n. 13), pp. 77–91.

31. Ibid., p. 84. Cf. (2, n. 51) where Newton used an almost identical argument.

totelian natural tendencies of water and earth to seek their separate spheres, a problem that later commentators frequently solved by invoking the axiom of the *horror vacui*: supposing that God had dilated the globe, the waters would have rushed in to fill the interstices, allowing dry land to appear.[32]

Though the purpose of the hexaemeral commentaries was not originally scientific in any sense, in them there appeared a cultural vehicle for doing science, a vessel, as it were, into which curiosity about the natural world could be poured in a religious age. Though the questions asked were framed as questions about how God did it, in the end the elaboration of the secondary procedures that came into play after the fiat passed insensibly over into a form of discourse that is recognizably scientific. The Renaissance period saw the climax of traditional hexaemeral literature, perhaps with the four folio volumes of Pererius, completed in 1598, in which not only contemporary science but all literature, history, mythology, and morality were encapsulated as comments on the book of Genesis.[33] As the seventeenth century waxed, less ambitious commentators, or those who followed new methods of exegesis, ceased to write books that were compendia of the entire culture and moved toward linguistic and historical analysis of the book of Genesis itself, whereas the business of reconciling Moses with philosophy passed into the hands of the natural philosophers. They took it up with great vigor. Mersenne's *Quaestiones celeberrimae in Genesim* (1623) was already top heavy with scientific information, but with Burnet's *Telluris theoria sacra* of 1681 the ongoing construction of the world after the fiat achieved open primacy. With two thousand years of tradition behind it, the problem of reconciling Moses with philosophy had taken on life in a new format and was fast becoming simply a natural history of the earth.

One comes close here to the intellectual core of the scientific revolution, one characteristic of which, it has been argued, was the recognition of an identity between knowledge and construction or reconstruction, a concept of "knowledge as making or doing."[34] Such a doctrine of knowledge recognizes the impact of broad developments in technology in West-

32. Williams, *Common Expositor* (3, n. 14), pp. 194–5.
33. Ibid., where Pererius is quoted and commented upon passim, but see esp. pp. 255–68 on "the waning of the tradition."
34. Paoli Rossi, "Hermeticism, rationality and the scientific revolution," in *Reason, Experiment, and Mysticism in the Scientific Revolution* (1, n. 5), pp. 247–73, esp. pp. 251–5.

ern society. It also suggests a reason for the intense interest in Hermetic cosmogonic documents in early modern Europe: they offered an entrée into the chemical–alchemical processes God had set in motion at certain stages of His creative work.

In doing this, alchemical and Hermetic materials made a unique contribution to hexaemeral literature, utilizing as they did an unusual application of microcosm–macrocosm analogies. Broadly speaking, a microcosm is some portion of the world which epitomizes in its structures or processes the essential features of the macrocosm, difference in scale being irrelevant or virtually so. In many early speculations of this type, man himself is a structural microcosm, and the physical or political universe is described as an analogous organic form. In other instances, all cosmic features are said to be reflected in the nature of man, or the divine is reflected in his soul.[35] Had not God said, "Let us make man in our image"?[36]

The alchemical and Hermetic materials, however, were much richer, for in them human being was not the only microcosm. The alchemical process, the great work of alchemy, epitomized in some way the coming-into-being of the larger cosmos. Alternatively, as one may see later, the coming-into-being of the cosmos sketched or prefigured the great work of alchemy and also the entire historical duration of the cosmos from beginning to end. The effective alchemical agency – the occult spirit named so variously by the alchemical philosophers – was thus seen as the exemplar of God's creative power. The alchemical miniature of creation in turn could offer parallels with the powerful transformations that take place in the human psyche upon occasion, the ferment of the soul dimly sensed as creative forces of cosmic dimensions.[37] Different alchemical writers emphasized different aspects of the various analogies, and probably few if any of them explicitly understood all the parallels they invoked, but a remarkable consensus emerged from this class of literature finally: an urgent description of the bursting vitality of creativity, of life

35. George Perrigo Conger, *Theories of Macrocosms and Microcosms in the History of Philosophy* (reprint of the 1950 ed.; New York: Russell & Russell, 1967).
36. Genesis 1: 26.
37. Conger, *Macrocosms* (3, n. 35), pp. 69–71; Marie-Louise von Franz, "The idea of the macro- and microcosmus in the light of Jungian psychology," *Ambix 13* (1965), 22–34; idem, *Patterns of Creativity Mirrored in Creation Myths* (Zürich: Spring Publications, 1972); S. Mahdihassan, "Imitation of creation by alchemy and its corresponding symbolism," *Abr-Nahrain (Melbourne)* 12 (1972), 95–117.

and growth, of the multiplication and ramification of life into the unexpected, the unpredictable, and the new as God's creative fiat worked His will upon the original chaos. Though in fact the alchemical sense of bursting vitality within the world probably had its ultimate roots in ancient Stoicism,[38] by Newton's time the cosmic creative forces of the alchemists had largely been amalgamated with the Genesis account of creation.[39] And from the conjunction of Moses and Hermes in the seventeenth century there arose vivid pictorial representations of cosmic creativity, such as one sees in Plate 4.[40]

Newton's Commentary on the Emerald Tablet

Newton's *Commentary* on the *Emerald Tablet* of Hermes Trismegistus belongs to that same conjunction of Moses and Hermes. If, as suggested

38. G. Verbeke, *L'evolution de la doctrine du pneuma du stoicisme à S. Augustin. Étude philosophique* (Bibliothèque de l'Institut Supérieur de Philosophie Université de Louvain; Paris: Desclée De Brouwer; Louvain: Éditions de l'Institut Supérieur de Philosophie, 1945), p. 37.
39. Christian assimilation of ancient light metaphysics lies behind this development of alchemical doctrine. Cf. David C. Lindberg, "The genesis of Kepler's theory of light: light metaphysics from Plotinus to Kepler," *Osiris*, Second Series, 2 (1986), 5–42; James McEvoy, *The Philosophy of Robert Grosseteste* (Oxford: Clarendon Press, 1982), pp. 51–222; Georges Duby, *The Age of the Cathedrals. Art and Society, 980–1420*, tr. by Eleanor Levieux and Barbara Thompson (Chicago: The University of Chicago Press, 1981), pp. 99–108.
40. "Janitor Pansophvs, Seu Figura AEnea Quadripartita Cunctis Museum hoc Introeuntibus, Superiorum ac Inferiorum Scientiam Mosaico-Hermeticum, analyticè Exhibens," Figura III, in *Musaeum Hermeticum reformatum et amplificatum, omnes sopho-spagyricae artis discipulos fidelissimè erudiens, quo pacto Summa illa veraque lapidis philosophici Medicina, qua res omnes qualemcunque defectum patienties, instaurantur, inveniri & haberi queat, Continens tractatus chimicos XXI. Praestantissimos, quorum Nomina & Seriem versa pagella indicabit* (Francofurti: Apud Hermannum à Sande, 1678). A modern facsimile reprint exists, introduced by Karl R. H. Frick (Graz: Akademische Druck-u. Verlagsanstalt, 1970), and Arthur Edward Waite's English translation of 1893 has been reprinted (2 vols,; New York: Samuel Weiser, 1973). See also Robert Fludd's numerous and imaginative illustrations of creation now conveniently gathered in Joscelyn Godwin, *Robert Fludd. Hermetic Philosopher and Surveyor of Two Worlds* (Boulder, CO: Shambhala, 1979). For the perennial sense of the ever ramifying living spontaneity of the cosmos restated in terms of modern cosmogonic theory, see J. McKim Malville, *The Fermenting Universe. Myths of Eternal Change* (New York: The Seabury Press, 1981).

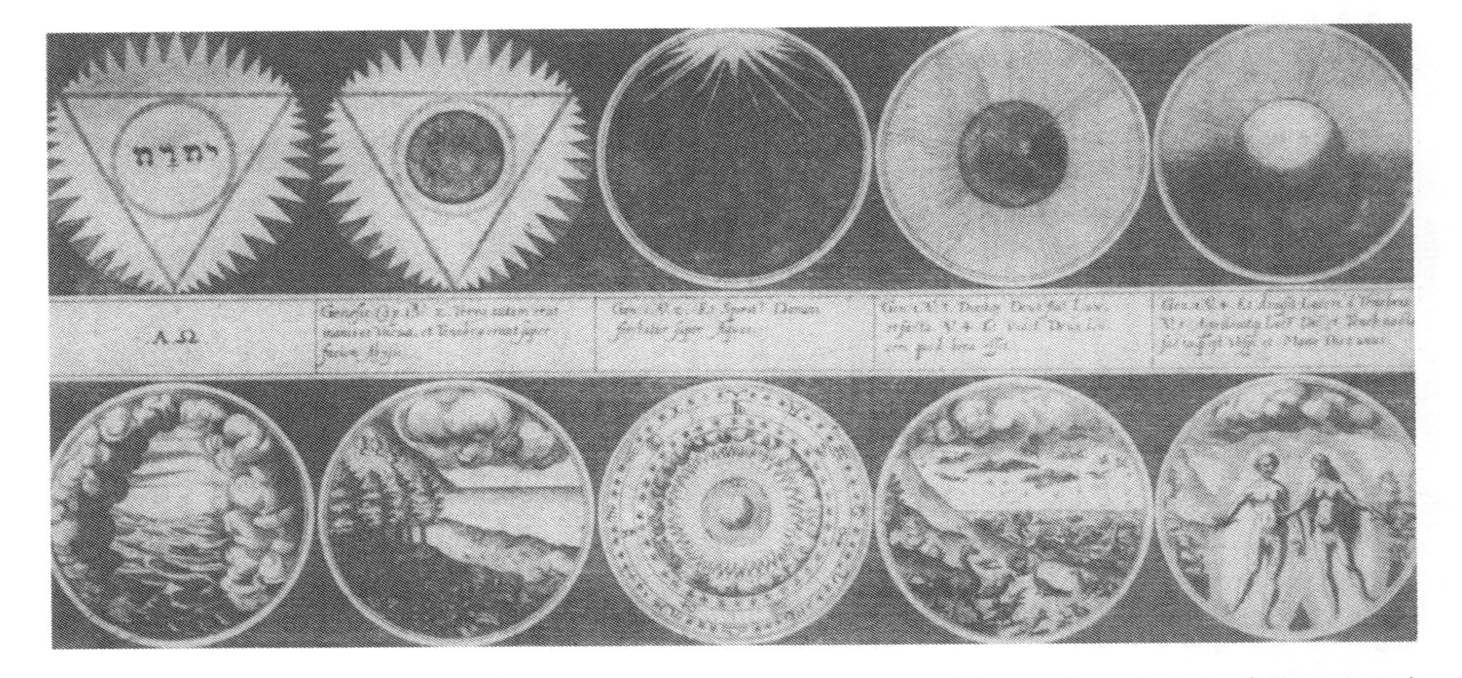

Plate 4. Janitor Pansophvs. Figura III, in *Musaeum Hermeticum reformatum et amplificatum* (Francofurti: Apud Hermannum à Sande, 1678). From the collections of the Library of Congress Rare Book and Special Collections Division, Washington, D.C.

above, its composition was stimulated by the arrival of Burnet's manuscript, then it was probably written early in the 1680s. In it Newton unselfconsciously poured much of the Genesis story into the empty mold provided by Hermes's "Thus was the world created," then provided an alchemical explication of God's actions.

> And just as the world was created from dark Chaos through the bringing forth of the light and through the separation of the aery firmament and of the waters from the earth, so our work brings forth the beginning out of black Chaos and its first matter through the separation of the elements and the illumination of matter. Whence arise the marvellous adaptations and arrangements in our work, the mode of which here was adumbrated in the creation of the world.[41]

Again, where Hermes offered a Neoplatonic cosmogony entailing the rise of multiplicity from an unnamed unity "by the mediation of one," Newton imposed first a Mosaic and then an alchemical interpretation: "And just as all things were created from one Chaos by the design of one God, so in our art all things... are born from this one thing, which is our Chaos...."[42]

Newton's study of Hermes Trismegistus extended over a period of at least twenty years, probably longer. The principal surviving evidence for it is comprised of two manuscripts, Keynes MSS 27 and 28, King's College, Cambridge. These manuscripts, though one should not suppose they represent the entirety of Newton's exploration of Hermetic materials, do show that he knew well the two primary alchemical tracts attributed to Hermes, the *Emerald Tablet* and the so-called *Seven Chapters* (or the *Golden Work*, the *Tractatus aureus de Lapidis Physici Secreto in Cap. 7 divisus*).[43]

Of these two principal Hermetic tracts with which Newton worked, the *Emerald Tablet* was by far the better known. It was, in fact, one of the best known tracts in all of alchemy. Of indeterminate but great antiquity, it was long supposed to encapsulate in its mysterious phrases all the occult wisdom of the ancients regarding divine actions in the creation of the world, phrases that properly interpreted would in turn

41. Keynes MS 28 (1, n. 35), f. 6v. Cf. Appendix B.
42. Ibid.
43. Cf. John Ferguson, *Bibliotheca Chemica: A Catalogue of the Alchemical, Chemical and Pharmaceutical Books in the Collection of the Late James Young of Kelly and Durris, Esq., Ll.D., F.R.S., F.R.S.E.* (2 vols.; Glasgow: James Maclehose and Sons, 1906; London: Derek Verschoyle Academic and Bibliographic Publications, 1954), vol. I, 389–94. For a reconstruction of Newton's work on the papers now in Keynes MSS 27 and 28, see the editorial note to Appendix B.

elucidate the alchemical work. In annotations to his own commentary, for example, Newton cited the work of Avicenna, who flourished in the eleventh century, and of "Senior" (Zadith ben Hamuel), who flourished in the thirteenth.[44] Though he did not, he might have cited Hortulanus, whose commentary on the *Emerald Tablet* was first published in 1541 and was often reprinted and translated.[45] Later Newton took note of an exposition by Gerhard Dorn, the sixteenth-century Paracelsian.[46] But there can be no doubt that the commentary on the *Emerald Tablet* in

44. Keynes MS 28 (1, n. 35), f. 7r. Cf. Appendix B. The works Newton cited were: "Avicennae Tractatulus de Alchimia, in *Artis avriferae, qvam chemiam vocant, Volumina duo, qvae continent Tvrbam Philosophorum, aliosqúe antiquissimos auctores, quae versa pagina indicat. Accessit nouiter volumen tertium, continens: 1. Lullij vltimum Testamentum. 2. Elucidationem Testam. totius ad R. Odoardum 3. Potestatem diuitiarum, cum optima expositione Testamenti Hermetis. 4. Compendium Artis Magicae, quoad compositionem Lapidis. 5. De Lapide & oleo Philosophorum. 6. Modum accipiendi aurum potabile. 7. Compendium Alchimiae & naturalis Philosophiae. 8. Lapidarium. Item Alberti Magni secretorum Tractatus. Abbreuiationes quasdam de Secretis Secretorum Ioannis pauperum. Arnaldi Quaest. de Arte Transmut. Metall. eiusqúe Testamentum. Omnia hactenus nunquam visa nec edita. Cum Indicibus rerum & verborum locupletissimis* (3 vols. in 1; Basileae: Typis Conradi Waldkirchii, 1610), vol. I, 260–79; "Senioris antiquissimi philosophi libellvs, vt Brevis ita artem discentibus & exercentibus utilissimus, & verè aureus, Dixit Senior Zadith filius Hamuel," in *Theatrum chemicum, praecipuos selectorum auctorum tractatus de chemiae et lapidis philosophici antiquitate, veritate, jure, praestantia, & operationibus, continens: In gratiam Verae Chemiae, & medicinae Chemicae studiosorum (ut qui uberrimam inde optimorum remediorum messem facere poterunt) congestum, & in Sex partes seu volumina digestum; singulis voluminibus, suo auctorum et librorum catalogo primis pagellis: rerum verò & verborum Indice postremis annexo* (6 vols.; Argentorati: Sumptibus Heredum Eberh. Zetzneri, 1659–61), vol. V, 193–239. Newton owned both titles: Harrison, *Library* (1, n. 6), items 90 and 1608, pp. 91 and 249, now, respectively, Trinity College, Cambridge NQ. 16. 121, and missing.
45. Cf. Ferguson, *Bibliotheca Chemica* (3, n. 43), vol. I, 419–22. Speculative dates for Hortulanus (or John Garland?) range from the late twelfth to the fourteenth century.
46. Keynes MS 28 (1, n. 35), f. 2v. Cf. Appendix B. Reference was to Dorn's "Tractatus de naturae luce physica ex Genesi desumpta, Iuxta Sapientiam Theophrasti Paracelsi," in *Theatrum chemicum* (3, n. 44), vol. I, 326–457, which contains: (1) "Physica Genesis," (2) "Physica Hermetis Trismegisti," (3) "Physica Trithemij," (4) "Philosophia Meditativa," (5) "Philosophia Chemica." Cf. Debus, *Chemical Philosophy* (3, n. 2), vol. I, 77–8.

Keynes MS 28 is Newton's own composition, for in it one finds the intersection of some of his deepest concerns.

One has already noted that Hermetic cosmogonic materials were especially attractive to early modern philosophers because they seemed to offer hints about the chemical–alchemical processes God had utilized in the creation of the world. As the hexaemeral commentators struggled to elaborate their treatments of the six days as stages in a concrete physical process, the chemical philosophers among them focused on questions about God's techniques in generating organized forms of matter from the original unformed chaos.

Alchemy had always concerned itself with the various manifestations and transformations of matter, and the *Emerald Tablet* was no exception to that rule despite its obscurity. One will not find in it the atoms associated with modern matter theory, nor even the structured hierarchies of particles that had taken near-final form in Newton's mind before the 1680s. But even a casual reading of the *Emerald Tablet* will reveal several pairs of related material opposites: sun–moon, father–mother, earth–fire, subtle–gross, things superior–things inferior. To Newton, pairs of that sort, and their union, had come to represent the most fundamental, the most basic, manifestations of matter as it arose in organized forms from a primitive chaos. He commented on and explained the pairs from the *Emerald Tablet* as follows.

> Inferior and superior, fixed and volatile, sulfur and quicksilver have a similar nature and are one thing, like man and wife. For they differ one from another only by the degree of digestion and maturity. Sulfur is mature quicksilver, and quicksilver is immature sulfur; and on account of this affinity they unite like male and female, and they act on each other, and through that action they are mutually transmuted into each other and procreate a more noble offspring to accomplish the miracles of this one thing.[47]

Newton might have used a large variety of paired symbols to name the two materials. Similar duos often occur in alchemical literature and in the illustrations for it, as in Plate 5.[48] It is possible to give a psycho-

47. Keynes MS 28 (1, n. 35), f. 6r, v. Cf. Appendix B.
48. Zentralbibliothek, Zürich, Codex rhenovacensis 172, *Aurora consurgens*, f. 10. See also the red and green lions from Fitzwilliam Museum, Cambridge, Fitzwilliam MS 276*, *George Ripley, Canon of Bridlington, The Emblematical Scroll*, reproduced in Dobbs, *Foundations* (1, n. 1), p. 33. I have suggested in my *Alchemical Death & Resurrection* (1, n. 30) that the significance of the lions on the Ripley Scrolls may have been otherwise for the original artists, having a meaning related to the sacred lions that guarded

Plate 5. MS Rh. 172, f. 10v, Zentralbibliothek Zürich. Reproduced by permission of Zentralbibliothek Zürich, Zürich, Switzerland.

logical explanation of their prevalence,[49] yet one has no real evidence that Newton participated in the psychological dialectic suggested by that explanation.

Similar duos still occur in modern chemistry (e.g., acid–base, electron–proton, anion–cation), not to mention biology (male–female), and it seems rather more likely that Newton had recognized in alchemical literature a valid symbolization of rather fundamental observational dichotomies. He was interested in the structure of matter and in what alchemy could teach him about its forms and changes and about the universal spirit that animated the changes and molded the forms. He continued to argue for many years for a matched pair of opposites as basic constituents of matter, speaking of them in many of his alchemical papers.[50] In the climactic "Praxis," probably written in the 1690s, his first chapter, "De materiis spermaticis," lists a number of the matched pairs he had garnered from the older literature,[51] and in his small tract

the Great Goddess in some of the ancient Mediterranean civilizations, but the lions do not seem to have held that meaning for Newton.

49. Dobbs, *Foundations* (1, n. 1), pp. 26–35, and the literature cited there.
50. Some of these are examined in Dobbs, "Newton's copy" (1, n. 5).
51. Babson MS 420 (1, n. 39), pp. 3–4. Cf. Appendix E.

On the Nature of Acids he observed "that what is said by chemists, that everything is made from sulphur and mercury, is true, because by sulphur they mean acid, and by mercury they mean earth."[52]

Of even greater interest than the pairs themselves is the marriage between them that produces "a more noble offspring," for it is in the begetting of the offspring that the activating spirit comes into play. The activating spirit was in some sense divine, for the mode of its acting was prefigured or adumbrated (*adumbratus*) in the creation of the world. It was the spirit that gave life, and it is impossible to mistake the vitalistic nature of the process as Newton described it: "this generation is similar to the human, truly from a father and mother...."[53] But Newton never assumed that two related forms of matter came together and generated "a more noble offspring" by the volition and powers of matter itself. He was quite clear that matter was passive and that only the spiritual realm could initiate activity.[54] In that general understanding he stood within both Neoplatonic and Stoic traditions, but that position led him inexorably to search for the spiritual agent that acted on matter to give it life as organized forms arose from chaos. For, as he later observed, "all matter duly formed is attended with signes of life." In that understanding he also stood within the general alchemical tradition.[55]

The persistent theme of the activating spirit runs through all of Newton's alchemical work. One has already seen him pursuing it with the greatest diligence, calling it at different times "a fermental virtue," "the vegetable spirit," and, later, "the force of fermentation." Knowledge about this occult spirit was one of the most carefully cloaked secrets of alchemy, and Newton constantly attempted to decode the alchemical literature for the meaning hidden in its various names for the active agent. In his commentary on the *Emerald Tablet* Newton identified the term "mercury of the philosophers" with the activating spirit and concluded, interestingly enough, that Hermes himself was a symbol for it.

> On account of this art [alchemy] Mercurius [Hermes] is called thrice greatest [Trismegistus], having three parts of the philosophy of the whole world, since he signifies the Mercury of the philosophers... and has dominion in the mineral kingdom, the vegetable kingdom, and the animal kingdom.[56]

52. Newton, *Correspondence* (2, n. 14), vol. III, 206 and 210.
53. Keynes MS 28 (1, n. 35), f. 6v. Cf. Appendix B.
54. Cf. McMullin, *Newton on Matter and Activity* (2, n. 61).
55. Dixon, *Alchemical Imagery* (2, n. 72), pp. 15–17, 24, and Fig. 7.
56. Keynes MS 28 (1, n. 35), f. 7r, my brackets. Cf. Appendix B.

The activating spirit "signified" by Hermes was thus universal in scope and operation, having "dominion" in all three kingdoms, where "all matter duly formed is attended with signes of life."

Alchemy as prophecy

Newton's use of the concept of adumbration or prefiguration in his commentary leads perforce into a consideration of typological exegesis and of Judeo–Christian perceptions of historical process.

From very early times the Jews had conceived of history as theophany, as revelations of the divine will and human responses to them. Each epiphany was unique, so even its details were of supreme importance, which gave an emphasis to historical particularity in sharp contrast to archaic and classical conceptualizations of history as endless cyclical repetition. Only once did God create the world; only once did He deliver Israel from Egypt. Only once did Christ die for our sins, as Augustine thundered at the classicists; the Atonement cut across all cycles. In Augustine's adaptation of the Jewish vision for Christian edification, the world is not a treadmill but a schoolhouse: humanity is not bound forever to a pattern of eternal rise and fall. History has a beginning, a middle, and an end. Time's motion is of the arrow, not the wheel.[57]

There is thus inherent in the Judeo–Christian tradition a sense of history as a progressive motion, not meaning that the future will get better and better as in some modern secular distortions, but meaning that more and more of God's plan will be revealed to humanity with the passage of time. Although God's plan is presumed to be constant, human understanding of it is not. God's plan is to bring humanity to ultimate salvation, and both Christian and Jewish communities embrace some

57. Mircea Eliade, *The Myth of the Eternal Return or, Cosmos and History*, tr. from the French by Willard R. Trask (Bollingen Series XLVI; reprint of the Princeton/Bollingen paperback ed. of 1971; Princeton: Princeton University Press, 1974); Charles Norris Cochrane, *Christianity and Classical Culture. A Study of Thought and Action from Augustus to Augustine* (reprint of the 1944 ed.; London: Oxford University Press, 1977), esp. pp. 456–516. For a cogent analysis of the origin of modern views on geologic time as the interplay and ultimate fusion of these two ancient and antagonistic views of history, see Stephen Jay Gould, *Time's Arrow, Time's Cycle. Myth and Metaphor in the Discovery of Geological Time* (The Jerusalem–Harvard Lectures Sponsored by the Hebrew University of Jerusalem and Harvard University Press; Cambridge, MA: Harvard University Press, 1987).

form of *Heilsgeschichte* – salvation history – in which Creation and Paradise are followed by Fall, Redemption, Beatitude (the Millennium, the New Jerusalem, a new heaven and a new earth in which all the tragedy of sin has been washed away). It is not the end that is in doubt but rather human ability to comprehend it and respond appropriately within the process of history. Faith in *Heilsgeschichte* is "faith enacted as history," and it assumes a progressive revelation of God's plan within the historical process itself; it sees history as the progressive epiphany of an eternal design.[58]

Typology becomes possible precisely because of these ideas on progressive revelation, for types are faint sketches or adumbrations of what is to be more perfectly revealed in the fullness of time. Yet as one may be able to envision some features of the finished portrait from its first penciled outline, so a prophet may catch a glimpse of the future antitype by recognizing that events of the past are prefigurations of events to come. Typology is thus a double movement in which a historical event or person is taken as the basis for prediction of a future one. The future revelation will be fuller, but since the overall plan remains the same, the type will necessarily bear resemblances and analogies to the antitype yet to come. The major Hebrew prophets were already typologists, using descriptions of Paradise (the past) to sketch the lineaments of the future messianic age, but Christian apologists developed a riotous flexibility in typology that enabled them firmly to incorporate the Old Testament into the Christian canon by the discovery of types of Christ in it. Adam was a type of Christ as were Moses and Noah. The brazen serpent lifted up in the desert and healing those who looked at it became a type of Christ crucified, as did the rock that gushed forth living water when struck. By such means the New Testament was defined as the fuller revelation of all that the Old had shadowed forth.[59] The tradition of typology built

58. Will Herberg, "Biblical faith as 'Heilsgeschichte.' The meaning of redemptive history in human existence," in Will Herberg, *Faith Enacted as History. Essays in Biblical Theology*, ed. with intro. by Bernhard W. Anderson (Philadelphia: The Westminster Press, 1976), pp. 32–42; Eliade, *Myth of the Eternal Return* (3, n. 57), pp. 102–37.
59. An influential Puritan publication in this genre was Thomas Tailor, *Christ Revealed: Or The Old Testament Explained. A Treatise of the Types and Shadowes of our Saviovr contained throughout the whole Scriptvre: All opened and made usefull for the benefit of Gods Church* (London: printed by M. F. for R. Dawlman and L. Fawne at the signe of the Brazen serpent in Pauls Churchyard, 1635), now reprinted as Thomas Taylor, *Christ Revealed. A Facsimile Reproduction with an Introduction by Raymond A. Anselment* (Delmar, NY: Scholars' Facsimiles & Reprints, 1979). See also:

up by the early fathers, and then reemphasized in Augustine's *Contra Faustum*, became so well established that the assumptions and presuppositions of typological analysis could be applied to pagan literature as well. Pagan gods and heroes were assimilated to Christianity as the patriarchs had been, by the discovery of their typological significance for the Judeo–Christian drama.[60]

As with the practice of typology itself, the notion of the week of creation as a type of the full course of *Heilsgeschichte* may be traced back to Jewish origins. Embraced by the fathers, Augustine, Isidore, and many medieval theologians, the doctrine received a humanistic restatement in the *Heptaplus* of Giovanni Pico della Mirandola in 1489. Although there were other types thought to adumbrate the complete career of history (e.g., the four empires in the prophecies of Daniel), the seven days of the original creation most commonly signified the totality of time. Since the seventh day (of rest) so clearly signified the Millennium toward which the entire process was tending, six epochs or eras were left for Creation, Paradise, Fall, Incarnation, Resurrection, and Redemption, and empirical history was ordered and periodized in various ways into those epochs.[61]

Furthermore, Augustine had argued that the book of God's words and the book of God's works were literally identical in structure, and this parallel texts metaphor came ever more to the fore in early modern Europe after the invention of printing. Therefore if the week of creation was a type or prophecy of the whole course of history, it was no less a type or prophecy of the whole course of the natural world. Moses had

Jean Daniélou, *From Shadows to Reality. Studies in the Biblical Typology of the Fathers* (London: Burns & Oates, 1960); M.-D. Chenu, "The Old Testament in twelfth-century theology," in M.-D. Chenu, *Nature, Man, and Society in the Twelfth Century. Essays on New Theological Perspectives in the Latin West*, preface by Etienne Gilson; selected, ed., and tr. by Jerome Taylor and Lester K. Little (Chicago: The University of Chicago Press, 1968), pp. 146–61; Northrop Frye, *The Great Code: The Bible and Literature* (New York: Harcourt Brace Jovanovich, 1982), esp. pp. 78–138.

60. Paul J. Korshin, *Typologies in England 1650–1820* (Princeton: Princeton University Press, 1982).

61. M.-D. Chenu, "Theology and the new awareness of history," in Chenu, *Nature, Man, and Society* (3, n. 59), pp. 162–201; Maren-Sofie Røstvig, "Structure as prophecy: the influence of biblical exegesis upon theories of literary structure," in *Silent Poetry. Essays in Numerological Analysis*, ed. by Alastair Fowler (London: Routledge & Kegan Paul, 1970), pp. 32–72; Giovanni Pico della Mirandola, *Heptaplus or Discourse on the Seven Days of Creation*, tr. with intro. and glossary by Jessie Brewer McGaw (New York: Philosophical Library, 1977).

been given a vision of the cosmic order imposed as a type upon the cosmos in the beginning, and he had incorporated the cosmic order into his narrative in such a way that the natural events described adumbrated not only cosmic history but the parallel *Heilsgeschichte*. Thus, for Pico, the creation of the sun on the fourth day was a prophecy of the birth of Christ in the midst of the time allotted for all history.[62] The moral Fall had had its parallel in the physical world as well: all nature had fallen with man and required to be redeemed.

Burnet's frontispiece offers a pictorial representation of all those themes, where the state of the physical globe is represented in seven successive images arranged in a circle. First comes the original chaos, then the perfectly smooth realm of Paradise, followed by a globe covered with the waters of the Flood, upon which rides the tiny Ark of Noah. Our present broken, corrupted, and irregular earth comes next, then the prophesied purgation by conflagration, followed by the perfect sphere of the Millennium, and, finally, the new heavens and new earth of Scripture. Christ stands above, completing the circle of globes as alpha and omega, beginning and end, with one foot on the original chaos and the other on the final representation of beatitude. Burnet thus captured in one complex image the idea of the seven-day week of creation as the type of the entire course of history from beginning to end, and also succeeded in representing the physical state of the earth at its different stages as correlative with the moral condition of mankind.[63]

62. Røstvig, "Structure as prophecy" (3, n. 61); Pico, *Heptaplus* (3, n. 61), VII, 4, pp. 98–104. See also the elaboration of Pico's typological identification of the creation of the sun with the birth of Christ, published by Cardinal de Bérulle in 1622 and reflecting his knowledge of the work of Copernicus, in Clémence Ramnoux, "Héliocentrisme et Christocentrisme (sur un texte du Cardinal de Bérulle)," in *Le soleil à la Renaissance. Sciences et mythes. Colloque international tenu en avril 1963 sous les auspices de la Fédération Internationale des Instituts et Sociétés pour l'Étude de la Renaissance et du Ministère de l'Éducation nationale et de la Culture de Belgique. Publié avec le concours du Gouvernement belge* (Bruxelles: Presses universitaires de Bruxelles; Paris: Presses universitaires de France, 1965), pp. 447–61.
63. Burnet's frontispiece was not in the original Vol. I published in 1681 (3, n. 3) but appeared in both volumes of the English "translation," to the first volume of which Burnet added much new material: Thomas Burnet, *The Theory of the Earth: Containing an Account of the Original of the Earth, and of all the General Changes Which it hath undergone, or is to undergo, Till the Consummation of all Things. The Two First Books Concerning The Deluge, and Concerning Paradise* [1684]. *The Two Last Books, Concerning the Burning of the World, and Concerning the New Heavens and*

Christian alchemy, with its ancient affinities with a gnostic philosophy where matter was inherently evil and in need of redemption, also restated the redemption theme as a natural parallel to *Heilsgeschichte*. Matter had indeed fallen with Adam and Eve but was redeemed by the philosopher's stone, the natural analogue of Christ. The alchemical process ended in beatitude as did the salvation history of humankind, a joy expressed in Newton's translation of the *Emerald Tablet* thus: "By this means you shall have ye glory of ye whole world & thereby all obscurity shall fly from you."[64] The doctrine of the parallel

New Earth [1690] (2 vols. in 1; London: printed by R. Norton, for Walter Kettilby, at the Bishop's Head in St. Paul's Church-Yard, 1684–90). It was also in Vol. II of the Latin edition: idem, *Telluris Theoria Sacra: Orbis Nostri Originem & Mutationes Generales, quas Aut jam subiit, aut olim subiturus est, complectens. Libri duo posteriores De Conflagratione Mundi, et De Futuro Rerum Statu* (Londini: Typis R. N. Impensis Gualt. Kettilby, ad Insigne Capitis Episcopi in Coemeterio Paulino, 1689), and has recently been reproduced in Gould, *Time's Arrow, Time's Cycle* (3, n. 57), p. 20. It has also been published in the modern reprint of the "second" edition of the English treatise (1690/91), introduced by Basil Willey (Carbondale, IL: Southern Illinois University Press, 1965).

64. Thomas Willard, "Alchemy and the Bible," in *Centre and Labyrinth. Essays in Honour of Northrop Frye*, ed. by Eleanor Cook, Chaviva Hošek, Jay Macpherson, Patricia Parker, and Julian Patrick (Toronto: University of Toronto Press in association with Victoria University, 1983), pp. 115–27; H. J. Sheppard, "Gnosticism and alchemy," *Ambix 6* (1957), 86–101; John Warwick Montgomery, *Cross and Crucible. Johann Valentin Andreae (1586–1654), Phoenix of the Theologians. Vol. I. Andreae's Life, World-view, and Relations with Rosicrucianism and Alchemy. Vol. II. The Chymische Hochzeit with Notes and Commentary* (International Archives of the History of Ideas, no. 55; The Hague: Martinus Nijhoff, 1973), vol. I, 17; J. B. Craven, *Count Michael Maier. Doctor of Philosophy and of Medicine, Alchemist, Rosicrucian, Mystic, 1568–1622. Life and Writings* (Kirkwall: William Peace & Son, Albert Street, 1910), pp. 13–30; C. G. Jung, *Psychology and Alchemy*, tr. by R. F. C. Hull (Bollingen Series XX), in *The Collected Works of C. G. Jung*, Vol. 12, ed. by Herbert Read, Michael Fordham, Gerhard Adler, and William McGuire (reprint of the 1968 ed.; Princeton/Bollingen Paperback; Princeton: Princeton University Press, 1980), pp. 345–431; Robert M. Schuler, "Some spiritual alchemies of seventeenth-century England," *Journal of the History of Ideas 41* (1980), 293–318; Stanton J. Linden, "Alchemy and eschatology in seventeenth-century poetry," *Ambix 31* (1984), 102–24; Dixon, *Alchemical Imagery* (2, n. 72), pp. 10–11, 65, 68–9; Karl Hoheisel, "Christus und der philosophische Stein. Alchemie als über- und nichtchristlicher Heilsweg,"

texts, by only a small extension, turned alchemical operations and the alchemical exegesis of Moses into suitable vehicles for the interpretation of prophecy, for natural and moral histories ran together from beginning to end. An early eighteenth-century alchemist whom Newton was to study closely espoused precisely that idea in speaking of the philosophical (alchemical) chaos:

> ... *'Tis a Matter as Ancient as the World it self; ... 'tis that one thing, whence all things proceed... therefore Operation on it shews what the World was, what it is, and what it shall be.*[65]

Newton, in his commentary on the *Emerald Tablet* in the early 1680s, recognized the beatitude at the end of the alchemical process by elaborating on the Hermetic statement: "Thus will you have the glory of the whole world and all obscurities and all need and grief will flee from you."[66] As the beginning of the world in dark chaos adumbrates the beginning of the great work, so the prophesied end of history in a new paradise finds its parallel in the successful conclusion of the alchemical redemption of matter. Although one does not wish to suggest that Newton thought a crude, mechanical transfer of knowledge from one parallel text to the other was ever possible, there do seem to be two ways in which his full acceptance of the parallelism between natural and human history are reflected in his work.

One is in the effort to interpret Moses alchemically. Hints of that effort appear in Newton's correspondence with Burnet and in his commentary on the *Emerald Tablet*, both from the early 1680s, and also in Newton's early and long continued interest in "illumination" as a part of the alchemical process. In 1687 – the same year he published his own *Principia* – a book appeared in Paris that offered the "true theory" of the philosopher's stone based on the resemblance of the work to the original creation: *La lumière sortant par soy même des tenebres.*[67] The book presented the usual analogies based on a primal chaos in which the four elements were confusedly mixed, but in addition it had interesting comments on the activating spirit that was assumed to be the same in both alchemy and cosmogony and to have the broad ongoing cosmological role of vivification. In Newton's abstracts from *La lumière* one recognizes

Wolfenbütteler Forschungen, 32, (1986) 61–84; Keynes MS 28 (1, n. 35), f. 2r. Cf. Appendix B.

65. Cleidophorus Mystagogus, *Mercury's Caducean Rod* (2, n. 55), p. 4. Cf. Linden, "Alchemy and eschatology" (3, n. 64).
66. Keynes MS 28 (1, n. 35), f. 6v. Cf. Appendix B.
67. *La lumière* (1, n. 37).

substantial affinities with the "vegetable spirit" of his treatise on the vegetation of metals of about 1672. For example,

> The vapor of the elements which is very pure and almost insensible and contains in it the spirit of fire or light, which is the form of the universe and being so impregnated with the spirit of the universe represents the first chaos containing all things necessary to the creation, that is the matter universal and form universal. And in descending and becoming sensible it first puts on the body of the air which we breathe and becomes enclosed in it to nourish and vivify all nature.[68]

Again, the activating spirit is called the "sulfur or fire of gold," engendered in the mines of the philosophers,

> fixed, pure, balsamic; a corporal spirit diffused through all nature; the principle of all vegetation, life, attraction, sympathy, and motion; a composite of salt, sulfur, and mercury; the fire of mercury and most digested part thereof; the form informing all things; the innate heat of the elements; the lawful son of the sun and the true sun of nature. . . .[69]

The presupposition in these passages is always that the activating spirit of alchemy may be identified with the spirit of God that shaped the original chaos and continues to govern the world of vegetation. No doubt a fuller understanding of that mysterious agent would assist in the interpretation of Moses, and there is reason to believe that Isaac Newton discussed just this sort of alchemical exegesis with his young friend Nicolas Fatio de Duillier sometime before 1692, as they worked on the interpretation of prophecy together.[70]

The other way in which Newton's acceptance of the parallel texts metaphor informed his work was in his perennial search for evidence of the activity of the spirit in the world. The attempt Newton made to interpret Moses alchemically was relatively narrow in scope, but his attempt to find evidence of spiritual activity was as broad as might be. It included alchemy, where, as argued in Chapter 2, Newton sought evidence for spiritual activity in micromatter in the "active" processes of vegetation and fermentation. But Newton's search also included his biblical and historical studies, where he sought evidence for the activity of the God of history, especially in the interpretation of prophecy – where

68. Yahuda MS Var. 1, Newton MS 30 (1, n. 36), f. 2v. In this and subsequent quotations from Newton's abstracts of the book I have modernized spelling and punctuation. For Newton's original English, cf. Appendix C.
69. Babson MS 414B (1, n. 36), f. 1r. Cf. Appendix C.
70. This possibility will be explored in more detail in Chapter 6.

divine inspiration and historical fact seemed to him to intersect. As one may see in the next section, the natural and moral worlds found for him their deepest relationship not in the details of their parallel movements but in the fact that both were theaters for the activity of the spirit. Prophecy, fulfilled and correctly interpreted, provided for Newton exactly the same sort of evidence for the divine governance of the world as did "active" alchemical processes, and it served precisely the same metaphysical purpose. There is also reason to suppose that he considered the divine agent in both worlds to be the same.

Cosmogony, alchemy, and history

Dr. Cudworth was "much mistaken" in declaring them atheistical, Newton said, those oldest pagan cosmogonists who said the world arose from chaos. Cudworth had said in his *True Intellectual System of the Universe* that they had created everything, gods, and all, "out of Senseless and Stupid Matter." But Newton knew better, for by "night" they meant an invisible deity and by "Love" the spirit that moved on the face of the waters. For a world to arise from chaos at all, divine action was required, and so it was implicit in the accounts of Homer and Aristophanes even when they were not explicit about the ancient beliefs.[71] Moses of course was explicit, and his account adumbrated the alchemical work in which a "philosophical chaos" was illuminated and given life by the activating principle. Everything comes "out of black Chaos and its first matter," Newton said, "through the separation of the elements and the illumination of matter."

"Illumination" was rather a technical term in alchemy, as one has already seen, relating to the activation of matter. Since for Newton matter per se was always passive, and activity was the province of divinity, the "illuminating" or activating agent in the alchemical work

71. The William Andrews Clark Memorial Library, Los Angeles, CA, Isaac Newton, "Out of Cudworth," p. 1, published in James E. Force and Richard H. Popkin, *Essays on the Context, Nature, and Influence of Isaac Newton's Theology* (International Archives of the History of Ideas, 129; Dordrecht: Kluwer Academic Publishers, 1990), Appendix, pp. 207–13, quotation from p. 208; Ralph Cudworth, *The True Intellectual System of the Universe: The First Part; Wherein, All the Reason and Philosophy of Atheism is Confuted; and Its Impossibility Demonstrated* (London: printed for Richard Royston, bookseller to His most Sacred Majesty, 1678; Stuttgart-Bad Connstatt: Friedrick Frommann Verlag [Günther Holzboog], 1964), pp. 120–1.

was in some sense divine, just as the agent that acted on chaos in the original creation had been. One must now ask in what sense it was divine, this creative, illuminating principle that acted to shape and inform all the forms of matter as they arose from chaos, and for an answer one must turn to Newton's unorthodox religious convictions, his acceptance of Arian doctrines that had been considered heretical since the fourth century.

It is by no means clear from his manuscript remains that Newton had any closely reasoned views on the nature of the divine active principle when he first began to study alchemy in the late 1660s. However, he probably had become a convinced Arian by early in 1673, and it is in passages from Newton's theological papers of the 1680s and later that one finds suggestions that he had satisfactorily situated the vegetable spirit in God's cosmogonical and cosmological scheme of things. There can be little or no doubt that it was his Arian theology that enabled him to do so.[72]

The Arian position involved a subtle interpretation of the divinity of Christ that denied Christ full equality and coeternity with God the Father. In fact, Arius had argued that there was a time "when he [Christ] was not," when there was only the unique and almighty God who could not even properly be called Father before the advent of the Son. The Arian divine Son had certain creaturely qualities, and though he was and remained first among the creatures and was a mediator between God and man, he was not of the same essence as the Father. The essentialist, Trinitarian opponents of Arius carried the day against him, and the Nicaean Creed of 325 defined the Son as "being of one substance with the Father," a definition that became Christian orthodoxy. Most subsequent interpretations of the Arian position have emphasized its horrid heretical nature in so downgrading the Son of God, but it is not impossible occasionally to glimpse a vast cosmic dignity in the Arian Son. Brought into being by the Father before all worlds and all ages, the Son in his turn became the framer of the cosmos, the agent of God's creativity. Subordinate he was, certainly, but

72. Westfall, *Never at Rest* (1, n. 4), pp. 309–34; Dobbs, "Newton's alchemy and his theory of matter" (1, n. 5); Frank E. Manuel, *Isaac Newton, Historian* (Cambridge, MA: The Belknap Press of Harvard University Press, 1963), pp. 156–9; idem, *The Religion of Isaac Newton. The Fremantle Lectures 1973* (Oxford: Clarendon Press, 1974); Isaac Newton, *Sir Isaac Newton Theological Manuscripts*, selected and ed. with an intro. by H. McLachlan (Liverpool: At the University Press, 1950); Popkin, "Newton's biblical theology" (3, n. 29).

central to God's purposes in everything, for the Arian God was a transcendent and wholly sovereign Deity whose will was His primary attribute. The will of God for creation and guidance was worked through the agency of the Son, the creative Word or Logos.[73] As Newton put it, "God and his Son cannot be called one God upon account of their being consubstantial," but they may and should be called one God through a "unity of dominion,"

> the Son receiving all things from the Father, being subject to Him, executing His will, sitting in His throne and calling Him his God, and so is but one God with the Father as a king and his viceroy are but one king.[74]

Thus Christ is the viceroy, the spiritual being that acts as God's agent in the world.

Newton identified Christ with the Word and argued that he was with God before his incarnation, "even in the beginning," and that he was God's active agent throughout time, speaking to Adam in paradise and appearing to the patriarchs and Moses "by the name of God": "For the Father is the invisible God whom no eye hath seen nor can see." Christ wrestled with Jacob and gave the law on Mount Sinai; after his resurrection his testimony was "the spirit of prophecy."

> He is said to have been *in the beginning with God* and that *all things were made by him* to signify that as he is now gone to prepare a place for the blessed so in the beginning he prepared and formed this place in which we live, and thenceforward governed it. For the supreme God doth nothing by himself which he can do by others.[75]

73. Edward Peters, " 'The Heretics of Old': The definition of orthodoxy and heresy in Late Antiquity and the Early Middle Ages," in *Heresy and Authority in Medieval Europe. Documents in Translation*, ed. with intro. by Edward Peters (Philadelphia: University of Pennsylvania Press, 1980), pp. 13–56; Robert C. Gregg and Dennis E. Groh, *Early Arianism – A View of Salvation* (Philadelphia: Fortress Press, 1981).
74. Yahuda MS Var. 1, Newton MS 15.7 (2, n. 50), f. 154r; I have omitted all indications of Newton's deletions and interlineations and modernized spelling and punctuation. See also David Castillejo, *The Expanding Force in Newton's Cosmos As Shown in His Unpublished Papers* (Madrid: Ediciones de Arte y Bibliofilia, 1981), p. 74 (where this passage is also quoted), and James E. Force, *William Whiston, Honest Newtonian* (Cambridge University Press, 1985), esp. pp. 105–13 (where Force convincingly argues that one may better understand Newton's secret Arianism by way of Whiston's openly stated Arianism).
75. Yahuda MS Var. 1, Newton MS 15.5 (2, n. 50), ff. 96r–97r; I have omitted all indications of Newton's deletions and interlineations and modernized spelling and punctuation, but the italics are Newton's. Cf. Castillejo, *Ex-*

Newton's Christ is a very unorthodox Christ indeed but one whose many duties keep him engaged with the world throughout time. A part of his function is to insure God's continued relationship with His creation. Even though Newton's God is exceedingly transcendent, He never loses touch with His creation, for He always has the Christ transmitting His will into action in the world.

The duties assigned to Newton's Christ would seem to place him in charge of – or perhaps even identify him as – such active natural entities as the vegetable spirit, which as argued above, Newton saw as exercising God's providential care in shaping "[a]ll that diversity of natural things which we find suited to different times and places," God's alchemical "Vicegerent." Newton apparently thought that when organized matter first arose from chaos, Christ, as God's executive, directed the vegetative, nonmechanical processes between the most minute primordials ("in the beginning he prepared and formed this place in which we live"), then continued to direct the vegetative operations of nature ("and thenceforward governed it"). For, Newton said in the General Scholium, natural diversity "could arise from nothing but the ideas and will of a Being necessarily existing." It was the Christ, united with God in a "unity of dominion" though not of substance, that put the ideas into effect, and one is reminded here by his use of the word "dominion" that in his commentary on the *Emerald Tablet* Newton had declared "the Mercury of the philosophers," signified by Hermes, to have "dominion" in all three kingdoms of nature. Hermes was thus a pagan type of Christ himself, the Christ that shared in the Godhead through a "unity of dominion." As the Christ acted in his capacity of assistant to the Father to frame the cosmos in the beginning, and since the creation of the world prefigured the alchemical work, the active agent in alchemy is thus identified as the Logos – still acting as God's creative agent in the framing of the world of matter. Alchemy, and the nonmechanical processes with which it dealt, was the story of God's ongoing activity in the world of matter. Although Newton was perhaps rather an isolated figure in his Arianism,[76] in his acceptance of the Christus-*lapis* parallel he stood well

panding Force (3, n. 74), pp. 61–62; also King's College, Cambridge, Keynes MS 3, f. 42: "He is called the God who was in the beginning with God to signify that he was that God who walked in Paradise in the cool of the day & sentenced Adam & Eve & the Serpent, & by whom God the father made all things in the beginning & gave the promisses to the Patriarchs."

76. Newton may not have been so isolated in his Arianism as has commonly been assumed. Although John Milton's *De Doctrina Christiana* was not

within a broad spectrum of contemporary thought.[77] And in Plate 6[78] one may see a late-fifteenth-century representation of Christ at his cosmic work of the separation of the elements, reminiscent of Newton's *Commentarium* in which "our work" of alchemy "brings forth the beginning out of black Chaos and its first matter through the separation of the elements and the illumination of matter... the mode of which... was adumbrated in the creation of the world."

As alchemy was the story of God's ongoing activity in the world of matter for Newton, so history was the story of God's ongoing activity in the moral world, and as such it was a key for the interpretation of prophecy. Prophecy in the Bible was divinely inspired, and Newton spent untold hours on the writings of Daniel and the Apocalypse of St. John. But human beings could fully understand prophecy only after it had been fulfilled, for it was written in "mystical" language that was not readily accessible. In any event, a person was not to presume to interpret it with an eye for concrete prediction of the future. Only after the prophesied events had occurred could one see that they had been the fulfillment of prophecy. Then God's action in the world was demonstrated.

> The folly of Interpreters has been, to foretel times and things by this Prophecy [John's], as if God designed to make them prophets. By this rashness they have not only exposed themselves, but brought the Prophecy also into contempt. The design of God was much otherwise. He gave this and the Prophecies of the Old Testament, not to gratify men's curiosities by enabling them to foreknow things,

published until 1825, and there is some scholarly doubt that Milton should properly be termed an Arian in any case, some knowledge of Milton's ideas perhaps circulated among the Cambridge Platonists, a group that overlapped chronologically at Cambridge with both Milton and Newton: William B. Hunter, Jr., General Ed., *A Milton Encyclopedia*, with John T. Shawcross and John M. Steadman (Co-editors) and Purvis E. Boyette and Leonard Nathanson (Associate Editors) (8 vols.; Lewisburg, PA: Bucknell University Press; London: Associated University Presses, 1978–80), s.v. "Arianism." One of the Cambridge Platonists, Cudworth, was actually accused in 1693 of a "moderate Arianism": Redwood, *Reason, Ridicule and Religion* (1, n. 24), pp. 165 and 258, n. 132.

77. Cf. esp. Linden, "Alchemy and eschatology" (3, n. 64).
78. Bartholomaeus Anglicus, *De proprietatibus rerum*, tr. by John Trevisa (Westminster: Wynken de Worde, 1495), sig. eiii. This illustration has been reproduced in S. K. Heninger, Jr., *The Cosmographical Glass. Renaissance Diagrams of the Universe* (San Marino, CA: The Huntington Library, 1977), p. 100, as an illustration of Christ as Plato's Demiurge.

Plate 6. Bartholomaeus Anglicus, *De proprietatibus rerum*, tr. by John Trevisa (Westminster: Wynken de Worde, 1495), sig. e iii. Reproduced by permission of The Henry E. Huntington Library, San Marino, CA.

> but that after they were fulfilled they might be interpreted by the event, and his own Providence, not the Interpreters, be then manifested thereby to the world. For the event of things predicted many ages before, will then be a convincing argument that the world is governed by providence.[79]

79. Isaac Newton, *Observations upon the Prophecies of Daniel, and the Apocalypse of St. John. In Two Parts* (London: printed by J. Darby and T. Browne in Bartholomew-Close. And sold by J. Roberts in Warwick-lane, J. Tonson in the Strand, W. Innys and R. Manby at the West End of St. Paul's Church-Yard, J. Osborn and T. Longman in Pater-Noster-Row, J.

Newton's methodology in prophetic interpretation was undoubtedly influenced by the methods of others: Mede, More, Alsted.[80] Yet there was in addition something peculiarly Newtonian about it. In Newton's mind history seemed to bear a direct correspondence with experimental or even mathematical demonstration. Just as an experiment might enable the investigator to decide between alternative theories of natural phenomena, so historical facts might enable the interpreter to choose between possible interpretations of prophecy. For Newton it was only the firm correspondence of fact with correctly interpreted prophecy that provided an adequate demonstration of God's providential action. What had been adumbrated by divine agency in the prophecy had been fulfilled by divine agency. What God had said He would do, He had done. That, and only that, provided for Newton a "convincing argument" for God's providential governance of the moral world. When actual historical developments exactly matched predicted ones in "the event of things predicted many ages before," one hears an echo of that universally satisfying geometrical conclusion, *quod erat demonstrandum*: Q. E. D.

It was of course just exactly that that Newton wanted to prove with his biblical and historical studies – God's providential action in the moral world – just as he desired by his alchemical studies to demonstrate God's providential action in the natural world. The divine agent was the same too, for the Christ was the channel through which God spoke to the prophets. True religion, though now corrupted, had consisted of:

> believing, adoring, and obeying one invisible God or supreme monarch of the universe whose dominion is boundless and irresistible: and in this monarchy one visible Lord Jesus Christ by whom God governs the world, the Word or Grace of God [whose voice is to be

Noon near Mercers Chapel in Cheapside, T. Hatchett at the Royal Exchange, S. Harding in St. Martin's lane, J. Stagg in Westminster-Hall, J. Parker in Pall-mall, and J. Brindley in New Bond-street, 1733), pp. 251–2.

80. Charles Webster, "Prophecy," in *From Paracelsus to Newton. Magic and the Making of Modern Science. The Eddington Memorial Lectures Delivered at Cambridge November 1980* (Cambridge University Press, 1982, pp. 15–47; Westfall, *Never at Rest* (1, n. 4), pp. 319–30; Popkin, "Newton's biblical theology" (3, n. 29). For Newton seen primarily as an interpreter of prophecy and as part of a growing, developing intellectual tradition on interpretation, see also LeRoy Edwin Froom, *The Prophetic Faith of our Fathers. The Historical Development of Prophetic Interpretation* (4 vols.; Washington, D.C.: Review and Herald, 1946–54), esp. vol. II, 506–669.

obeyed because the name of God is in him, and] who by his messenger spake to the Prophets.[81]

God's activity

Human beings had once known how to worship God properly, and had understood the true religion. But humanity had slipped into idolatry, had begun to attribute the powers of the one true God to false gods, and the knowledge of true religion had been corrupted or lost. However, true religion might be restored. One comes finally to that powerful fountain of motivation that kept Newton at his furnace year after year, at his desk through tome after tome, patiently nursing his vegetating chemicals, patiently matching Germanic invasions with the horns of the prophetic beasts, patiently searching for what had been lost – the knowledge of God's activity in the world.

In a sermon against idolatry in all its forms Newton argued that it is for His *actions* that God wants to be worshipped, His actions in "creating, preserving, and governing all things." To celebrate God for His essence – His "eternity, immensity, omnisciency, and omnipotency" – is "very pious" and indeed "is the duty of every creature to do it according to his capacity." But those attributes spring not from the freedom of God's will "but from the necessity of His nature."

> And as the wisest of men delight not so much to be commended for their height of birth, strength of body, beauty, strong memory, large fantasy, or other such gifts of nature, as for their wise, good, and great actions, the issues of their will: so the wisest of beings requires of us to be celebrated not so much for His essence as for His actions, and the creating, preserving, and governing of all things according to His good will and pleasure. The wisdom, power, goodness, and justice which He always exerts in His actions are His glory....[82]

His actions are His glory, and He is jealous of His honor regarding them, "even to the least tittle." His actions are the glory celebrated from one end of Scripture to the other, by Prophets, Apostles, and especially the Psalmist. His actions, the "issues" of His will, are indeed

81. Yahuda MS Var. 1, Newton MS 15 (2, n. 50), p. 99, quoted in Castillejo, *Expanding Force* (3, n. 74), p. 64.
82. Isaac Newton, Jewish National and University Library, Jerusalem, Yahuda MS Var. 1, Newton MS 21, f. 2. I have omitted all indications of Newton's deletions and interlineations in this quotation and in the following one from this same manuscript, and also modernized spelling and punctuation.

> the only glory by which God manifests Himself to His creatures, and which His creatures are able to behold in Him, the reason why His creatures worship Him, and the life and soul of all the worship we can give Him. . . .[83]

The worship of God for His *activity* in the world – in creating it, preserving it, and governing it according to His will – that is the true religion. That is what Isaac Newton intended to restore through his study of God's activity in cosmogony, alchemy, and history.

83. Ibid., ff. 2–3.

4

Modes of divine activity in the world: before the Principia

Introduction

Newton's conviction that God acted in the world never wavered, but his explanations of the manner of God's acting underwent startling and dramatic changes as he examined certain questions. Which events does God manage by "mechanical principles" and which ones by "active principles"? Into which class – "mechanical" or "active" – should gravity fall? Chemical changes? The phenomena of life? Did God employ a special agent to put His will into effect at the beginning of the world, before He established the laws of His ordinary concourse? How much activity in the world results from God's ordinary concourse, His *potentia ordinata*, and how much results directly from His absolute will, His *potentia absoluta*? Is the law of gravity evidence of His *potentia ordinata* or His *potentia absoluta*? Will gravity play a role in the final conflagration of this world promised by Scripture? If the alchemical spirit is God's vice-regent in the phenomena of life, is it to be identified directly with the Logos, or is it to be identified with the "electric and elastic spirit" that is itself presumably an agent of the Logos? Have prophecy and miracles ceased in these latter days now that they have accomplished their part in the providential design?

Newton did not of course ask all those questions from the beginning of his work in natural philosophy though there is every reason to think that he always considered the world to have been created by God and to continue under divine guidance. In that he was not alone: much natural philosophy in the seventeenth century was done under that rubric. To doubt those particular cultural assumptions was to court charges of atheism, as Hobbes, for example, had done, and Newton, so far from challenging the assumptions, always strove to find evidence to sustain them.

The changes in Newton's thought seem not to have come from any questioning of God's creativity or providence but rather from the impact of several different traditions upon his various partial syntheses as well as from the impact of his own discoveries. One can hardly hope now to unravel the full complexity of his mental processes. Nevertheless, there are significant passages in many of Newton's published and unpublished writings that serve as indicators of his development and suggest the differential influence of various ways of attacking the problems associated with divine activity during different periods of his life. The next four chapters will explore the impact upon Newton's thought of some of those influences, as he continually strove to establish the parameters of a system that encompassed both the natural and the divine.

Of the almost limitless variety of systems generated or recovered during the Renaissance, some forms of Neoplatonism and Stoicism certainly were influential upon him. Both provided severe challenges to the orthodox mechanism that held all motion in the world to come from the impact of particles of matter upon each other. For the Neoplatonists matter was totally passive and required the guidance of spiritual principles for many classes of phenomena. But where the Neoplatonists might postulate a hierarchy of spiritual entities between matter and the divine, the Stoics offered a rather sharp dichotomy between "passive principle" and "active principle," equating the two principles with matter and God. Even so, the Stoics related activity very closely to certain tenuous forms of matter. Both approaches had been incorporated into alchemical literature and were readily available elsewhere as well, as one has already seen, and there is evidence for Newton's consideration of first the one and then the other at different times.

The pure Cartesian mechanism that Newton early encountered was soon modified in some of his papers by the addition of active principles, but at the same time Newton probed Scriptural and theological traditions. His decision in favor of the Arian heresy was made on the basis of his own researches into the polemics of fourth-century Christianity, and, so far as one can tell from the surviving manuscripts, the decision was made independently of any issues relating to God's activity in the world. Nevertheless, the Christology of the Arians had important implications for natural philosophy, for the Arian God was so transcendent that He required an intermediate agent to put His will into effect both in the creation of the world and in its governance. The Christ, in his capacity as creative Logos, was that agent, and Newton was not unaware of the implications of Arian Christology. As one has already seen, the Logos came to be intimately connected in his mind with the alchemical spirit. Furthermore, Newton adhered to the theological framework of *potentia dei ordinata et absoluta*. That was the common structure of thought that guided many of his contemporaries when they inquired into the relationships between God and the world, and manuscript evidence declares Newton's acceptance of it.

To the welter of traditions that Newton encountered and considered, his own mathematical and experimental discoveries must be added. The determinative influence of his own mathematics has seldom been lost to view in considerations of his conceptual development. In the predominant strand of Newtonian historiography, mathematics has often been treated as the only factor worthy of consideration in Newton's characterization of the force of gravity, which force, however, Newton finally considered to be one mode of divine activity, as one may see in Chapters 6 and 7. When it was supposed that only scientific laws of lasting importance need be considered by historians of science, and not the cultural context in which they were generated, the close connection between Newton's mathematics and his law of universal gravitation seemed so obvious as to exclude other factors as irrelevant. The disciplinary maturation of the history of science, plus the increasingly intensive study of Newton's unpublished papers during the last twenty years, have, however, produced challenges to the earlier view. Newton changed his mind about the nature of the gravitational force so many times, as the papers show, that it became reasonable to argue that other factors were involved in his formulation of universal gravity. Both Westfall and I have been forward in suggesting that the active principle in alchemy must have served Newton as a conceptual guide in his treatment of gravity as an active principle.[1]

The religious dimension of this historiographic debate has been somewhat neglected on both sides, however, and it is precisely that deficiency that has made the "mathematical" and the "alchemical" points of view seem so irreconcilable. As is often the case in such arguments, the removal of issues to a new and more comprehensive plane of analysis helps to resolve their apparent contradictions. By treating both the action of gravity and the action of the alchemical spirit as modes of divine activity in the world, one may hope to effect such a resolution here. It will certainly be seen that both mathematics and alchemy played substantial roles in Newton's conceptualization of gravity, but it will also be seen that the relationship between gravity and the active alchemical principle was never a simple one.

In this chapter one may see that Newton at first considered gravity to operate strictly by a mechanical principle, in contradistinction to the active principle of alchemy or vegetable chemistry, but for a short time he partially fused the different speculative material agents responsible respectively for gravitation and vegetation. In Chapter 5 one finds, on the other hand, that Newton was forced to abandon his mechanical definition of gravitation through a combination of mathematical and

1. Cf. the discussion supra (1, pp. 3–5).

observational–experimental evidence. Prior to the monumental mathematical and experimental discoveries of 1684, it is doubtful that Newton had ever entertained the notion that gravity might be strictly an active principle, much less that gravitation might be due to an incorporeal cause. He had long since decided, however, that all natural operations were effected through either mechanical or active principles; so, by his own dichotomy, if gravity was not mechanical, it must be active. Mathematics, observation, and experiment were thus at least partially responsible for Newton's later conceptualization of gravity as an active principle.

But, as will appear in Chapter 6, Newton's penultimate characterization of gravity as an active principle was almost certainly mediated by the generalized Stoic dichotomy between the passive and the active (in which the ultimate active principle was fully equivalent to the Stoic Deity) rather than stemming directly and exclusively from the active principle of alchemy itself. Finally, though, Newton's last speculative cosmic aether, suggested in the 1717/18 *Opticks* as a cause for gravity, probably was a conceptual extrapolation from the microcosmic "electric and elastic spirit" that was closely associated with alchemical vegetability, as one may see in Chapter 7.

It is only when one holds firm to the realization that Newton was searching for Truth and for the modes of divine activity in the world that the logic of his many explanatory shifts and turns becomes apparent. In what is to follow, one may most profitably analyze Newton's conceptual development by considering his views on *both* gravity and alchemy from the 1660s until the end of his remarkable life.

Gravity and alchemy in the 1660s: mechanical and Neoplatonic influences

At first in the 1660s Newton considered gravity to be a mechanical mode of action. He began to ponder the problem about 1664, and the editors of the undergraduate notebook that records his early ideas have suggested influences from Descartes, Digby, and Boyle on Newton's first mechanical theory, which was derivative and nonmathematical.

> The matter causing gravity must pass through all the pores of a body. It must ascend again, (1) for either the bowels of the Earth must have had large cavities and inanities to contain it, (2) or else the matter must swell the Earth. . . . For it must descend very fast and swift as appears by the falling of bodies and by the great pressure toward the Earth. It must ascend in another form than it descends,

> or else it would have a like force to bear bodies up as it has to press them down.... The stream descending will grow thicker as it comes nearer the Earth; but it will not lose its swiftness until it finds as much opposition as it has help from the flood following behind it. But when the streams meet on all sides in the midst of the Earth, they must needs be compressed into a narrow room, closely pressed together, and thus very much opposing one another either turn back the same way they came, or crowd through one another's streams with much difficulty and pressure, and so be compacted and the descending stream will keep them compacted, by continually pressing them to the Earth until they arise to the place from whence they came. There they will attain their former liberty.[2]

Though the precise form is Newton's own, his theory is a restatement of impact physics, and it is conventional, orthodox mechanical philosophy: in short, an aether theory of gravitation. Bodies descend through the impulsion of fine material particles; it is a mechanical stream of aethereal matter causing gravity.

No hint exists in Newton's earliest statement of what gravity is to become for him: an active principle directly or indirectly dependent upon God's will. Yet even in the student notebook there is a recognition of God's omnipresence, the omnipresence that is later to subsume universal gravity. Violent motion in a plenum requires that some matter be crowded out of the way of a moving body, Newton said; therefore the motion meets with resistance. Not so with violent motion in a vacuum.

> But *in vacuo* it meets with nothing impenetrable to stay it. It is true God is as far as vacuum extends, but he, being a spirit and penetrating all matter, can be no obstacle to the motion of matter; no more than if nothing were in its way.[3]

Newton is to repeat much later his conviction that God is present where there is no body, as well as present where body is also present.[4] There, as here, God penetrates all matter.[5] But whereas later the omnipresence of God and His ability to penetrate matter have the utmost significance with respect to gravity, that is not the case in the student notebook, where

2. McGuire and Tamny, *Certain Philosophical Questions* (1, n. 22), "Gravitation, attraction, and cohesion," pp. 275–95; Newton's "Of gravity and levity," pp. 362–5, 426–31, quotation from pp. 363, 365, 427.
3. Ibid., p. 409.
4. This is discussed in detail in Chapter 6.
5. Cf. Edward Grant, *Much Ado about Nothing. Theories of Space and Vacuum from the Middle Ages to the Scientific Revolution* (Cambridge University Press, 1981), pp. 227–8 on Henry More's belief in an "All-penetrating" Deity.

gravity is caused by the mechanical motion of small particles of matter to which God's presence simply constitutes "no obstacle."

Some five years later the first active principle appeared in Newton's papers, in his "Propositions" of about 1669.[6] In the "Propositions" Newton argued for the existence of a vital agent diffused through all things. The agent is "the mercurial spirit, most subtle and wholly volatile." Though it does have a "general method of operating in all things," it also has "particularities" of operation according to its subject matter and is responsible for both putrefaction and generation. The mercuriality, subtlety, and volatility of the vital "spirit" seem to imply that it is material, though barely so. It has nothing to do with gravity, only with phenomena of life and death.

Though Newton's synopsis in the "Propositions" came entirely from the alchemical tradition, as his own annotations in the manuscript indicate, his interest in the alchemical vital spirit was probably stimulated by Neoplatonic critiques of Descartes's mechanical system.[7] Descartes had attributed too much efficacy to brute matter moving by ordinary mechanical laws. If the original common matter of the universe only moved "naturally," there could be no differentiation, neither sun, nor sky, nor earth, nor vortices, nor any heterogeneous thing, as Henry More had been quick to observe to Descartes.[8] More, in his later publications, as well as Ralph Cudworth in his *True Intellectual System of the Universe*, forced the issue of the inadequacy of a purely mechanical system. The realm of matter was the realm of necessity, of fate, of nature. Matter to the Platonists had the lowest possible ontological standing and could not by itself perform the higher operations associated with providential design. For design, spiritual guidance of material particles was a necessity. A "spirit of nature" or a "soul of the world," intermediate between God and matter, provided such guidance and organized matter according to divine design.[9] Newton fully accepted the program of the Cambridge

6. Keynes MS 12A (1, n. 29). Cf. the discussion supra (2, pp. 22–5).
7. Dobbs, *Foundations* (1, n. 1), pp. 100–5.
8. Henry More to René Descartes, 23 July 1649, in René Descartes, *Correspondance avec Arnauld et Morus. Texte latin et traduction. Introduction et notes par Geneviève Lewis* (Bibliothèque des textes philosophiques, Directeur: Henri Gouhier; Paris: Librairie Philosophique J. Vrin, 1953), pp. 170–81, esp. pp. 176–9; also in René Descartes, *Oeuvres de Descartes publiées par Charles Adam & Paul Tannery* (11 vols.; Paris: Librairie Philosophique J. Vrin, 1964–74), vol. V, 376–83, esp. 381.
9. Cassirer, *The Platonic Renaissance in England* (1, n. 25); Serge Hutin, *Henry More. Essaie sur les doctrines théosophiques chez les Platoniciens de Cambridge* (Studien und Materialien zur Geschichte der Philosophie, Herausge-

Platonists, not rejecting mechanical philosophy per se but desiring to modify it to include spiritual principles capable of initiating motion in passive matter and then guiding it into providentially designed configurations. But not content with a philosophical statement of that program, Newton turned to a study of alchemy in order to implement it, for alchemy not only discussed active principles but also claimed to provide experimental evidence of their operation.

McGuire has argued that the more general metaphysical tradition of Neoplatonism provides an adequate explanation of Newton's use of active principles in his later work and that there is thus no need to see alchemy as playing an essential role.[10] From the strictly philosophical point of view, McGuire perhaps makes a legitimate point, but his argument is not tenable from the historical point of view. As a matter of historical record, Newton's papers demonstrate that he used alchemical literature and alchemical experimentation to search for the facts regarding active principles. With those facts he hoped to found a more adequate natural philosophy that included not only mechanical principles but spiritual ones as well.

There is in any event one way in which Newton's earliest work on active principles differs from that of the Neoplatonic philosophers, a difference having to do with the materiality of the alchemical spirit. The most likely source of Newton's philosophical stimulus in the 1660s on the needed addenda to the mechanical philosophy was Henry More's *Immortality of the Soul* (1659); the guiding spirit of nature More postulated there was decisively incorporeal.[11] But, as one has seen, Newton's alchemical spirit had elements of materiality with its subtlety, mercuriality, and volatility. He called it a "spirit," as he had called God a "spirit" in the student notebook, but seventeenth-century "spirits" were notoriously ambiguous, existing in a broad gray area between solid matter and the Deity. They could be either corporeal or incorporeal as required

geben von Heinz Heimsoeth, Dieter Henrich, und Giorgio Tonelli, Band 2; Hildesheim: Georg Olms Verlagsbuchhandlung, 1966); J. E. McGuire, "Force, active principles, and Newton's invisible realm," *Ambix 15* (1968), 154–208; idem, "Neoplatonism and active principles" (2, n. 52).

10. McGuire, "Neoplatonism and active principles" (2, n. 52), pp. 140–1, n. 84.
11. Henry More, *The Immortality of the Soul, So farre forth as it is demonstrable from the Knowledge of Nature and the Light of Reason* (London: printed by J. Flesher, for William Morden bookseller in Cambridge, 1659), p. 450; Edwin Arthur Burtt, *The Metaphysical Foundations of Modern Science* (reprint of the 2nd rev. ed.; Doubleday Anchor Books; Garden City, NY: Doubleday & Co., 1954), pp. 135–43.

by the exigencies of the problem to be explained.[12] Most of the seventeenth-century ambiguity can probably be traced ultimately to the initial ambiguity incorporated into the active principle of the Stoics, which, over the course of centuries, had produced systems in which "spirit" or *pneuma* ranged from complete corporeality to total incorporeality.[13] Evidence will be advanced later to show that for Newton God was the only purely incorporeal spirit; all created spirits had some degree of corporeality. That had not been the case with More, for whom created spirits as well as the Deity were incorporeal. Had Newton deliberately broken with More on this point by the end of the 1660s? That seems unlikely; probably Newton was only recording in the "Propositions" the characteristics of the vital spirit as they were reported in the alchemical literature. Although degrees of corporeality were later to be of great concern to him, there is no evidence that he attempted to resolve such issues in the 1660s.

Gravity and alchemy through 1675: Stoic and theological influences

In the 1670s Newton began to weave together those two concepts that had originated in totally different areas of his studies: the concept of the material, mechanical aether that was the agent of gravity, and the active, yet still material "mercurial spirit" that was the agent of vitality. He seems to have made the relatively simple assumption that the more "spiritual" an agent is, the finer the matter of which it is composed, or, conversely, the smaller the material particle, the greater its "spirituality" and activity. That appears to be the assumption that allowed him to treat his mechanical and nonmechanical agents in a comparable format during most of the ensuing decade and even sometimes to combine their functions in the economy of nature.

Among the Stoics an assumption similar to Newton's was made, relating decreasing materiality with increasing activity. In ascending from earth through water and air to fire in the conventional ranking of the elements, the Stoics considered elementary matter to become more tenuous but at the same time more tense and active. Even more significantly,

12. Pagel, *Paracelsus* (1, n. 25), passim; McMullin, *Newton on Matter and Activity* (2, n. 61), pp. 54–6; Norma E. Emerton, *The Scientific Reinterpretation of Form*, foreword by L. Pearce Williams (Cornell History of Science Series; Ithaca, NY: Cornell University Press, 1984), pp. 179–93.
13. Verbeke, *L'évolution de la doctrine du pneuma* (3, n. 38), pp. 21–4.

in Stoic philosophy all elementary bodies continued to bear within themselves, in their innermost constitution, traces of the divine and almost totally immaterial creative fire that had formed the universe and that was the most tense and active substance of all. Distinguished from the elementary fire that can only burn and destroy (fire conceived as an element or qualified body), the hidden creative fire of the Stoics (fire conceived as the active principle) was the material agent for the shaping of matter but was also identified with the Stoic Deity.[14]

The alchemists were often very close to that Stoic distinction between fiery principle and fiery element, insisting that their occult fire, their hidden active principle, was by no means to be equated with ordinary fire, for the latter could only burn and destroy while "their" fire was a creative one.[15] Newton spoke of the "secret fire" in his "Vegetation" manuscript, and among his later alchemical notes one finds just such a concept embodied in an alchemical description of the creation. Following the descending order for the creation of the elements from fiery creative principle to air, and so forth, Newton recorded the following scheme.

> The vapor of the elements ... is very pure and almost insensible and contains in it the spirit of fire or light, which is the form of the universe.... And in descending and becoming sensible it first puts on the body of the air which we breathe and becomes enclosed in it to nourish and vivify all nature. And that it may act more easily upon the grosser bodies of vegetables and minerals, it becomes still denser and insinuates itself into the water.[16]

Newton's assumption that "spirituality" and activity are inversely proportional to materiality thus could have been rooted in alchemy, in Stoic

14. David E. Hahm, *The Origins of Stoic Cosmology* (n.p.: Ohio State University Press, 1977); Michael Lapidge, "*Archai* and *Stoicheia*: a problem in Stoic cosmology," *Phronesis 18* (1973), 240–78; Paul Hager, "Chrysippus' theory of the pneuma," *Prudentia 14* (1982), 97–108.
15. See, for example, an alchemical paper of Newton's in which he is attempting to understand "The three mysterious fires," published in Samuel Devons, "Newton the alchemist?" *Columbia Library Columns 20* (1971), 16–26. See also F. Sherwood Taylor, "The idea of the quintessence," in *Science, Medicine, and History. Essays on the Evolution of Scientific Thought and Medical Practice Written in Honour of Charles Singer*, collected and ed. by E. Ashworth Underwood (2 vols.; London: Oxford University Press, Geoffrey Cumberlege, 1953), vol. I, 247–65; Lapidge, "*Archai* and *Stoicheia*" (4, n. 14); Hager, "Chrysippus' theory" (4, n. 14).
16. Yahuda MS Var. 1, Newton MS 30 (1, n. 36), f. 2v. I have modernized spelling and punctuation in this and subsequent quotations from this manuscript. For Newton's original English, cf. Appendix C.

philosophy, or in both. Whatever its source, it was an assumption that served him not only in the 1670s but beyond. Translated into the terminology of force, the same assumption appeared finally in one of the Queries to the *Opticks* in Newton's public presentation of his last late speculative aether for the explanation of gravity: "The exceeding smallness of its Particles may contribute to the greatness of the force by which those Particles may recede from one another...,"[17] and McGuire has noted a number of closely related draft Queries to which the same idea is fundamental.[18]

Utilizing the assumption, then, that the smaller the material particles the greater their activity, Newton was able to treat together in one physical system both mechanical and active processes. In the early 1670s the mechanical gravitational aether and the nonmechanical alchemical spirit appeared partially fused in Newton's treatise "Of Natures obvious laws & processes in vegetation."[19] In this important manuscript the matter causing gravity was thoroughly "entangled" with active spirits, yet also in this same set of papers Newton distinguished with some care the natural processes he thought were "mechanical" from those he thought were "vegetable." That Newton took up the problem of gravity at all in this primarily alchemical document is of supreme significance, and one must follow his thoughts on it in detail.

Newton was apparently led to a reconsideration of his earlier gravitational aether through a discussion of its "grosser" counterpart of atmospheric air. Air is generated by several processes, he said, the chief of which are "mineral dissolutions and fermentations" in the earth. Being then "protruded" by more air ascending under it, the air rises higher and higher until it "straggle into the aethereal regions," carrying with it vapors, exhalations, and even whole clouds, the whole mass losing its gravity in the ascent.[20] The air ascending in this speculative scenario is evidently that matter in a "grosser consistence" than aether of which Newton spoke in his student notebook. It needs to be grosser so it will not press bodies upward with as great a force as the finer aether presses them down.

In the "Vegetation" manuscript Newton proceeded to a quantitative estimate of the air thus being continuously generated, then noted that when air reaches the aethereal regions it presses on the aether.

17. Newton, *Opticks* (2, n. 15), p. 352.
18. McGuire, "Force, active principles, and Newton's invisible realm" (4, n. 9), p. 160.
19. Dibner Collection MSS 1031 B (1, n. 30), ff. 3v–4r. Cf. Appendix A.
20. Ibid., f. 3v.

> This [the air] constantly crowding for room, the aether will become pressed thereby and so forced continually to descend into the earth from whence the air came; and there it is gradually condensed and interwoven with bodies it meets there and promotes their actions, being a tender ferment.[21]

In that brief passage there are three points of special interest. First, particles of air and aether change places with each other by "crowding." The motion is a circulatory one, not unlike that of the Cartesian vortices. Second, the limits for the rising air are set by the "aethereal regions," which are at no great distance from the earth. Thus the whole system seems to be earthbound, an interpretation reinforced by the recollection of the frequent references to the earth in the student description, where "the bowels of the Earth" must have many cavities or else swell with the aethereal flood or extrude the aethereal matter again in another form. Such an interpretation is further suggested by a related passage in the "Vegetation" manuscript in which the earth "draws in aethereal breath ... and transpires again with gross exhalation."[22] Third, and perhaps most important for the understanding of Newton's evolving thought, is the fact that the "tender ferment," the vital and vegetable principle, has come to be associated with the gravitational aether. The descending aether becomes "interwoven" with bodies and "promotes their actions," that is, it initiates or supports activity in ordinarily passive matter.

Not that the gravitational function of the aether had been displaced. By no means, for Newton continued,

> But in its descent it [the aether] endeavors to bear along what bodies it passes through, that is, makes them heavy, and this action is promoted by the tenacious elastic constitution, whereby it takes the greater hold on things in its way, and by its vast swiftness. So much aether ought to descend as air and exhalations ascend, and therefore the aether being by many degrees more thin and rare (as air is than water), it must descend so much the swifter and consequently have so much more efficacy to drive bodies downward than air has to drive them up.[23]

Newton still thought at that time that his gravity was mechanical in its operation, for he explicitly observed later in the manuscript that "Nature's actions are either vegetable or purely mechanical (gravity, flux, meteors, vulgar chemistry)."[24] But in fact, though his aethereal particles still acted primarily by impact, they had also acquired a "tenacious elastic

21. Ibid.
22. Ibid.
23. Ibid.
24. Ibid., f. 5r.

constitution" that enabled them to take a "greater hold" on objects. Those new properties added an element to gravitational action that was not strictly mechanical, one that was later to become more prominent in Newton's speculations.

But in the early 1670s gravity remained primarily mechanical for Newton and it also still seemed to be earthbound. Newton continued his discussion of it in a way that apparently referred it only to the earth.

> And this is very agreeable to nature's proceedings, to make a circulation of all things. Thus this earth resembles a great animal or rather inanimate vegetable, draws in aethereal breath for its daily refreshment and vital ferment and transpires again with gross exhalations. And, according to the conditions of all things living, ought to have its time of beginning, youth, old age, and perishing.[25]

Newton almost certainly derived that animistic conception of the earth from the Stoics.[26]

Perhaps as a result, however, of restating the vital processes of the earth-animal so explicitly for himself, Newton then abruptly dropped his discussion of the gravitational function of the aether and resumed a discussion of its vegetative function. Indeed, it is not clear in these sections of the manuscript that the two functions were really quite as distinct in his mind as they had once been. Although parts of the following passages were quoted before to demonstrate the nature of Newton's alchemical "vegetable spirit," they will bear repeating here to emphasize the partial fusion of the mechanical and vegetative functions of the aether in Newton's thought during the transitional period of the early 1670s.

> [This is the subtle spirit which searches the most hidden recesses of all grosser matter, which enters their smallest pores and divides them more subtly than any other material power whatever. (Not after the way of common menstruums by rending them violently asunder, etc.) This is nature's universal agent, her secret fire, the only ferment and principle of all vegetation. The material soul of all matter, which being constantly inspired from above, pervades and concretes with

25. Ibid., f. 3v.
26. Cf. Seneca, *Natural Questions*, VI, 16.1, in *Seneca in Ten Volumes, Volumes VII and X*, tr. by Thomas H. Corcoran, ed. by E. H. Warmington (The Loeb Classical Library; Cambridge, MA: Harvard University Press; London: William Heinemann, 1971–72). I am indebted to David E. Hahm for this reference to Seneca. See also *Plato's Cosmology* (2, n. 32), p. 332. I am indebted for this reference to Plato and his commentators to Shmuel Scolnicov. See also the arguments for the animation of the world put forward by the sixteenth-century Christian Stoic Justus Lipsius: Jason Lewis Saunders, *Justus Lipsius, The Philosophy of Renaissance Stoicism* (New York: The Liberal Arts Press, 1955), pp. 196–9.

> it into one form; and then, if incited by a gentle heat, actuates and enlivens it; but so tender and subtle is it withal as to vanish at the least excess and (having once begun to act) to cease acting forever and congeal in the matter at the defect of heat, unless it receive new life from a fresh ferment. And thus perhaps a great part if not all the moles of sensible matter is nothing but aether congealed and interwoven into textures whose life depends on that part of it which is in a middle state, not wholly distinct and loose from it like the aether in which it swims as in a fluid, nor wholly joined and compacted together with it under one form; in some degree condensed, united to it yet remaining of a much more rare, tender, and subtle disposition; and so this seems to be the principle of its acting, to resolve the body and be mutually condensed by it, and so mix under one form, being of one root, and grow together till the compositum attain the same state which the body had before solution. Hence 1 the earth needs a constant fresh supply of aether, 2 Bodies are subtiliated by solution.][27]

In its descent aether made bodies heavy, and it condensed to form sensible matter, yet the rarer, tenderer, and more subtle parts of the aether were those that contributed life. The rarer parts were thus more active than the general aether in which they swam "as in a fluid," yet they were "united" to it. Thus were the gravitational and vegetative functions conjoined so that the mechanical gravitational aether became the bearer of the life-giving spirit.

Newton, however, set brackets around that entire passage, an act that often may be taken to mean a reconsideration on his part of the bracketed material, perhaps with an eye to cancellation. Certainly in this instance he seems to have made an effort to differentiate again the mechanical (passive) and vegetative (active) aethereal functions, for immediately after the bracketed passage he began a new paragraph by saying: "Note that it is probable the aether is but a vehicle to some more active spirit." That statement seems to place the main body of the aether – presumably that responsible for gravitation – firmly in the category of mechanical action again. Yet, if that was Newton's intention, he was not entirely successful in his effort, for in the remainder of the paragraph, the vegetable spirit remains "entangled" in the aether, and generating and growing bodies imbibe aether as well as air and spirit, which seems again to link even the gravitational aether to the process of vegetation and to give it a more than mechanical function.

> Note that it is probable the aether is but a vehicle to some more active spirit. The bodies may be concreted of both together; they may imbibe aether as well as air in generation, and in that aether

27. Dibner Collection MSS 1031 B (1, n. 30), ff. 3v–4r. Cf. Appendix A.

> the spirit is entangled. This spirit is perhaps the body of light because both have a prodigious active principle, both are perpetual workers.[28]

By 1675 Newton's thoughts had matured to the extent that he was willing to launch them into the world in "An Hypothesis explaining the Properties of Light, discoursed of in my severall Papers." When one compares the "Hypothesis," which he sent to the Royal Society, with the private manuscript on "Vegetation," one sees that three notable shifts have taken place. Two of these shifts have to do with the nature of the gravitational principle: it becomes both more active and more universal. The third shift reflects the revolution in theological principles Newton had undergone by about 1673.

First, the gravitational principle had been further spiritualized. By 1675 it was *not* the relatively phlegmatic bulk of the aether that was responsible for gravitation, but a special thin part of it, something especially subtle that was similar to the vital spirit in air. Though still material, Newton's aether had become extraordinarily complex and was filled with vapors, exhalations, and effluvia. After treating electrical matter as aethereal and suggesting the condensation of aethereal electrical matter to be the cause of what one ordinarily calls electrical attraction, Newton made a parallel suggestion for gravitational attraction.

> [S]o may the gravitating attraction of the Earth be caused by the continuall condensation of some other such like aethereall Spirit, not of the maine body of flegmatic aether, but of something very thinly & subtily diffused through it, perhaps of an unctuous or Gummy, tenacious & Springy nature, and bearing much the same relation to aether, wch the vitall aereall Spirit requisite for the conservation of flame & vitall motions . . . does to Air.[29]

So a state of corporeality even finer than that of the main body of the aether had definitely been assigned to the gravitational "spirit" by 1675. In addition the operation of this "aethereall Spirit" by mechanical impact had been further eroded by a stronger emphasis on its unctuous, gummy, tenacious, and springy qualities. And even more significant, the gravitational function and the vegetative function of the aether have been even more closely fused than in the "Vegetation" manuscript, for the passage just quoted continued, without a break, with a discourse on what is transparently the vegetative function even though the topic as Newton introduced it had been concerned with gravitation only.

> For if such an aetherall Spirit may be condensed in fermenting or burning bodies, {or otherwise inspissated in ye pores of ye earth to

28. Ibid., f. 4r.
29. Newton, *Correspondence* (2, n. 14), vol. I, 362–82 (Newton to Oldenburg, 7 Dec. 1675), quotation from p. 365.

> a tender matter wch may be as it were ye succus nutritious of ye earth or primary substance out of wch things generable grow} [or otherwise coagulated, in the pores of the earth and water, into some kind of humid active matter for the continuall uses of nature, adhereing to the sides of those pores after the manner that vapours condense on the sides of a Vessell subtily set]; the vast body of the Earth, wch may be every where to the very center in perpetuall working, may continually condense so much of this Spirit as to cause it from above to descend with great celerity for a supply.[30]

With the gravitational and vegetative functions thus joined, the gravitational aether fully shared in the activity of the vegetable spirit. The gravitational particles were not only thin and subtle; when condensed they became a "tender matter," the "succus nutritious," the "primary substance out of wch things generable grow," or a "humid active matter." By conflation with the active vegetable spirit, gravity itself had become in many ways an active principle.

Yet the general motion of the gravitational aether was still essentially the same as it had been for ten years, since the original entry in his student notebook, for Newton explained, again without a break, how that mechanical descent of the aether "with great celerity" contrived to bring bodies down with it.

> In wch descent it may beare downe with it the bodyes it pervades with force proportionall to the superficies of all their parts it acts upon; nature makeing a circulation by the slow ascent of as much matter out of the bowells of the Earth in an aereall forme wch for a time constitutes the Atmosphere, but being continually boyed up by the new Air, Exhalations, & Vapours riseing underneath, at length, (Some part of the vapours wch returne in rain excepted) vanishes againe into the aethereall Spaces, & there perhaps in time relents, & is attenuated into its first principle.[31]

So the movement up of air and the movement down of aether continue in this still earthbound system. "For nature is a perpetuall circulatory worker ..., Some things to ascend & make the upper terrestrial juices, Rivers and the Atmosphere; & by consequence others to descend for a Requitall to the former."[32] In the "Hypothesis" of 1675 one finds at first only a slightly refined restatement of the "crowding" of the aether by

30. Ibid., vol. I, 365–6; curly and square brackets both inserted by the editor of the *Correspondence* to indicate variant manuscripts. Cf. ibid., vol. I, 386–7, nn. 1 and 9, also ibid., vol. I, 413–15 (Newton to Oldenburg, 25 Jan. 1675/6).
31. Ibid., vol. I, 366.
32. Ibid.

the air that one saw in the "Vegetation" manuscript. The motion is like that of the aether in its old purely mechanical days: the aether descends to the earth, circulates through the bowels of the earth, condenses, and reemerges as "grosser" forms of matter that then ascend in a great circulatory pattern. The larger particles of matter, constituted as they are only of condensed aether, are "attenuated" again in the aethereal regions just beyond the atmosphere, then descend once again as aether. However, even though the movement is the same, the association of the gravitational function with the vegetative function has done its work. Gravity no longer arises from "the maine body of flegmatic aether" but from a portion of it that is thin, subtle, active, and spiritual, a portion of it that seems in fact to be identical with the vegetable spirit.

But with increased activity, gravity also suddenly moved toward universality, and that is the second important difference to be found in Newton's "Hypothesis." Appearing in the published paper almost as an afterthought at the end of his description of the great circulation down to the earth, through it, and back up to the aethereal regions, the operations of the gravitational–vegetable "Spirit" were expanded to include the solar system.

> And as the Earth, so perhaps may the Sun imbibe this Spirit copiously to conserve his Shineing, & keep the Planets from recedeing further from him. And they that will, may also suppose, that this Spirit affords or carrys with it thither the solary fewell & materiall Principle of Light; And that the vast aethereall Spaces between us, & the stars are for a sufficient repository for this food of the Sunn & Planets.[33]

The concept there enunciated bears little resemblance to the "real" Newton's "real" universal gravity, operating by mathematical law and applicable to all bodies in the universe. In that passage there has only been an equivocal extension of the animistic concept of the "Vegetation" manuscript to include the sun and the other planets. Earlier it had been the earth only, that "great animal or rather inanimate vegetable," that drew in "aethereal breath for its daily refreshment and vital ferment." By 1675 what had been food only for the earth had become food for the whole solar system, while the aethereal "repository" had been moved out of its place just beyond earth's atmosphere to the "vast aethereall Spaces" between the solar system and the fixed stars.

33. Ibid. Newton later told Halley that he had "interlined" this material in his own draft of the letter he sent to the Royal Society, confirming the impression that his expansion of the operations of the "Spirit" to the solar system was something of an afterthought. Cf. ibid., vol. II, 435–41 (Newton to Halley, 20 June 1686), esp. pp. 439–40.

Newton's speculative aethereal system was enormously expanded in its scale of application when he thus extended it to the sun and other planets, but in no way did that expansion affect the mode of operation of the system, for both gravitational and vegetative functions were still attributed to the "Spirit." The "Spirit" had the stated gravitational function of keeping the planets in their orbits: the sun imbibed the "Spirit" to "keep the Planets from recedeing further from him." The vegetative function of the "Spirit," on the other hand, is readily apparent in other phrases: this spirit provided "food" and "fewell." It furthermore carried with it the "materiall Principle of Light," a sharp and unmistakable echo of Newton's identification of the vegetable spirit with "the body of light" in the "Vegetation" manuscript.

The entire conceptual structure of Newton's aethereal hypothesis of 1675 is so closely affiliated with that of the alchemical "Vegetation" manuscript that one would hesitate to suggest a relationship between it and his later concept of universal gravitation had not Newton himself claimed in 1686 those very words quoted from the "Hypothesis" as evidence that he had the concept of universal gravity in 1675. The occasion for Newton's claim was Hooke's attempt to establish his own priority, and Newton in white anger looked through his manuscripts until he found the exact passage, then begged Halley to search for the Royal Society's copy.[34] After quoting to Halley the section of the "Hypothesis" beginning "And as the Earth, so perhaps may the Sun...," Newton said,

> In these & ye foregoing words [i.e., the discussion of his earthbound system] you have ye common cause of gravity towards ye earth Sun & all the Planets, & that by this cause ye Planets are kept in their Orbs about ye Sun. And this is all ye Philosophy Mr Hook pretends I had from his letters some years after, the duplicate proportion only excepted. The preceding words contein ye cause of ye phaenomena of gravity as we find it on ye surface of the earth without any regard to ye various distances from ye center: For at first I designed to write of nothing more. Afterwards, as my manuscript shews, I interlined ye words above cited relating to ye heavens, & in so short & transitory an interlined hint of things, the expression of ye proportion may well be excused. But if you consider ye nature of ye Hypothesis you'l find that gravity decreases upward & can be no other from ye superficies of ye Planet then reciprocally duplicate of ye distance from the center, but downwards that proportion does not hold.[35]

The "duplicate proportion" was not an insignificant aspect of New-

34. Ibid., vol. II, esp. p. 439.
35. Ibid., vol II, 439–40.

ton's gravity a decade after 1675, when he wrote the *Principia*, but did he really have it in mind when he wrote the "Hypothesis"? Perhaps he did. The weight of the evidence seems to indicate that he had formulated the inverse square relationship even though he had mentioned it to no one.[36] Perhaps he had even done so with respect to objects gravitating toward the earth in the 1660s, as he later claimed he had done, but he can hardly have applied it universally even by 1675, for the papers show that he did not arrive at a concept of the *mutual* attraction between gravitating bodies until he was actually writing the *Principia* in the 1680s.[37] But whether he had the "duplicate proportion" or not, what he did have in 1675, or at least believed himself to have, was the *cause* of gravity. Later, he was to say the cause of gravity was what he did not pretend to know,[38] but in 1686, looking back to his aethereal hypothesis of 1675, he was eager to claim that knowledge: the words of the "Hypothesis" "contein ye cause of ye phaenomena of gravity as we find it on ye surface of the earth" and they also contain "ye common cause of gravity towards ye earth Sun & all the Planets."

The third shift in emphasis to be discovered through a comparison of the "Vegetation" manuscript with the "Hypothesis" occurs in Newton's usage of the word "protoplast," a word with two distinctive meanings that are dependent upon differential etymologies.

The first, in which the meaning derives from the Greek πρωτο- and πλαστ-υζ (first molded, first formed), is that used by Newton in the treatise on vegetation: "Of protoplasts, that nature can only nourish, not form them; that's God's mechanism; that these nature's."[39] Newton quite evidently meant "protoplasts" to be understood as the models, the ex-

36. L. Rosenfeld, "Newton's views on aether and gravitation," *Archive for History of Exact Sciences* 6 (1969), 29–37; E. J. Aiton, "Newton's aether-stream hypothesis and the inverse square law of gravitation," *Annals of Science* 25 (1969), 255–60; Derek T. Whiteside, "Newton's early thoughts on planetary motion: a fresh look," *British Journal for the History of Science* 2 (1964), 117–37; idem, "Before the *Principia*: the maturing of Newton's thoughts on dynamical astronomy, 1664–1684," *Journal for the History of Astronomy* 1 (1970), 5–19; Richard S. Westfall, "Newton's development of the *Principia*," in *Springs of Scientific Creativity. Essays on Founders of Modern Science*, ed. by Rutherford Aris, H. Ted Davis, and Roger H. Stuewer (Minneapolis: University of Minnesota Press, 1983), pp. 21–43.
37. Cohen, *Newtonian Revolution* (1, n. 8), pp. 258–71, 354–57; Westfall, *Never at Rest* (1, n. 4), pp. 394–5, 408–9, 421–3.
38. Newton, *Correspondence* (2, n. 14), vol. III, 238–41 (Newton to Bentley, 7 Jan. 169²⁄₃), esp. p. 240.
39. Dibner Collection MSS 1031 B (1, n. 30), f. 1r. Cf. Appendix A.

emplars, the originals of all created things (first formed), that could only have been made by God. Nature cannot form protoplasts. Only God can do that, though nature can then nourish them.

At issue in Newton's mind was the balance of power between God and nature. So much can nature do by the powers delegated to her by God; so much, but not all. Some operations within the economy of the universe are beyond the powers of nature and can only be done by God. One has encountered a similar idea in Henry More's objection to an all-mechanical system of nature. More, however, wanted to rectify Cartesian mechanism by the introduction of spiritual intermediaries between God and nature, whereas Newton at first assigned the creation of the protoplasts to God directly. In the early 1670s Newton had drawn the line between God and nature quite differently from More. The nourishing of the protoplasts, which is spiritual and nonmechanical – a primary function of Newton's vegetable spirit – had in fact been assigned by Newton to nature herself. This point constitutes an important difference between Newton and More and shows Newton to have been at that time aligned rather more closely with traditional Christian theology than with More's Neoplatonism, and one must soon consider in more detail Newton's affiliation with the theology of God's ordained and absolute powers.

Newton took up his thoughts about the originals of all the various forms in the natural world again in the "Hypothesis," but in so doing he used the word "protoplast" in its second meaning of first former or creator, a meaning derived from the Greek πρωτο- and πλαστηζ (first agent). Arguing that "the whole frame of Nature" may be nothing but condensed "aethereall Spirits," Newton suggested that, after condensation, the "Spirits" had been

> wrought into various formes, at first by the immediate hand of the Creator, and ever since by the power of Nature, wch by vertue of the command Increase & Multiply, became a complete Imitator of the copies sett her by the Protoplast.[40]

The use of the word "protoplast" in this second sense has never been common in any period. Most dictionaries do not carry it at all, and though the *Oxford English Dictionary* gives two earlier seventeenth-century examples, Newton's use of it in the "Hypothesis" is its most significant citation.[41] No examples prior to the seventeenth century appear, nor any from the eighteenth century, when even that great word specialist Dr. Samuel Johnson did not know the second meaning.[42] A

40. Newton, *Correspondence* (2, n. 14), vol. I, 364.
41. *Oxford English Dictionary* (2, n. 79), s.v. "protoplast."
42. Samuel Johnson, *A Dictionary of the English Language in which The Words*

single nineteenth-century example in the *OED* comes from the poet Robert Browning.

In all of the other examples cited by the *OED* the word was used as a synonym for the Deity, and that is evidently Newton's general meaning as well. But the lack of context might make it difficult to determine whether he had a more precise signification in mind if one did not know that Newton had become a convinced Arian during the first half of the 1670s. Newton's shift from the common meaning of "protoplast" as first prototype or exemplar of created beings to the rare second meaning of "protoplast" as first agent in the creative process can only be a reflection of his recent shift from orthodox Trinitarianism to the Arian heresy, and in fact this entire passage from the "Hypothesis" must be subjected to an Arian interpretation.

When Arius rejected the argument that the Second Person of the Triune God (the Son of God, Jesus Christ) was of the same "essence" as the First Person of the Trinity (the Father), he was perhaps attempting to return to a stricter monotheism. Pagan critics of fourth-century Christianity sometimes accused the Christians of reinstituting a form of polytheism with their characterization of God as Three Persons. Arius reemphasized the absolute supremacy of the original uncreated One, insisting that there was a period before the creation when the Son had not yet been called into existence. The status of the Arian Son was necessarily diminished as the status of the Arian Father was elevated: Father, Son, and Holy Ghost did not share as co-equals in the Arian Godhead. The Son was the first and most important of created beings, even though he was not granted full equality with the Father, and in Arian doctrine he became the mediator or intermediary between God and the created world. One result of Arius's emphasis on the supremacy of the Father was to set Him wholly apart from the creation, to make Him entirely "other" and transcendent. Consequently, the cosmological role of the Arian Christ was also given emphasis. The Christ was the intermediary through whom and by whom God created the world and

are deduced from their Originals, and Illustrated in their Different Significations by Examples from the best Writers. To which are prefixed, A History of the Language, and An English Grammar (2nd ed.; 2 vols.; London: printed by W. Strahan, for J. and J. Knapton; T. and T. Longman; C. Hitch and L. Hawes; A. Millar; and R. and J. Dodsley, 1755–6), s.v. "Protoplast." Johnson gives a single signification: "Original; thing first formed as a copy to be followed afterwards." His illustrative quotation is from Harvey: "The consumption was the primitive disease, which put a period to our *protoplasts*, Adam and Eve."

interacted with it.[43] That was what Newton had come to believe in the first half of the 1670s, even though Arian doctrines had been declared heretical in the fourth century and adherence to them in seventeenth-century England was still fraught with danger.[44]

Newton was silent in public regarding his heretical beliefs, and, except for a few intimate friends, no one knew about them until his papers were examined after his death. But those papers have in them exceedingly clear statements on a transcendent Arian Deity who interacts with the world through the active agency of the Logos, the Arian Christ. As already seen in Chapter 3, for Newton it was the Christ who "prepared and formed" the cosmos and continues to govern it according to God's will. The Father in His transcendent perfection is "the invisible God whom no eye hath seen nor can see," yet all is done according to His omnipotent will, "the Son receiving all things from the Father" and subject to Him in all his activity. "For the supreme God doth nothing by Himself which He can do by others."

Given the commitment Newton had made to the principles of Arius before 1675, one must look again at his statement in the "Hypothesis" regarding the origin of natural variety. All comes from a common aether that has been condensed and shaped into various forms "at first by the immediate hand of the Creator." Newton must surely be read metaphorically there. The transcendent Arian Deity does not Himself reach directly into the created world in such a manner; indeed it is the Arian Christ that constitutes His "immediate hand" in all the contacts between God and man reported in the Bible. As Newton made quite plain in his theological papers, those encounters were always through the Christ acting "by the name of God." So also must it have been in Newton's mind with respect to the primordial creation of all natural forms, for the Christ also "prepared and formed this place in which we live." Within this Arian context then the "Protoplast" that has fashioned the copies for nature to imitate must be interpreted not as the supreme Deity but

43. Cf. the discussion and references supra (3, pp. 81–4).
44. An anti-Trinitarian, Matthew Hamont, had been burned at the stake as recently as 1578 (John Stow and Edmund Howe, *The Annales or Generall Chronicle of England* [London, 1615], p. 685, cited in Frank Luttmer, "Enemies of God: Atheists and Anxiety about Atheists in England, 1570–1640" (Northwestern University: Ph. D. dissertation, 1987), p. 68. William Whiston was expelled from the University of Cambridge in 1710 for his Arianism (probably learned at least in part from Newton himself): Force, *Whiston* (3, n. 74), pp. 3, 14–19, 105–13. See also, Redwood, *Reason, Ridicule and Religion* (1, n. 25), pp. 156–72.

rather as the Christ acting in his capacity as cosmological agent. Newton perhaps chose the second and unusual meaning of the word "protoplast" for its etymological exactitude: "first agent."

Nevertheless, in Newton's papers it is always the will of the supreme God that is put into effect, both in the beginning of the creation and in its governance, and one must now examine in more detail Newton's place in the strong current of Christian theology that emphasized the primacy of God's will, the theological tradition of voluntarism. As Newton argued, it is not for His essence that God desires to be worshipped but for His actions, the "issues" of His will, especially in "creating, preserving, and governing all things" in which the Christ acts as His executor. Newton's heavy emphasis upon the will of God places him firmly within the camp of voluntarist theology, in which God's will is considered to be His primary attribute. The power of God's will is complete and even the laws of nature are subsumed under it, for natural laws – both physical and moral – were ordained by God.

With Newton, as with other voluntarists, there was a concern to define the realm of natural law, which represents God's ordained power, his *potentia ordinata*. Though God can abrogate His natural laws whenever He so wills, ordinarily He does not do so because of His goodness and beneficence, but rather maintains the world in an orderly fashion by His ordinary concourse. The laws of nature act regularly and naturally because God in His goodness has so ordained it; the laws of nature function smoothly through the powers God has placed in the natural world. But the voluntarist never forgets that with His absolute power, His *potentia absoluta*, God may at any time use His laws in an extraordinary way, abrogate the laws completely, or decree that they be other than what they now are. For the voluntarist the world is always contingent upon the will of God.

The theological framework of *potentia dei ordinata et absoluta* guided Newton and many of his contemporaries when they inquired into the relationships between God and the world, and it is now widely recognized that Newton and Newtonian science were deeply indebted to voluntarist theology and to the struggle to define God's ordained and absolute powers. *Potentia* doctrines had comprised important strains of thought in Western Christendom at least since the thirteenth century, when the absolute power of Jehovah collided openly with the necessitarianism of Greek and Arabic philosophy.[45] Several articles in the Condemnations

45. Edward Grant, "The condemnation of 1277, God's absolute power, and physical thought in the Late Middle Ages," *Viator 10* (1979), 211–44; idem, *Much Ado about Nothing* (4, n. 5); Étienne Gilson, *History of Chris-*

of 1277 by the Bishop of Paris forbade Christian philosophers to limit God's ability to act and so reemphasized the *potentia dei absoluta*, God's ability to do whatever He willed (short of a logical contradiction). Concessions were made to natural law: because God's will was absolute it was not therefore capricious. But after 1277 even natural law was imposed by the will of God and was not necessitated by the nature of things, as it had been for some of the Greeks. As an external imposition, natural law was valid for the ordinary course of events, but it could always be abrogated or supervened if God so willed in extraordinary circumstances. God was not bound by what He had ordained. The laws of nature represented God's ordained power, for He had ordained and established both natural and moral laws at the time of creation; nevertheless God's absolute power remained just that – absolute.

Late medieval voluntarist theology, viewing creation itself as the ultimate example of *potentia absoluta* and maintaining a consistent emphasis on the primacy of God's will, was especially sensitive to issues of ordained and absolute power. Restated by Luther, Zwingli, and Calvin among the great reformers, elaborated in England by both Puritan and Anglican divines and by members of the Royal Society as well, Reformation voluntarism produced discussions of *potentia dei ordinata et absoluta* that continued throughout the seventeenth century, often under the rubric of "ordinary" and "extraordinary" providence. Descartes, Gassendi, Charleton, Boyle, and Newton were only a few of the figures of the scientific revolution to concern themselves with the theology of God's power in and over nature.[46]

tian Philosophy in the Middle Ages (New York: Random House, 1955), esp. 387–545; Francis Oakley, *Omnipotence, Covenant, & Order. An Excursion in the History of Ideas from Abelard to Leibniz* (Ithaca, NY: Cornell University Press, 1984).

46. Eugene M. Klaaren, *Religious Origins of Modern Science. Belief in Creation in Seventeenth-Century Thought* (Grand Rapids, MI: William B. Eerdmans Publishing Co., 1977); Daniel O'Connor and Francis Oakley, Eds., *Creation. The Impact of an Idea* (Scribner Source Books in Religion; New York: Charles Scribner's Sons, 1969), esp. the following: "General introduction," pp. 1–12, Daniel O'Connor, "Introduction: two philosophies of nature," pp. 15–28, M. B. Foster, "The Christian doctrine of creation and the rise of modern science," pp. 29–53, Francis Oakley, "Christian theology and the Newtonian science: the rise of the concepts of the laws of nature," pp. 54–83, and Hans Jonas, "Jewish and Christian elements in the western philosophical tradition," pp. 241–58; Amos Funkenstein, "Descartes, eternal truths, and the divine omnipotence," *Studies in History and Philosophy of Science* 6 (1975), 185–99; J. E. McGuire, "Force, active principles, and

Among so many thinkers there were inevitably shades of opinion. Descartes considered himself to be safely positioned within the mainstream of orthodoxy in saying that his mechanical universe could continue to operate only through the constant application of God's will, a position sometimes identified as a "radical voluntarism." But laws of matter and motion, as well as the laws of mathematics, even though created and maintained by God's will in Descartes's system, assumed the character of "eternal truths" because of their very constancy. Such a position edged Descartes toward necessitarianism and toward an emphasis on the primacy of God's intellect rather than His will; it tended to obliterate the sense of the radical contingency of the universe that had long been predominant among the voluntarists, a universe in which even ordained laws could vary if God so willed it. Though Descartes was never overtly opposed to that voluntarist opinion, because of the emphasis in Descartes's system on God's perfection and immutability (characteristics of God's essence rather than of His will), some other voluntarists resisted the Cartesian formulation. Descartes's Deity constantly "maintains... by His normal participation" the same amount of motion with which He endowed the universe in the beginning and so guarantees that nature will always operate in a regular manner.[47]

The Cartesian system thus emphasized *potentia ordinata* to a greater extent than was customary in voluntarist theology. Within the theological context of *potentia dei ordinata et absoluta* Descartes's mechanical laws of impact were acceptable as a part of God's ordination, however, and were so understood. It was not on that point that Gassendi, for example, mounted his criticism. It was rather on the inadequacy of Descartes's total picture, in which the *potentia ordinata* was not balanced by the *potentia absoluta*. Gassendi's position was that of course God uses secondary causes in His general providence that establishes the ordinary course of nature, but that God did not abandon the world to run by

Newton's invisible realm" (4, n. 9); Margaret J. Osler, "Descartes and Charleton on nature and God," (2, n. 42); idem, "Providence and divine will in Gassendi's views on scientific knowledge," ibid., 44 (1983), 549–60; idem, "Eternal truths and the laws of nature: the theological foundations of Descartes' philosophy of nature," ibid., 46 (1985), 349–62.

47. Rene´ Descartes, *Principles of Philosophy*, II, 36 in Rene´ Descartes, *Principles of Philosophy*, tr., with explanatory notes, by Valentine Rodger Miller and Reese P. Miller (Collection des Travaux de l'Acade´mie Internationale d'Historire des Sciences, No. 30; A Pallas Paperback; Dordrecht: D. Reidel, 1984), pp. 57–8, quotation from p. 58; Funkenstein, "Descartes" (4, n. 46); Osler, "Descartes and Charlton" (2, n. 42); idem, "Eternal truths and the laws of nature (4, n. 46).

itself after He created it. Scope must also be allowed for His special (extraordinary) providence; Descartes's constantly operating mechanical laws cannot be the whole story.[48]

Gassendi's assimilation of secondary causes to God's general providence or His *potentia ordinata* was not an isolated instance. The English Puritan William Perkins, for example, similarly equated general providence with secondary causes and contrasted God's general providence with His special providence that "worketh without means."[49] One may see here that the thrust of traditional theological arguments relating God to the natural world was different from that of the Neoplatonists. Although Gassendi and Henry More attacked the Cartesian account at virtually the same weak point – on the overall inadequacy of a closed system operating *only* by constant mechanical law – their attacks were mounted from quite distinctive positions. Whereas the Neoplatonists suggested spiritual agents intermediary between God and the world as a corrective to Cartesianism, the more traditional theologian might suggest, as Perkins did, that in cases of special providence, when God exercised his *potentia absoluta*, He operated on the natural world directly and "without means."

Newton's Arianism should have precluded Perkins's solution to the problem, though Newton perhaps later wavered on this point, as one will see in Chapter 6. In the 1670s, his struggle with these issues incorporated elements from the Neoplatonic stance, as one has already seen, but it also drew heavily upon the theology of *potentia dei absoluta et ordinata* and upon Stoic doctrines as well as Arian ones. He attempted to find a balance that would satisfy all his religious, metaphysical, experimental, and mathematical concerns, but the problems involved grew ever more complex. One may attempt to evaluate first the balance Newton tried to establish in the first half of the 1670s, then return in subsequent chapters to his later efforts.

In his treatise on "Vegetation" Newton entered unequivocally into the lists of the voluntarists with a succinct discussion of God. His first point seems to have been that God can do anything that does not involve a logical contradiction.

> Of God. Whatever I can conceive without a contradiction, either is or may be made by something; that is, I can conceive all my own

48. Osler, "Providence and divine will" (4, n. 46).
49. William Perkins, *A Resolution to the Countrey-man*, in *The Workes of That Famous and Worthy Minister of Christ... M. W. Perkins* (3 vols.; Cambridge, 1608–31), vol. III, 657, quoted in Oakley, *Omnipotence* (4, n. 45), p. 139, n. 54.

> powers (knowledge, activating matter, etc.) without assigning them any limits. Therefore such powers either are or may be made to be.
>
> Example. All the dimensions imaginable are possible. A body by accelerated motion may transcend all distance in any finite time assigned. Also it may become infinitely long. This, if thou deniest, it is because thou apprehendest a contradiction in the notion, and, if thou apprehendest none, thou wilt grant it in the power of things.[50]

Presumably communing with himself alone as he committed the intimate "thou" to paper in his lonely study, Newton reworked the classical statement of God's *potentia absoluta*. God can do any conceivable thing as long as it involves no contradiction. Then exploring the implications of that proposition for God's cosmic creativity, Newton argued that since God could have created any sort of world, our world is as it is not by necessity but by God's "voluntary and free determination that it should be thus."

> Argument 2. The world might have been otherwise than it is (because there may be worlds otherwise framed than this). It was therefore no necessary but a voluntary and free determination that it should be thus. And such a voluntary [cause must be a God]. Determination implies a God. If it be said the world could be no otherwise than it is because it is determined by an eternal series of causes, that's to pervert not to answer the first proposition. For I mean not that the world might have been otherwise notwithstanding the precedent series of causes, but that the whole series of causes might from eternity have been otherwise here, because they may be otherwise other places.[51]

The "whole series of causes" that makes this world what it is could have been different, because God's power is absolute and His choice of any particular set of causes is "voluntary and free," that is, an issue of His unlimited free will.

Newton's purpose in inserting these passages into an alchemical treatise, however, cannot simply have been to rehearse the traditional arguments for a voluntarist theology. On the contrary, he was concerned in the "Vegetation" manuscript to find the laws of vegetable

50. Dibner Collection MSS 1031 B (1, n. 30), f. 4v. Cf. Appendix A.
51. Ibid. A restatement of this idea eventually appeared in Query 31 of the *Opticks* (2, n. 15), pp. 403–4: "And since Space is divisible *in infinitum*, and matter is not necessarily in all places, it may be also allow'd that God is able to create Particles of Matter of several Sizes and Figures, and in several Proportions to Space, and perhaps of different Densities and Forces, and thereby to vary the Laws of Nature, and make Worlds of several sorts in several Parts of the Universe. At least, I see nothing of Contradiction in all this."

action and to distinguish them from the laws of mechanical action. Mechanical action, by which Newton's predecessors in the mechanical philosophy meant action by contact or by impact, was readily comprehensible to the human mind, which was surely one reason for the seventeenth-century popularity of mechanical systems. By reminding himself of God's unlimited power to institute *any* series of causes and not just mechanical ones, Newton was making room within the natural world for the nonmechanical laws of vegetation. One may recall the initial phrase of Newton's treatise: "Of Natures obvious laws & processes in vegetation." One may also recall that in distinguishing between mechanical and vegetable chemistry Newton said that the "mechanical coalitions or separations of particles" may explain many chemical changes. "But, so fast as by vegetation such changes are wrought as cannot be done without it, we must have recourse to some further cause." Mechanical causation in the seventeenth-century sense may be more easily understandable by limited human minds, but man's mind is not the measure of God's unlimited power. God could have instituted any sort of causal laws He wanted, and the only way human beings have of grasping those laws is to investigate the world as God actually instituted it. That there is thus a close linkage in practice between voluntarism and empiricism has long been recognized.[52] In rehearsing his voluntarist theology in an alchemical context Newton justified to himself his empirical investigations into the laws of vegetation (i.e., his alchemical experimentation) that so radically contravened the contemporary consensus on mechanical philosophy.

Furthermore, Newton gave an additional clue to his purpose in inserting the *potentia dei absoluta* doctrine into this manuscript with the phrases that immediately precede the theological discourse: "Of the contrivance of vegetables and animals. Of sensible qualities. Of the soul's union."[53] None of those topics could be adequately explicated by contemporary mechanical philosophy. They were to Newton examples of just that sort of natural process for which "we must have recourse to some further [nonmechanical] cause" and to which his alchemical studies were to provide the key.

However, the nonmechanical laws of vegetation were in no sense above or outside of nature in Newton's conception but were on the contrary a part of nature. The whole thrust of Newton's "Vegetation" manuscript was to place whatever active, nonmechanical principles

52. Hooykaas, *Religion and the Rise of Modern Science* (3, n. 25), esp. pp. 29–53; Osler, "Providence and divine will" (4, n. 46).

53. Dibner Collection MSS 1031 B (1, n. 30), f. 4v. Cf. Appendix A.

there might be firmly within the realm of nature. It was "nature's" laws and processes in vegetation for which he searched. He noted that "nature" begins with putrefaction or fermentation; vegetation is the "natural" work of metals; it is agreeable to "nature's" proceedings to make a circulation of all things; the vegetable spirit is "nature's" universal agent; "nature's" actions are either vegetable or purely mechanical; it is "nature" that makes the strange transmutations of vegetation. One cannot escape the conclusion that Newton conceived the operation of his active principles under the rubric of natural law.

Within the tradition of *potentia dei ordinata et absoluta* natural laws represented God's ordained power. But if all the apparently spontaneous or spiritually guided processes of vegetation were to be brought under the rule of law, and secondary causes be found to account for them, they would, it would seem, no more demonstrate God's continuing activity in the world than does the regular operation of mechanical law. To the extent then, that Newton achieved success in his alchemical enterprise of determining the laws of vegetation, to that same extent he defeated his original, fundamental purpose of directly demonstrating divine activity in the world. Or so it might seem to us, whose concept of matter itself encompasses the "active principle" electromagnetically, whose "laws of vegetation" have been given a physico-chemical basis, and who are only interested in efficient causes.

But Newton's matter remained passive; his active principles operated only between and among its particles and simply constituted the natural vehicle by which God's will was enacted. The "contrivance" of vegetables and animals – the so obviously designed concatenation of the particles that fitted the forms of plants and animals so perfectly to their habitats – was the issue of God's will. God Himself had formed the "protoplasts" in the "Vegetation" manuscript (or the "Protoplast," the Christ acting as God's executor, had done so in the "Hypothesis"); then the vegetative spirit within "nature" nourished them and organized inchoate matter into the designed forms, regularly sustaining their lives in the lawful manner God had ordained. Though the "spirit" had some corporeal qualities, it was defined in opposition to the passive matter upon which it worked. Partaking as it did of the fiery creativity of the divine Stoic *pneuma*, it carried divinity directly into the world. The cosmic Stoic *pneuma* from the beginning had represented a combination of efficient and formal causes,[54] and so Newton used it. The formal causes within the *pneuma*, that carried the

54. Verbeke, *L'évolution de la doctrine du pneuma* (3, n. 38), p. 39.

divine plans for organization, pointed for Newton directly to the willed activity of the Deity through the argument from design.

Alchemy and gravity in the later 1670s

When Boyle pressed Newton for his thoughts about "Physicall qualities" in 1678/9, Newton responded with two speculative aethereal systems that omitted all mention of active principles.[55] Not too long before that, in 1676, Newton had thought Boyle to have been much too free with alchemically sensitive information, and Newton had seriously protested Boyle's apparent indiscretion in a letter to Oldenburg.[56] Perhaps Newton thought Boyle to be less than fully trustworthy with respect to Hermetic secrets; perhaps Boyle had only asked for Newton's opinion on the class of phenomena Newton considered mechanical, but in fact no vital, vegetative phenomena were discussed in Newton's response.

The first of the systems Newton offered Boyle operated by means of a density gradient in which the aether stood rarer in the pores of large bodies than in the spaces between them. Newton utilized this new static aethereal system to explain some aspects of cohesion at a gross level, capillary attraction, surface tension, the diffraction of light by a knife edge, and several chemical phenomena of the sort he thought belonged to mechanical, common, or vulgar chemistry.

The second system partially transferred the gradient to aethereal particle size and postulated that more of a finer type of aether (and less of a grosser) stood within the pores of bodies than in free spaces. With respect to the earth there was in addition an overall aethereal gradient from the upper atmosphere down to the center of the earth in which the particles of aether became finer and finer as one descended. Any body suspended in the air or lying on the earth participated in the overall gradient in such a way as to have grosser aether in its upper pores and finer in its lower. Since (by the hypothesis) the large bodies were to have more of the finer type throughout, the bodies dislodged their grosser internal aether by descending to the region of finer aether below, and thus Newton offered his second system as a "cause of gravity."

Newton's new "cause of gravity" no longer invoked an aethereal shower of particles to move heavy bodies kinetically, but relied on a

55. Newton, *Correspondence* (2, n. 14), vol. II, 288–96 (Newton to Boyle, 28 Feb. 1678/9).
56. Ibid., vol. II, 1–3 (Newton to Oldenburg, 26 April 1676).

tensional stasis in the aether. No great circulatory pattern appeared, though aerial substances were still being generated in the bowels of the earth and rising up to constitute the atmosphere. In his letter to Boyle, Newton went to some length to establish that true permanent air came from metals and, being heaviest and lowest in the atmosphere, buoyed up the vapors, mists, clouds, and smoke. "The [permanent metallic] air also is ye most gross unactive part of ye Atmosphere affording living things no nourishment if deprived of ye more tender exhalations & spirits yt flote in it...."[57] The vital parts of the air were not further mentioned, nor was there any mention of a vital spirit floating in the aether. Newton had defined the aether in his usual way by its relationship to air: "an aethereal substance capable of contraction & dilatation, strongly elastick, & in a word much like air in all respects, but far more subtile."[58] But he omitted the remainder of his customary proportion: the vital vegetable spirit in the aether that was much like the vital spirit in air although far more subtle.

In the "Hypothesis" of 1675 he had made the full proportion quite explicit. As aether is to air, so the gravitational–vegetable "Spirit" of the aether is to "the vitall aereall Spirit requisite for the conservation of flame & vitall motions." Even though he omitted the alchemical spirit in his letter to Boyle, one should not assume that he had abandoned his belief in it any more than he had abandoned his belief in the vital aerial spirit that he barely mentioned to Boyle. The multitude of alchemical manuscripts he wrote after 1678/9 testify to Newton's continued faith in his active vegetative principle. The purpose of the letter to Boyle probably was simply to discuss possible aethereal mechanisms for phenomena Newton classified as mechanical, and, if so, Newton had no need to invoke the vegetable spirit in his explanations. In any case, what he gave Boyle was evidently a set of mechanisms for which "the maine body of flegmatic aether" might be held responsible.

So the vegetable spirit had by no means been eliminated from Newton's thinking by 1678/9, but a significant change had certainly occurred in the status of gravity. In the letter to Boyle Newton completely detached gravity from the vital aethereal spirit and made it again fully mechanical. The mechanism of its operation was different, for his earlier postulate of a shower of aethereal particles bringing bodies down with it had evidently been abandoned, but certainly no activity was invoked – only the presence of the static and apparently

57. Ibid., vol. II, 294.
58. Ibid., vol. II, 289.

rigid aether gradient coupled with a gradation of aethereal particle size within bodies.

Newton told Boyle that that particular gravitational conjecture came into his mind only as he was writing the letter, and, though one will meet a variant of it again, it did not have a very long life in its original form. For a few months later, toward the end of 1679, Hooke's challenges to Newton regarding the motion of bodies set Newton on a course of development that changed forever the conditions for aethereal speculation.[59] The correspondence with Hooke provided the stimulus for Newton's first solution to the problem of celestial dynamics in the terms later to appear in the *Principia*, where planetary orbital motion was understood to be compounded of an inertial component and an inverse-square attraction toward a center of force located at one focus of the elliptical orbit. Newton's prior analyses had been cast in terms of other dynamical principles and complicated by the presence of aethereal mechanisms; indeed, having been very busy with his alchemical and theological studies during the 1670s, he had barely considered celestial dynamics quantitatively and had done no original work in that area since the 1660s. Hooke diverted him from his other studies, and irritated him by correcting some errors and by what he called Hooke's "dogmaticalnes," so Newton was "inclined" to try Hooke's mode of analysis and then found the theorem by which he "afterward examined ye Ellipsis."[60] It is possible that Newton made his trial of the new method late in 1679 or early in 1680, but even if so he once more quickly put his calculations aside for other studies.[61]

Conclusion

Newton ended the decade of the 1670s, then, with some rather well-settled convictions and with other convictions in a state of flux.

59. Ibid., vol. II, 297–300 (Hooke to Newton, 24 Nov. 1679); ibid., vol. II, 300–4 (Newton to Hooke, 28 Nov. 1679); ibid., vol. II, 304–7 (Hooke to Newton, 9 Dec. 1679); ibid., vol. II, 307–8 (Newton to Hooke, 13 Dec. 1679); ibid., vol. II, 309–12 (Hooke to Newton, 6 Jan, 1679/80); ibid., vol. II, 312–13 (Hooke to Newton, 17 Jan. 1679/80).
60. Ibid., vol. II, 446–8 (Newton to Halley, 27 July 1686).
61. John Herivel, *The Background to Newton's Principia. A Study of Newton's Dynamical Researches in the Years 1664–84* (Oxford: Clarendon Press, 1965); Whiteside, "Before the *Principia*" (4, n. 36); Westfall, *Never at Rest* (1, n. 4), pp. 381–8.

Alchemy had seemed to him from the first to be a way of studying the active principles at work in the world, principles that were the natural vehicles of divine will, principles that activated and guided the small particles of passive matter into the variegated contrivances of minerals, vegetables, and animals. In the context of his theological convictions regarding God's ordained and absolute powers, one may say that Newton's vegetative principles represented the divinely ordained powers of natural laws and processes, yet they were not mechanical in the seventeenth-century sense. Operating by some cause other than mechanical impact, they constituted the channel by which divine ideas were given shape in the natural world. With the argument from design, they brought divinity directly into nature, and a demonstration of their mode of operation would be a demonstration of the divine governance of the world. At some level, God's absolute powers also were in evidence in active vegetative phenomena, as well as His ordained powers, for the original exemplars of all living creatures sprang directly from the ideas and will of the supreme God or from His "first agent" who always put His will into effect in the world. "Nature" could not form those first patterns, though she could nourish them afterwards. None of Newton's convictions in this area ever suffered substantial change.

Gravitational principles, on the other hand, had a more fluid career in this period of Newton's life, as well as later. He accepted the basic premise of mechanical philosophy at first, that gravitational effects are due to the impacts of particles of an imperceptible aether acting mechanically. God had ordained the laws of impact "at the beginning," when He set everything in motion – few major thinkers in the seventeenth century would have wished to deny that. But since in the context of a strict mechanism, mechanical operations continued more or less automatically (with general divine concourse but without further direct divine action), one could hardly use the laws of gravity to argue for God's continued providential governance of the world. At best, they represented only *potentia dei ordinata*; at worst, they set the natural philosopher on the slippery slope that led toward deism or even atheism, for a world without divine providential care was nothing more than fate and nature. If in fact Newton had the inverse-square relationship in hand already, as he later claimed he did, he would have been fully aware of the *regularity* of the natural laws of gravitation and would have thought of them as evidence of God's ordained powers operating in the natural world through the secondary cause of the mechanical gravitational aether, but he would not have conceived them to be potential demonstrations of divine providence.

In some ways Newton's conceptualization of the causal principle of gravity shifted in the course of the 1670s, however, from the mechanical to the active. Within the speculative kinetic systems he generated from the mid-1660s through 1675, the mechanical gravitational aether became more and more closely assimilated to the active, nonmechanical vegetable spirit. As it was so assimilated, Newton apparently came to think of the gravitational cause as only a small portion of the main body of phlegmatic aether and comprised of smaller and more active particles. Its motion nevertheless continued to be primarily that of a great mechanical circulation that swept bodies down to the earth's surface (that is, made them heavy) or swept the planets in orbits around the sun. It seems unlikely that Newton thought he might be able to demonstrate God's providential activity in the world through the operations of gravity, even at that stage in his development; in any case, the assimilation of the gravitational principle to the nonmechanical alchemical principle was relatively short-lived, for by the end of the decade gravity had again become fully mechanized in its operations. There is thus no continuity between the quasi activity Newton assigned to the causal principle of gravitation in the early to mid-1670s and the full activity the gravitational principle achieved after 1687.

5

Modes of divine activity in the world: the Principia *period*

Introduction

With gravity again fully mechanized, as in his letter to Boyle of 1678/9, Newton concentrated his efforts to find evidence of divine activity in the world in his alchemical and theological work. He renewed his study of Eirenaeus Philalethes, whose several treatises he already knew intimately.[1] He again worked over the ideas of Basilius Valentinus[2] and Hermes Trismegistus.[3] He discovered John de Monte

1. (1) "Philalethes. Notes out of Philalethes, beginning: 'Mercury vulgar is prepared for conjunction with sol vulgar by frequent cohobations of Reg of [Iron],' " Sotheby lot no. 69, now in the Isaac Newton Collection (M132), Department of Special Collections, Stanford University, Stanford, CA; (2) "Ripley (Sir George) 'Ripley expounded' [The Twelve Gates]...; 'Notes upon Ripley' " (in actuality this manuscript represents Newton's study of Philalethes' comments on Ripley), Sotheby lot no. 94, now King's College, Cambridge, Keynes MS 54; (3) annotations in his own copy of Eirenaeus Philalethes, *Secrets Reveal'd*, Sotheby lot no. 121, now in the Duveen Collection, University of Wisconsin, Madison, WI. On the last, see Dobbs, "Newton's copy" (1, n. 5).
2. (1) "Valentine (B.) Abstracts from the Works of Basil Valentine on Minerals; Transmutation of Metals; Vitriol, etc.," Sotheby lot no. 110, now British Library, London, Add. MS 44,888; (2) "Philosopher's Stone. 'Regulae seu canones aliquot Philosophici de Lapie Philosophico. Authore docto quodam Anonymo. Impress in fine Curationem Paracelsi,' followed by 'Mayer's Figures praefixed to Basil Valentine's Keys,' " Sotheby lot no. 71, now King's College, Cambridge, Keynes MS 43; (3) "Valentine (B.) Verses at the end of B. Valentine's Mystery of the Microcosm," Sotheby lot no. 111, now King's College, Cambridge, Keynes MS 63. Newton's earliest study of Basilius Valentinus had resulted in "Valentine (B.) Currus Triumphalis Antimonii [Notes and Abstracts]," Sotheby lot no. 112, now King's College, Cambridge, Keynes MS 64, discussed in Dobbs, *Foundations* (1, n. 1), pp. 129–30, 150–2, 191–2.
3. Keynes MS 28 (1, n. 35). Cf. Appendix B.

Synders[4] and selectively but systematically made notes on Ashmole's *Theatrum Chemicum Britannicum.*[5] He made transcripts of a number of unpublished alchemical manuscripts[6] and of four published alchemical books by Michael Maier.[7] In paper after paper he attempted to bring order out of the chaos of alchemical theory and practice: *De secreto solutione*;[8] *Notanda chemica* and *Sententiae notabiles*;[9] *Quomodo metalla generantur, De radice semine spermate et corpore mineralium, De Mineralibus ex quibus desumitur*, and *De Conjunc-*

4. (1) "Snyders (John de Monte) 'Commentatio de Pharmaco Catholico... donata per Authorem Chymicae Vannus' [Notes and Abstracts]," Sotheby lot no. 35 (part), now Jewish National and University Library, Jerusalem, Yahuda MS Var. 259.10; (2) "Snyders (John de Monte) The Metamorphosis of the Planets, That is A Wonderfull Transmutation of the Planets and Metallique formes into their first Essence..." (Newton's autograph transcript of the book plus four pages of notes on it), Sotheby lot no. 102, now in the Medical Historical Library, Cushing/Whitney Medical Library, Yale University, New Haven, CT.
5. (1) Annotations in his own copy of Elias Ashmole, *Theatrum Chemicum Britannicum*, now in the Van Pelt Library, University of Pennsylvania, Philadelphia, PA (cf. Harrison, *Library* [1, n.6], item 93, pp. 91–92); (2) "Bloomfield. 'Out of Bloomfield's Blossoms,'... 'A short work that beareth the name of Sr. George Ripley,'... and 'Fragments,'" Sotheby lot no. 8, now King's College, Cambridge, Keynes MS 15; (3) "Copper. 'The Hunting of ye Green Lyon,'... 'The Standing of ye Glass for ye time of Putrefaction & Congelation of ye medicine,'" Sotheby lot no. 13, now King's College, Cambridge, Keynes MS 20; (4) "Norton, "Out of Norton's Ordinal,'... 'Out of Chaucer's Tale of ye Chanon's Yeoman,'... and a Poem by Richard Carpenter," Sotheby lot no. 55, now King's College, Cambridge, Keynes MS 37; (5) "'Pearce the black Monck upon ye Elixir,'" Sotheby lot no. 68, now King's College, Cambridge, Keynes MS 42.
6. (1) "Epistola. 'Anno 1656. Serenissimi Principio Frederici Ducis Holsatiae et Sleswici, etc.,'" Sotheby lot no. 21, now King's College, Cambridge, Keynes MS 24; (2) "Philosopher's Stone. Several Questions concerning the Philosopher's Stone," Sotheby lot no. 72, now King's College, Cambridge, Keynes MS 44; (3) "Thesaurus Thesaurorum sive Medicina Aurea," Sotheby lot no. 107, now King's College, Cambridge, Keynes MS 61.
7. "Maier (M.) Symbola aurea mensae duodecim nationum... Atalanta Fugiens, hoc est Emblemata nova, etc....; Viatorium, hoc est De Montibus Planetarum septem...; Septimana Philisophica [Days 1–6 only]," Sotheby lot no. 43, now King's College, Cambridge, Keynes MS 32.
8. "Solution. 'De Secreto Sol[utions]'... [on Diana's doves, etc.]," Sotheby lot no. 101, now King's College, Cambridge, Keynes MS 59.
9. "Note-Book: Containing 'Notanda Chemica'... and 'Sententiae notabiles,'" Sotheby lot no. 58, now King's College, Cambridge, Keynes MS 38.

tione in hora nativitatis;[10] *De igne sophorem et materia quam cale facit;*[11] on "The Three Fires";[12] on the regimens of the work;[13] on the operations of the work.[14] Between the time he wrote Boyle in 1678/9 and the day that Halley's visit to Cambridge in August of 1684 redirected his thoughts to celestial dynamics,[15] Newton re-

10. "Metals. 'Quomodo metalla generantur,'... Minerals. 'De radice semine spermate et corpore mineralium,'... 'De Mineralibus ex quibus desumitur,' ... 'De Conjunctione in hora nativitatis,' " all sold as parts of Sotheby lot no. 50, now parts of King's College, Cambridge, Keynes MS 35.
11. "Pyrotechny. De igne sophorem et materia quam cale facit," Sotheby lot no. 76, now Jewish National and University Library, Jerusalem, Yahuda MS Var. 1, Newton MS 38.
12. "Pyrotechny. The Three Fires," Sotheby lot no. 77, now King's College, Cambridge, Keynes MS 46.
13. (1) "Regimen. Notes on the Regimen, three sets," Sotheby lot no. 85, now Dibner Library of the History of Science and Technology of the Smithsonian Institution Libraries, Washington, D.C., Dibner Collection MSS 1032 B (formerly Burndy MS 15); (2) "[Regimens. A Treatise on the Regimens, etc.]... under the following headings: 'Lapidis Compositio'; 'Elementorum Conversio Conjunctio et Decoctio'; 'Regimen Ignis'; 'Materia'...; 'Decoctio Regimen Mercurii'; 'Regimen Jovis'; 'Regimen Lunae'; 'Regimen Veneris, Martis et Solis,' with an Early Draft, much altered of the 'Decoctio Regimen Mercurii,' " Sotheby lot no. 84, now King's College, Cambridge, Keynes MS 48; (3) "Regimen [a Series of Seven Aphorisms... followed by Annotations...]," Sotheby lot no. 86, now King's College, Cambridge, Keynes MS 49.
14. (1) "[Operations: Notes on the Operations, 1, 2, and 6–9, some fragmentary ...]," Sotheby lot no. 66, now Dibner Library of the History of Science and Technology of the Smithsonian Institution Libraries, Washington, D.C., Dibner Collection MSS 1070 A (part); (2) "Operations. Opus Primum; Opus Quintum, Two sets of Notes; Opus Sextum, Two sets of Notes; Opus Octavum; Extractio auri vivi et conjunctio ejus," Sotheby lot no. 64, now King's College, Cambridge, Keynes MS 40; (3) "[Operations: 1–6, beginning with 'Extractio et Rectificatio Spiritus'; ending with 'Solutio sicca et humida metallorum vulgi eorumque purgatio et multiplicatio infinita mercurii sophici et extractio auri vivi,' " Sotheby lot no. 65, now King's College, Cambridge, Keynes MS 41.
15. Halley certainly visited Newton in August 1684; there is some evidence that Halley also visited Newton in May of that year: summarized in Herivel, *Background to Newton's Principia* (4, n. 61), pp. 96–7; disputed by Whiteside in Isaac Newton, *The Mathematical Papers of Isaac Newton*, ed. by Derek T. Whiteside with the assistance in publication of M. A. Hoskin (8 vols.; Cambridge University Press, 1967–80), vol. IV, 17–18, n. 52.

corded at least 15,000 words on his alchemical experimentation.[16] Since each brief laboratory report represented untold hours of labor invested at the furnace, it is impossible to estimate the amount of time he spent on his experimentation, but it must have been simply immense. Yet, in addition, during that period he wrote and rewrote many essays on the history of the church and occasionally compiled papers on Judaism and on the role of Jewish ceremony as a figure or type in prophecy; he assembled variant readings of the book of Revelation, made notes on interpretative methods, and attempted his own interpretations of the prophecies.[17] He composed and often revised a treatise on the philosophical origins of gentile theology.[18] He demonstrated a vast interest in the comet that appeared in November of 1680,[19] the comet he later came to regard as God's agent in the fiery destruction of the world prophesied in Scripture.[20] Small wonder that Newton's papers reflect no further pursuit of the dynamical analysis Hooke had suggested to him in late 1679 and early 1679/80, until Halley arrived in 1684; small wonder that by then Newton could not even find his original paper on the subject. He had busied himself with other things.

Small wonder also if Newton's conceptualization of gravity remained unsettled for the few years immediately following his interchange with Hooke, as the work he edited as well as his correspondence and papers between 1680 and 1684 seem to indicate it did. In his exchange with Burnet in 1680–1680/1 Newton suggested a mechanism of vortical pressure for gravity, discussed the *vis centri-*

16. University Library, Cambridge, Portsmouth Collection MSS 3973 and 3975.
17. Richard S. Westfall, "Newton's theological manuscripts," in *Contemporary Newtonian Research* (1, n. 10), pp. 129–43.
18. Richard S. Westfall, "Isaac Newton's *Theologiae Gentilis Origines Philosophicae*," in *The Secular Mind. Transformations of Faith in Modern Europe. Essays presented to Franklin L. Baumer, Randolph W. Townsend Professor of History, Yale University*, ed. by W. Warren Wagar (New York: Holmes & Meier, 1982), pp. 15–34. Westfall's arguments in this article for Newton's deistic tendencies have been effectively challenged by James E. Force,"Newton and Deism," in *Science and Religion/Wissenschaft und Religion. Proceedings of the Symposium of the XVIIIth International Congress of History of Science at Hamburg-Munich, 1.–9. August 1989*, ed. by Änne Bäumer and Manfred Büttner (Bochum: Universitätsverlag Dr. N. Brockmeyer, 1989), pp. 120–32.
19. Newton, *Mathematical Papers* (5, n. 15), vol. V, xiii and vol. VI, 15, n. 47; Westfall, *Never at Rest* (1, n. 4), pp. 391–7.
20. Dobbs, "Newton and Stoicism" (1, n. 5).

fuga of the planets (a component of his pre-Hookian dynamical analysis), and mentioned "gravitation towards a center" without offering any mechanism for it.[21] When conferring with Flamsteed about the comet of 1680, Newton mentioned the "attraction of ye earth by its gravity" but also mentioned the "motion of a Vortex." He was willing to "allow an attractive power" in the sun "whereby the Planets are kept in their courses about him from going away in tangent lines," which seems to presuppose Hooke's analysis, but, in refuting Flamsteed's notion that such an attraction might be magnetic, Newton utilized both the idea of the sun's vortex and the concept of the *vis centrifuga*. In a draft paragraph never actually sent to Flamsteed Newton hinted that he had a method "of determining ye line of a Comets motion (what ever that line be) almost to as great exactness as the orbits of ye Planets are determined" but for want of certain data he had not done so.[22] In his second (1681) edition of Varenius's *Geographia generalis* Newton maintained without comment Varenius's explication of the flux and reflux of the oceans that was based on Cartesian vortex principles and also kept unchanged the improved diagram of the earth's vortex he (Newton) had devised for the first edition he had edited in 1672, as one may see in Plates 7,[23] 8,[24] and 9.[25] About a year

21. Newton, *Correspondence* (2, n. 14), vol. II, 319–34 (Newton to Burnet, 24 Dec. 1680; Burnet to Newton, 13 Jan. 1680/1; Newton to Burnet, ? Jan. 1680/1), esp. 319, 322, 329, 331–2.
22. Ibid., vol. II, 340–56, 358–67 (Newton to Crompton for Flamsteed, 28 Feb. 1680/1; Flamsteed to Crompton for Newton, 7 March 1680/1; Newton to [? Crompton], ? April 1681; Newton to Flamsteed, 16 April 1681), esp. 342, 359–60, 366.
23. Bernhardus Varenius, *Geographia generalis. In qua affectiones generales Telluris explicantur Autore Bernh: Varenio Med: D.* (Amstelodami: Ex Officina Elzeviriana, 1671), p. 171. The diagram illustrated Propositions IX–XI of Chapter XIV of Book I, pp. 170–81 of the 1671 ed. Varenius's text and diagram remained unchanged from the first edition (Amstelodami: Apud Ludovicum Elzevirium, 1650) through the last Elzivier edition of 1671 (on this subject) although the pagination did change. For a discussion of the Cartesian principles involved and Newton's original encounter with them, see also McGuire and Tamny, *Certain Philosophical Questions* (1, n. 22), pp. 175–80.
24. Bernhardus Varenius, *Bernhardi Vareni Med. D. Geographia generalis, In qua affectiones generales Telluris explicantur, Summâ curâ quam plurimis in locis emendata, & XXXIII Schematibus novis, aere incisis, unà cum Tabb. aliquot quae desiderabantur aucta & illustra. Ab Isaaco Newton Math. Prof. Lucasiano Apud Cantabrigienses* (Cantabrigiae: Ex Officina

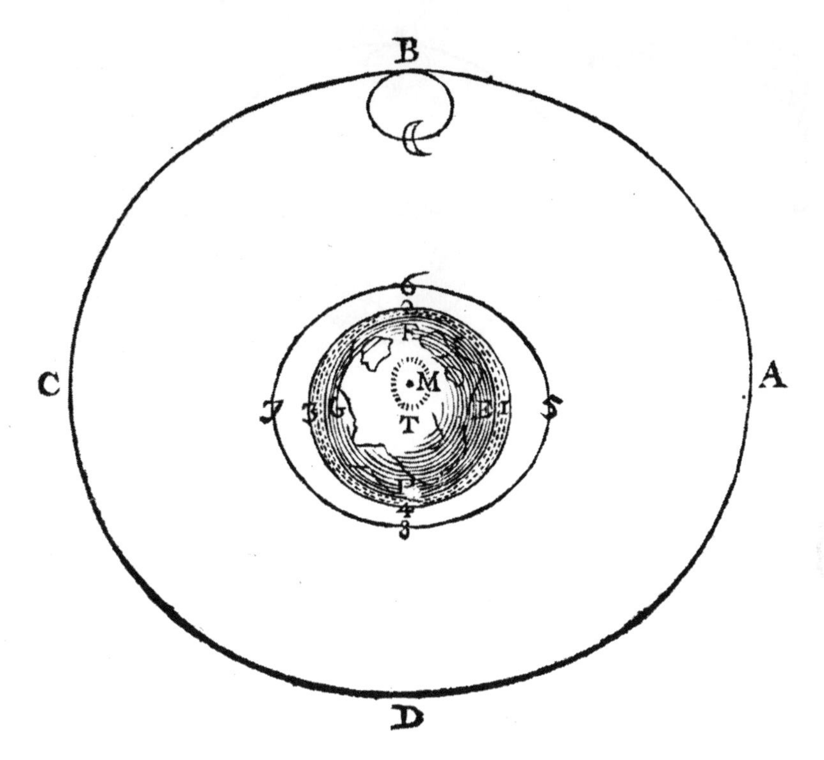

Plate 7. Bernhardus Varenius, *Geographica generalis* (Amstelodami Ex Officina Elzeviriana, 1671), p. 171. Reproduced by permission of The Henry E. Huntington Library, San Marino, CA.

later he referred to the material fluid of the heavens that gyrates around the center of the cosmic system according to the course of the

Joann. Hayes, Celeberrimae Academiae Typographi, Sumptibus Henrici Dickinson Bibliopolae, 1672), diagram between pp. 104 and 105, text pp. 120–6.

25. Idem, *Bernhardi Vareni Med. D. Geographia generalis, In qua Affectiones Generales Telluris Explicantur, Summâ curâ quam plurimis in locis Emendata, & XXXIII Schematibus Novis, AEre incisis, unà cum Tabb. aliquot quae desiderabantur Aucta & Illustrata, Ab Isaaco Newton Math. Prof. Lucasiano Cantabrigienses. Editio Secunda Auctior & Emendatior* (Cantabrigiae: Ex Officina Joann. Hayes, Celeberrimae Academiae Typographi, Sumptibus Henrici Dickinson Bibliopolae, 1681, diagram between pp. 298 and 299, pagination of text unchanged.

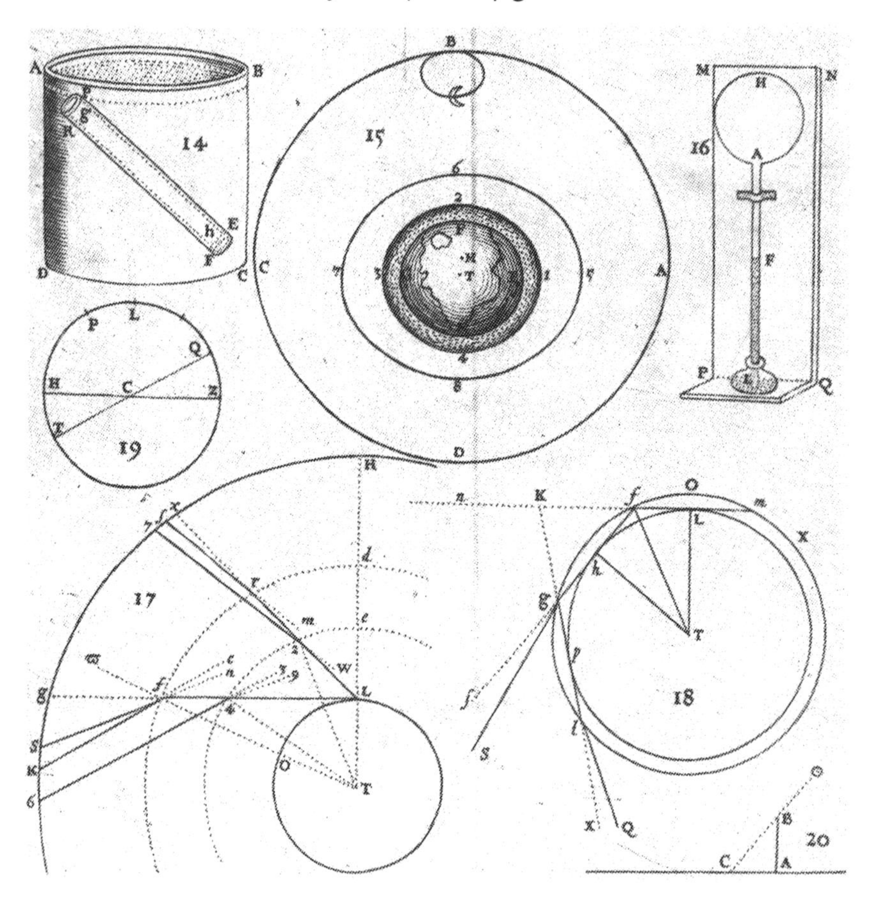

Plate 8. Bernhardus Varenius, *Geographica generalis. . . . Ab Isaaco Newton Math. Prof. Lucasiano Apud Cantabrigienses* (Cantabrigiae: Ex Officina Joann. Hayes, Celeberrimae Academiae Typographi, Sumptibus Henrici Dickinson (Bibliopolae, 1672), diagram between pp. 104 and 105. Reproduced by permission of The Henry E. Huntington Library, San Marino, CA.

planets.[26] Not until 1684 do Newton's papers reflect the clarity of thought on dynamical principles that enabled him to launch the *Prin-*

26. University Library, Cambridge, Portsmouth Collection Add. MS 3965.14, f. 613r, cited in Whiteside, "Before the *Principia*" (4, n. 36), pp. 14 and 18, n. 42. The dating is Whiteside's. The passage is also quoted and translated in McGuire and Tamny, *Certain Philosophical Questions* (1, n. 22), p. 169, n. 120.

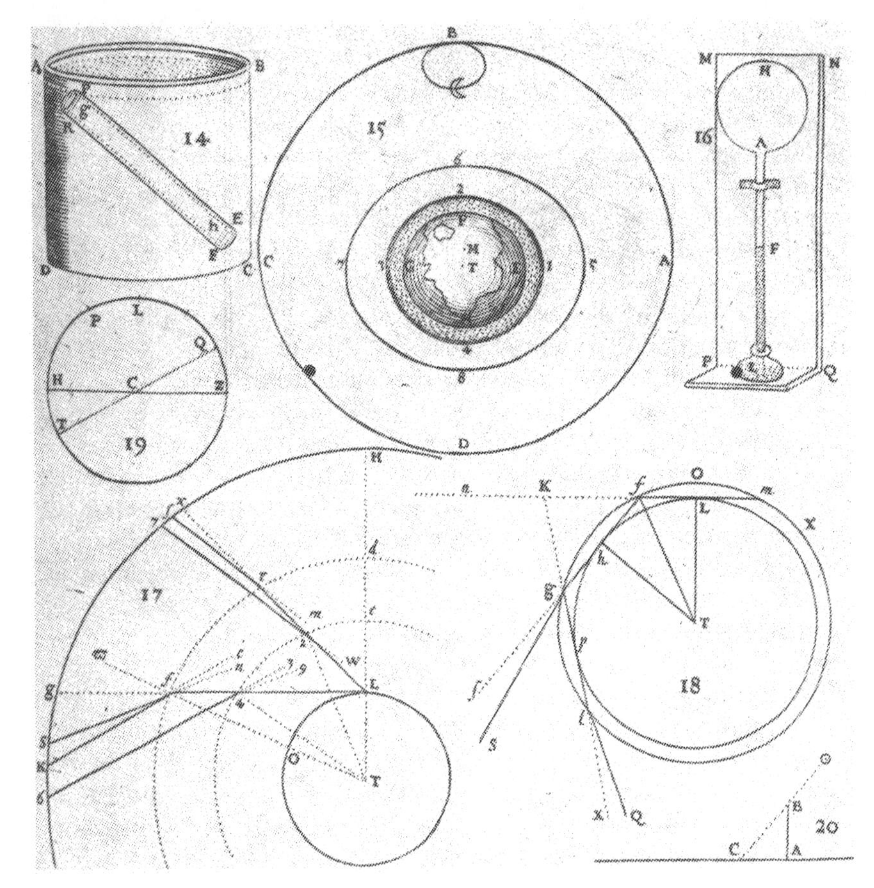

Plate 9. Bernhardus Varenius, *Geographica generalis . . . , Ab Isaaco Newton Math. Prof. Lucasiano Cantabrigienses. Editio Secunda Auctior & Emendatior* (Cantabrigiae: Ex Officina Joann. Hayes, Celeberrimae Academiae Typographi, Sumptibus Henrici Dickinson Bibliopolae, 1681), diagram between pp. 298 and 299. Reproduced by permission of The Henry E. Huntington Library, San Marino, CA.

cipia,[27] and only in the course of writing that work did Newton confront the problems that inhered in all his various early aethereal gravitational systems.

27. Whiteside, "Before the *Principia*" (4, n. 36); idem, "Introduction," in Newton, *Mathematical Papers* (5, n. 15), vol. VI, 3–19.

Gravity

The first fruits to be made public from the new dynamical work that Newton began at Halley's instigation in 1684, and that eventuated in the *Principia*, were sent to the Royal Society in November of that year. Entitled *De motu corporum in gyrum*, the first tract concentrated on central-force orbits but included also some work on terrestrial ballistics.[28] In it Newton defined "centripetal force" for the first time: "that by which a body is impelled or attracted towards some point regarded as its centre."[29] Although he also defined "resistance" – "that which is the property of a regularly impeding medium" – he immediately noted that in his first several propositions "the resistance is nil."[30] Of course that was not the case in the propositions on ballistics, in which the terrestrial atmosphere constituted an impeding medium, but Whiteside has observed that the ballistics problems were probably added as an afterthought and that regarding resistance Newton originally wrote: "Bodies are hindered neither by the medium nor by other external causes from yielding perfectly to their innate and to centripetal forces."[31] That general statement, presumably written when the tract consisted only of propositions on bodies in orbit, conclusively demonstrates that by November 1684 Newton knew the motions of celestial objects were not impeded by the medium through which they moved.

Theorem 1 of Newton's first tract on motion was a mathematical demonstration of a general area law for bodies revolving about an immovable center of force, a demonstration that remained substantially unchanged through all the versions of *De motu* and then became Proposition I, Theorem I of Book I of the *Principia*.

> The areas which revolving bodies describe by radii drawn to an immovable center of force do lie in the same immovable planes, and are proportional to the times in which they are described.[32]

The Keplerian area law for elliptical orbits was of course subsumed in Newton's general demonstration.[33]

28. Whiteside, "Introduction," in ibid., vol. VI, 18–19; Newton, ibid., vol. VI, 30–75.
29. Ibid., vol. VI, 30–1.
30. Ibid., vol. VI, 32–3.
31. Ibid., vol. VI, 32, n. 7.
32. Newton, *Principia* (Motte–Cajori: 2, n. 13), vol. I, 40–1 and (Koyré–Cohen: 1, n. 9), vol. I, 88–90; Westfall, *Never at Rest* (1, n. 4), pp. 411–14; Herivel, *Background to Newton's Principia* (4, n. 61), pp. 258–9, 278; Newton, *Mathematical Papers* (5, n. 15), vol. VI, 35–7, 539–42.
33. B. J. T. Dobbs, "Newton's rejection of a mechanical aether for gravitation:

The area law had first emerged in Kepler's work as an "approximation," based on an Aristotelian doctrine that made force proportional to velocity, though Kepler later realized that it was a deductive consequence of his hypothesis that an immaterial virtue emanating from the sun constituted the physical cause of planetary motion about the sun. In the Keplerian hypothesis the component of the planet's velocity at right angles to the radius vector was inversely proportional to the distance from the sun, so that the planet moved most rapidly at perihelion and most slowly at aphelion, which in turn implied the area law in which the radius vector sweeps out equal areas in equal times no matter where the planet is in its orbit. Combined with Kepler's even later discovery of the elliptical orbit for Mars, the area law received empirical verification (in that combined form), though for technical reasons the area law had been little used by post-Keplerian astronomers, even by those who accepted elliptical orbits.[34] Newton's new dynamical analysis, that included an inertial component, rendered the Aristotelian and Keplerian analyses obsolete; nevertheless, this convergence of mathematical and observational lines of investigation must have been immensely satisfying to Newton, for the verification of the first two Keplerian laws had been effected with the very best observational data assembled by Tycho Brahe.

But however satisfactory that may have been, within the context of the mechanical philosophy a problem immediately arose. The Keplerian area law that matched empirical observation should *not* fit so closely with the exact area law derived mathematically by Newton if the heavens are filled with a resisting medium. Unless the medium is somehow disposed to move with exactly the same variable speed that the planetary body exhibits, the planet should encounter enough resistance from the medium to cause an observed deviation from the mathematical prediction, just as projectiles in the terrestrial atmosphere are observed to deviate from mathematical prediction.

Newton's realization that no form of the hypothetical gravitational

empirical difficulties and guiding assumptions," in *Scrutinizing Science: Empirical Studies of Scientific Change*, ed. by Arthur Donovan, Larry Laudan, and Rachel Laudan (Synthese Library Studies in Epistemology, Logic, Methodology, and Philosophy of Science, Vol. 193; Dordrecht: Kluwer Academic Publishers, 1988), pp. 69–83.

34. Curtis Wilson, "Kepler's derivation of the elliptical path," *Isis* 59 (1968), 5–25; idem, "How did Kepler discover his first two laws?" *Scientific American* 226 (1972), No. 3, 93–106; idem, personal communication, 30 January 1989.

aether of mechanical philosophy could be reconciled with actual celestial motions must have been rather a shock to him. From the time of his introduction to mechanical philosophy in the 1660s until early in the 1680s he left an extensive record of his aethereal speculations, as one has seen in the preceding section and in Chapter 4. Even as he modified his schemata from time to time, even as "the matter causing gravity" became sometimes more active and then again more mechanical, Newton had never doubted that the "cause of gravity" was material. But if the heavens were filled with a material aether, then its presence should produce some notable retardation on the motions of bodies passing through it. Yet Newton said in the first draft of *De motu* that none was in evidence. Newton had had to rethink all of his aethereal mechanisms in order to make that statement, and one result of his rethinking is already in evidence in his definition of centripetal force. Whereas in earlier documents Newton had offered explanations of apparent attractions in terms of aethereal impulsions in the traditional fashion of mechanical philosophers, in the new definition he equivocated. Bodies are "impelled or attracted" by a centripetal force, he said. No causal mechanism was suggested nor any preference indicated between the two ways of describing the action of the force, a stance soon to be adopted in the *Principia* itself.

The paper to be discussed next is not overtly concerned with issues of gravitation. But even though it is not, it is probable that Newton composed his short but formal essay on air and aether when he first became aware of the problem of celestial retardation, or, rather, aware of the problem generated by the lack of celestial retardation, at some time in the months between August (or May) and November of 1684.[35] *De aere et aethere* is closely related to Newton's letter to Boyle of 1678/9 and covered that same set of phenomena that Newton considered to be mechanical: capillary attraction, surface tension, the diffraction of light by a knife edge, some forms of cohesion, and so forth. In the new essay, as in 1678/9, different types of aerial substances were generated by different terrestrial substances, with metals alone generating true permanent air of an unusually inactive sort. But whereas in his letter to Boyle Newton had proposed aethereal mechanisms for capillarity and the other phenomena, in *De aere et aethere* he undertook to explain them all by aerial density gradients.

When the Halls first published *De aere et aethere* in 1962, they argued

35. University Library, Cambridge, Portsmouth Collection Add. MS 3970, ff. 652–3, published and translated in Newton, *Unpublished Papers* (2, n. 51), pp. 214–28.

for a composition date for the manuscript between 1673 and 1675.[36] Certainly Newton could not have written it before 1673, the date of publication of some calcination experiments of Boyle's mentioned by Newton in the manuscript. The handwriting of the original document is ambiguous, however, and Newton himself left it undated, as he did most of his papers. Recognizing the similarities and differences between *De aere et aethere* and Newton's "Hypothesis" of 1675 and the letter to Boyle of 1678/9 as well, the Halls argued that the aerial mechanisms of *De aere et aethere* represent a less mature version of Newton's later aethereal mechanisms and so were written before the "Hypothesis" of 1675. More recent research in Newton's papers raises doubts regarding the terminal date suggested by the Halls. In the student notebook, in the "Vegetation" manuscript, in the "Hypothesis" of 1675, in the letter to Boyle of 1678/9, and in the general correspondence, one finds instead a strong continuity in Newton's use of aethereal mechanisms. It is of course possible that Newton flirted briefly with an aerial model during the 1670s, then returned to his favored aethereal one. The notable continuity found in all the other documents until 1684 argues against such a possibility, however, as does the fact that the changes that occur in Newton's thought seem better explained in terms of a slow evolution rather than in terms of abrupt and rapid oscillations between fundamentally different explanatory paradigms. Also, the general popularity of aethereal speculation among mechanical philosophers argues against a date in the 1670s for *De aere et aethere*. Aethereal speculation was so commonplace, and was indeed such a necessary concomitant of impact physics, that Newton would have needed an urgent reason to abandon it. That reason, one may suggest, struck him with its full force only in 1684, when he realized that the precision of planetary motion in effect precluded aethereal gravitational mechanisms.

In addition, there are other reasons for a later date for *De aere et aethere* than the Halls suggested. Much of Newton's speculation on aerial mechanisms in this document was couched in terms of a repulsive force that itself hardly constituted orthodox mechanism. In a paragraph that he ultimately canceled Newton struggled to articulate a "cause" for such repulsion.

> Many opinions may be offered concerning the cause of this repulsion. The intervening medium may give way with difficulty or not suffer itself to be much compressed. Or God may have created a certain incorporeal nature which seeks to repel bodies and make

36. Hall and Hall, "Introduction to part III, theory of matter," in Newton, *Unpublished Papers* (2, n. 51), pp. 187–9.

> them less packed together. Or it may be in the nature of bodies not only to have a hard and impenetrable nucleus but also [to have] a certain surrounding sphere of most fluid and tenuous matter which admits other bodies into it with difficulty. About these matters I do not dispute at all.[37]

Before canceling it all, Newton ended the paragraph with more equivocation and an additional disclaimer. As in the first draft of *De motu*, he recorded no personal preference. The tone of the entire paragraph contrasts sharply with the relative certainty with which he had put forward the earlier aethereal mechanisms. Against their background, the aerial mechanisms of *De aere et aethere* seem to be the production of a man who had unexpectedly been forced to doubt a favored and heretofore unquestioned explanatory device.

An additional, and quite substantial, reason for a date of 1684 for *De aere et aethere* lies in the nature of the abruptly terminated chapter on the aether, which consists only of a single paragraph. Newton began by defining the aether in terms of part of his customary proportional relationship (as air is to aether, so the vital principle of air is to the vegetable spirit).

> And just as bodies of this Earth by breaking into small particles are converted into air, so these particles can be broken into lesser ones by some violent action and converted into yet more subtle air which, if it is subtle enough to penetrate the pores of glass, crystal and other terrestrial bodies, we may call spirit of air, or the aether.[38]

Reminiscent of earlier passages in which air "relents" into its "first principle" when it reaches the aethereal regions above the atmosphere, this definition implied that aether differed from air only in the size of its particles. However, as already noted, their very small dimensions seemed to Newton to give aethereal particles some measure of activity.

In the remainder of the paragraph Newton marshaled such empirical evidence as he could for the existence of the aether. He cited the Boylean calcination experiments in which metals gained weight in hermetically sealed glasses when heated. "It is clear," Newton said, "that the increase is from a most subtle saline spirit" that came through the pores of the glass. He cited pendulum experiments in vacuo in which the motion of the bob was dampened almost as quickly as in air. He cited the case of iron filings arranged in curved lines by a lodestone: "I believe everyone who sees [that]... will acknowledge that these magnetic effluvia are of this [aethereal] kind." He began an argument that electrical attraction

37. Newton, *Unpublished Papers* (2, n. 51), pp. 216 and 223.
38. Ibid., pp. 220 and 227.

was "caused in the same way by a most tenuous matter of this kind..." And there he broke off, never to return.[39] Again, one sees in that brief paragraph the efforts of a man who had suddenly had his most fundamental explanatory device called into question. It is as if he had asked himself whether he could find evidence of any sort that would sustain his twenty-year-old belief in aethereal mechanisms.

It is, furthermore, well worth emphasizing that Newton never mentioned gravity in *De aere et aethere* even though in all his earlier aethereal systems the gravitational function of the aether had been a prominent component. It was of course precisely that function of the aether that had been called into question in 1684. It was the cosmic gravitational aether that seemed to be precluded by the lack of retardation in celestial motions. The other functions he had always assigned to the aether – chemical, alchemical, tensional, luminous, magnetic, and electrical – were all short-range phenomena. In the first chapter of *De aere et aethere*, on air, many of those short-range phenomena had been assigned to air and its density gradients or to repulsive force. Short-range phenomena had always been functionally separate from gravity in Newton's aethereal systems, even when they were quite closely assimilated to the gravitational aether in the early and middle 1670s. Nevertheless, even in 1684 he did not remove all of them to the air: chemical, magnetic, and electrical phenomena – the more active ones – remained situated in the aether and seemed to him indeed to offer evidence for the existence of the aether, as did the one item in Newton's evidentiary list that was directly related to the issue of celestial retardations: the dampening of pendulum motion even when the retarding air had been evacuated.

Newton had understood the in vacuo pendulum experiments to indicate the presence of aether within the glass even after the air was removed. By definition, aether had the ability to penetrate the pores of glass and all other terrestrial bodies. Therefore, the pendulum motion "ought not to cease unless, when the air is exhausted, there remains in the glass something much more subtle which damps the motion of the bob."[40] For Newton that experiment had indicated the presence of a gravitational aether that had acted in conjunction with the air to slow the motion of the bob when air was present, and had acted almost as effectively to dampen the motion when air was absent. But the experiment had only been designed to separate the effects of air and aether; with the very presence of the postulated aether in question Newton needed further evidence. Was there any way he could experimentally justify the hy-

39. Ibid., pp. 220 and 227–8.
40. Ibid.

pothesis that a gravitational aether remained within the evacuated container? Was there a property of the aether that could be detected by its varying reactions under experimentally varied conditions?

The answer was yes. Since, by the hypothesis, aether penetrated not only the pores of the glass container but also the pores of every other terrestrial body, it should interact with the internal parts of the bodies it permeated. By holding constant the retardation due to the presence of air (presumed to act only on the surface of the pendulum bob) and varying the quantity of matter within the interior of the bob (where the aether alone was presumed to interact with particles of matter), one should expect an increase in aethereal retardation with an increasing quantity of matter. If, that is, the gravitational aether really existed.

Newton reported just such a set of experiments in the *Principia*, in which he had first used an empty box as a pendulum bob, then filled the same box with lead or another metal.[41] The total resistance of the empty box was to the total resistance of the full box in the proportion of 77 to 78; therefore the increased quantity of matter made very little if any difference, certainly not enough to justify the existence of the gravitational aether. Westfall has made the suggestion that Newton performed these experiments not long after he had abruptly broken off the composition of *De aere et aethere*, and the logic of Newton's development makes that suggestion plausible in the extreme.[42] Although Westfall's dating of these events to 1679 is made doubtful by Newton's continued use of vortex terminology at least until 1682, these new and more refined pendulum experiments were exactly the sort of test Newton might have devised to shore up a wavering faith in an aethereal mechanism for gravity.

As he increased the quantity of matter within the box that served as the pendulum bob in these experiments, Newton had expected to observe increasing retardation of the pendulum motion because of the interaction of the aether with the increasingly numerous interior parts of the matter in the box, for by the common understanding of the mechanical philosophers, that was how the aether acted – by penetrating to and interacting with the interior parts of bodies. But that expectation was not met: retardation or resistance did not increase as he increased the quantity of matter in the bob, so he obtained no positive evidence in these experiments for the existence of the aether. In his own words, the new experiments showed him that "the resistance on the internal parts of the box

41. Newton, *Principia* (Motte–Cajori: 2, n. 13), vol. I, 325–6; (Koyré–Cohen: 1, n. 9), vol. I, 461–3.

42. Westfall, *Force* (1, n. 7), pp. 375–7, 410.

[was] either nil or wholly insensible,"[43] and that the gravitational aether as he had always envisioned it probably did not exist. If he wrote *De aere et aethere* and performed the new pendulum experiments shortly before he wrote the first draft of *De motu* in 1684, as seems likely, one has there a full and sufficient explanation for Newton's equivocal use of "impelled or attracted" in the first definition of centripetal force.

Newton did not, however, depart fully from aethereal systems as a result of the new pendulum experiments. In what seems to be his first written attempt to come to grips with the difficult information on nonresistance, he described the resistance of "pure aether" as "either nonexistent or extremely small." But one must note that it was the *resistance* of the aether that was nonexistent and not the aether itself. The passage was in a scholium in the "augmented" tract on motion, composed about December 1684. The new tract had the general title *De motu sphaericorum Corporum in fluidis* and was divided into two sections: "De motu corporum in mediis non resistentibus" and "De motu Corporum in medijs resistentibus." The former section contained the scholium of special interest for the present line of analysis.[44]

> Thus far I have explained the motions of bodies in nonresisting mediums, in order that I might determine the motions of the celestial bodies in the aether. For I think that the resistance of pure aether is either nonexistent or extremely small. Quicksilver resists strongly, water far less, and air still less. These mediums resist according to their density, which is almost proportional to their weights and hence (I may almost say) according to the quantity of their solid matter. Therefore the solid matter of air may be made less, and the resistance of the medium will be diminished nearly in the same proportion until it reaches the tenuousness of aether. Horsemen riding swiftly feel the resistance of the air strongly, but sailors on the open seas when protected from the winds feel nothing at all of the continuous flow of the aether. If air flowed freely between the particles of bodies and thus acted not only on the external surface of the whole, but also on the surfaces of the single parts, its resistance would be much greater. Aether flows between very freely, and yet does not sensibly resist. All those sounder astronomers think that comets descend below the orb of Saturn, who know how to compute their distances

43. Newton, *Principia* (Koyré–Cohen: 1, n. 9), vol. I, 463, note for line 22. This conclusion appeared only in the first edition, Newton removing it when he began to revive his aethereal speculations after 1706.

44. Newton, *Mathematical Papers* (5, n. 15), vol. VI, 74–81; idem, *Unpublished Papers* (2, n. 51), pp. 243–5, 247–70, 271–92.

> from the parallax of the Earth's orbit, more or less; these therefore are indifferently carried through all parts of our heaven with an immense velocity, and yet they do not lose their tails nor the vapour surrounding their heads, which the resistance of the aether would impede and tear away. Planets will persevere in their motion for thousands of years, so far are they from experiencing any resistance.[45]

The medium in which celestial bodies moved was a nonresisting one; nevertheless it was the aether still. The argument on diminishing resistances from quicksilver to water to air was designed not to argue the aether away completely but to highlight its extreme "tenuousness." Air interacts with the external surfaces of whole bodies to produce resistance to motion, and, if it acted also on "the surfaces of the single parts" internal to gross bodies, it would produce much more resistance. Yet aether does flow between the single parts internal to bodies "very freely" without producing resistance to motion – again an argument for the ultimate tenuousness of the aether. But undeniably the aether was still a medium, albeit a nonresisting one, and, since Newton discussed it under the general rubric of the motion of spherical bodies in fluids, it was to him presumably a fluid medium, comparable in some ways to the other fluids he mentioned.

Did Newton's tenuous, nonresisting, fluid medium still cause gravity at this stage of his thinking? One might think that the nonresisting aether had become much too tenuous to provide a mechanical "cause" of gravity by impact, as in his earlier sense of the aether as "the matter causing gravity." But if the cosmic aether was for Newton still corporeal, as the terms "fluid" and "tenuous" seem to imply it was, and if the heavens were completely filled with it, as Descartes had taught him to believe, probably he did still think in terms of a material substrate for gravity. All he said in the "augmented" version of *De motu* was that gravity was "one kind of centripetal force," such as he had been discussing. "Motion in the heavens, therefore, is ruled by the laws demonstrated."[46] The fact that those statements on gravity came in the same scholium as, and directly following, the long paragraph on the nonresistance of the aether just quoted is suggestive but far from conclusive. At most one may infer that gravity was still intimately connected in his mind with the aethereal medium, as it had always been.

Newton was soon to clarify his thinking by rejecting the corporeal nature of the medium "utterly and completely" in a formal essay of

45. Idem, ibid., pp. 261 and 285–6.
46. Ibid., pp. 261 and 286.

considerable length, left untitled, but with the incipit, *De Gravitatione et aequipondio fluidorum et solidorum in fluidis scientiam*:

> ...as water offers less resistance to the motion of solid bodies through it than quicksilver does, and air much less than water, and aetherial spaces even less than air-filled ones, should we set aside altogether the force of resistance to the passage of bodies, we must also reject the corporeal nature [of the medium] utterly and completely. In the same way, if the subtle matter were deprived of all resistance to the motion of globules, I should no longer believe it to be subtle matter but a scattered vacuum. And so if there were any aerial or aetherial space of such a kind that it yielded without any resistance to the motions of comets or any other projectiles I should believe that it was utterly void. For it is impossible that a corporeal fluid should not impede the motion of bodies passing through it, assuming that...it is not disposed to move at the same speed as the body....[47]

In the opinion of the present investigator this important manuscript on gravitation and on the equilibrium of fluids and solids in fluids has in the past been seriously misdated. When the Halls first published it in 1962 they argued that it showed so many signs of immaturity that it must belong to Newton's student days. They assigned it an approximate date between 1664 and 1668 but then frequently noted parallels with the published *Principia*.[48] Subsequent scholars have followed the Halls' dating despite the difficulties it causes. Koyré, who analyzed *De gravitatione* in his essay on "Newton and Descartes," found himself again and again comparing it to the *Principia*.[49] Herivel thought the handwriting probably restricted the date of *De gravitatione* to the period 1665–9 (certainly before 1673),[50] but then pointed out its relationships with the various versions of Newton's *De motu* tract, with his *De motu* lectures, and with certain other papers constituting preliminary drafts of the definitions and axioms in the lectures.[51] Westfall observed that the manuscript must have been written after 1668,[52] but treated it as written not long after Newton's undergraduate career,[53] only to return to it

47. Ibid., pp. 90–156, quotation from pp. 112–13 and 146–7, Halls' brackets.
48. Hall and Hall, "Introduction to part II," in ibid., pp. 75–85; idem, "Introduction to MS Add. 4003," in ibid., pp. 89–90.
49. Koyré, *Newtonian Studies* (2, n. 12), pp. 53–114.
50. Herivel, *Background to Newton's Principia* (4, n. 61), pp. 91–3.
51. Ibid., pp. 219–35 and 257–326.
52. Westfall, *Force* (1, n. 7), p. 403, n. 26.
53. Ibid., pp. 337–42.

frequently as the essential point of departure for Newton's mature dynamics.[54] Similarly, McGuire has found in it a close conceptual affinity with the theology and metaphysics of the *Principia* even though dating it between 1666 and 1670.[55] Grant, adhering to the Halls' dating, found that the manuscript demonstrated substantial resemblances to the first edition of the *Principia* in its spatial ideas although the manuscript focused more on the relationship of God to space while the *Principia* focused more on the physical properties of space.[56] Shapiro, who compared Newton's formulations of hydrostatics in *De gravitatione* with those in the *Principia*, found no change in fundamental principles in the later work even though the method of proof changed in some instances and suggested that the two documents might profitably be read together – even though Shapiro's following of the conventional dating would have meant that *De gravitatione* was written fifteen to twenty years earlier.[57] Palter, on the other hand, in doubt about the correct dating of *De gravitatione*, calls it Newton's "most substantial philosophical text" and points up the profundity of many of Newton's philosophical arguments in the text, arguments that have been treated dismissively by scholars who considered *De gravitatione* to be a bit of juvenilia,[58] and Biarnais, even though wishing to date it between 1662 and 1665, has recently found the foundational concepts of *De gravitatione* important enough to warrant a lengthy essay as well as a full translation of the original manuscript into French.[59] Similarly, Böhme, treating it as a text of importance for the philosophical foundations of classical mechanics, has provided a facsimile of the original manuscript and a German translation.[60]

54. Ibid., pp. 424–52 et passim.
55. J. E. McGuire, "Existence, actuality and necessity: Newton on space and time," *Annals of Science 35* (1978), 463–508.
56. Grant, *Much Ado about Nothing* (4, n. 5), pp. 233, 240–7, 407–11 (nn. 335–79).
57. Alan E. Shapiro, "Light, pressure, and rectilinear propagation: Descartes' celestial optics and Newton's hydrostatics," *Studies in History and Philosophy of Science 5* (1974), 239–96, esp. 266–84.
58. Robert Palter, "Saving Newton's text: documents, readers, and the ways of the world," *Studies in History and Philosophy of Science 18* (1987), 385–439.
59. Isaac Newton, *De la gravitation ou les fondements de la mécanique classique. Introduction, Traduction et Notes de Marie-Françoise Biarnais. Ouvrage publié avec le concours du CNRS* (Science et Humanisme; Paris: Les Belles Lettres, 1985).
60. Isaac Newton, *Über die Gravitation... Texte zu den philosophischen*

Some differentiation between the concepts of *De gravitatione* and those of the *Principia* obviously is in order and is quite appropriate, but it is striking that so many eminent Newton scholars have found it profitable to make a close comparison of the two documents and have in addition found *De gravitatione* to carry intrinsic significance. Those facts, when combined with internal evidence from *De gravitatione*, actually constitute an argument for a date of 1684 or early in 1684/5 for *De gravitatione*, for Newton introduced the manuscript by saying: "It is proper to treat the science of gravity and of the equilibrium of fluids and solid bodies in fluids by two methods." One of those methods is to demonstrate propositions "strictly and geometrically," he continued; the other is a "freer method of discussion, disposed in scholia, [so that it] may not be confused with the former which is treated in Lemmas, propositions, and corollaries."[61] Since the two studies mentioned – of gravitation and of the equilibrium of fluids and solid bodies in fluids – bear a strong resemblance to Books I and II of the published *Principia*, and since the published *Principia* was constructed "strictly and geometrically" except for a number of scholia utilizing a "freer method of discussion," the introductory paragraph of *De gravitatione* could almost serve for the published work. Koyré noted long ago that the two methods Newton described in *De gravitatione* match "exactly" the way he wrote the *Principia*.[62] Probably Newton did intend to write an introduction to the *Principia* as he envisioned it late in 1684 or early in 1684/5 when he wrote *De gravitatione*.

Then, as subsequent discoveries produced alterations in his planned discourse, the introductory essay in *De gravitatione* became outmoded and was abandoned as a unit, while at the same time many of its ideas that had been written in a "freer" mode were retained and ultimately appeared dispersed throughout the published work, often in scholia. The time period allowable for the composition of *De gravitatione* is indeed a very brief one, probably the few weeks falling between the "augmented" *De motu* of about December 1684 and the "initial revise" of *De motu* of winter/early spring 1684/5, as will be developed in more detail below. *De gravitatione* could not have been written later than the "initial revise," for by then Newton had recognized the mutuality of gravitational influences and had developed his third law of motion. Some signs of the germination of these new developments are already present in an insertion

Grundlagen der klassischen Mechanik. Text lateinisch-deutsch Übersetzt und erläutert von Gernot Böhme (Frankfurt/M.: Vittorio Klostermann, 1988).

61. Newton, *Unpublished Papers* (2, n. 51), pp. 90 and 121. I have modified the Halls' translation slightly.

62. Koyré, *Newtonian Studies* (2, n. 12), p. 82, n. 3.

in the "augmented" version; in the "initial revise" they are in flower; yet *De gravitatione* bears no evidence of them.[63]

The essay, in the form in which it has survived, is carefully and neatly entered in a calf-bound notebook with formal catchwords at the bottoms of the pages and with exact citations to the ideas of Descartes with which Newton takes issue – all of which, from Newton, implies prior drafts that are no longer extant, for it was his custom usually to belabor his compositions to the point of illegibility with deletions, additions, and corrections, and then, if the ideas still seemed worthwhile to him, to make fair copies and belabor them again. The polished surviving copy of this work in turn testifies to the importance Newton once attached to it, and the fact that most of the alterations in it occur toward the end suggests his relative satisfaction with most of the essay. Many ideas from the first part of *De gravitatione* do appear in later papers, but Newton abandoned the essay in the midst of an attempt to establish mathematically the properties of a "non-gravitating fluid."[64]

While Newton's interest in the properties of fluids may originally have stemmed from his long-standing belief in a fluid cosmic aether, in the course of writing *De gravitatione* he evidently came to realize that, as long as the aether did not retard the bodies passing through it, there was really no necessity for bringing it under mathematical law. He could ignore the presence of the aether in mathematizing the motions of the bodies in it since most of the aether was in fact empty space, as he said in the essay, and thus did not require to be treated as a fluid of any sort.

> But lest any doubt remain, it should be observed from what was said earlier that there are empty spaces in the natural world. For if the aether were a corporeal fluid entirely without vacuous pores, however subtle its parts are made by division, it would be as dense as any other fluid, and it would yield to the motion of bodies through it with no less sluggishness, indeed with a much greater, if the projectile should be porous, because then the aether would enter its internal pores, and encounter and resist not only the whole of its external surface but also the surfaces of all the internal parts. Since the resistance of the aether is on the contrary so small when compared with the resistance of quicksilver as to be over ten or a hundred thousand times less, there is all the more reason for thinking that by

63. I am indebted to I. Bernard Cohen who kindly called my attention to this fact in a private communication. See also Cohen, *Newtonian Revolution* (1, n. 8), pp. 258–71, esp. pp. 264–9; Newton, *Mathematical Papers* (5, n. 15), vol. VI, 74–81, esp. 78 (for the "augmented" tract) and 92–187 (for the "initial revise").
64. Newton, *Unpublished Papers* (2, n. 51), pp. 117–21 and 152–6.

> far the largest part of the aetherial space is void, scattered between the aetherial particles.[65]

The lack of retardation had simply forced Newton to conclude that the aether consisted mostly of "vacuous pores."

That passage from *De gravitatione* seems clearly to depend upon the results of the more sophisticated pendulum experiments that had shown the resistance of the aether to be null or very small, for when Newton described the hypothetical interaction of any corporeal aether with the surfaces of all the internal parts of bodies he described the rationale for those experiments. Applied to the heavens, the results of those pendulum experiments demonstrated that "by far the largest part of the aetherial space is void." The close dependence of *De gravitatione* upon those experiments has not generally been recognized. Though Cohen cited *De gravitatione* as containing the reasoning behind the pendulum experiments, he treated the example as an "early" one and so failed to connect *De gravitatione* directly with the revolutionary work in which Newton was engaged in 1684–5.[66] But there are many reasons for dating *De gravitatione* to this very period and to that very revolutionary work.

First, although one cannot rely too strongly on dating by the changing character of Newton's handwriting during different periods of his long life, in fact the handwriting of *De gravitatione* is indistinguishable from the drafts of *De motu* dated by Whiteside to 1684 and 1684/5. It is also indistinguishable from the handwriting of the alchemical *Commentarium* dated independently by the present writer to 1680–4.[67] Second, there are the conceptual affinities between *De gravitatione* and *De motu* as well as the affinities between *De gravitatione* and the published *Principia* found by the Halls, Koyré, Herivel, Westfall, McGuire, Grant, Shapiro, Palter, Biarnais, and Böhme, already noted in a general way.

One may now add to those general statements of affiliation the specific evidence from Newton's new pendulum experiments. Both Westfall and Cohen have argued that the pendulum experiments mark a critical change in Newton's assumptions regarding an aethereal medium.[68] Though Westfall and Cohen date and interpret that change somewhat differently, they agree that it was associated with the writing of the *Principia*. With

65. Ibid., pp. 113 and 147.
66. Cohen, *Newtonian Revolution* (1, n. 8), p. 314, n. 16.
67. I am grateful to Dr. Michael Halls, Modern Archivist of King's College, Cambridge, for providing a xerox copy of Newton's *Commentarium* from Keynes MS 28 for this direct comparison.
68. Westfall, *Never at Rest* (1, n. 4), pp. 376–7, 390, 455; Cohen, *Newtonian Revolution* (1, n. 8), pp. 116 and 314–15, nn. 16–7.

the evident dependence of *De gravitatione* upon the pendulum experiments, one must also thus associate *De gravitatione* with that critical change and with the writing of the *Principia.* All previous datings of *De gravitatione* would place it somewhere between 1662 and 1673, which would associate it closely in time with the mechanical aethereal gravitational system Newton recorded in his student notebook and with the vegetable–gravitational aethereal system of the early 1670s in the "Vegetation" manuscript (that in its gravitational component bore such a marked resemblance to the aether of Newton's student days). Conceptually, neither one of Newton's early systems could have been formulated with the results of the new pendulum experiments in mind. Nor could the aethereal systems of the "Hypothesis" of 1675, the letter to Boyle of $167^8/_9$, or *De aere et aethere.*

If, however, *De gravitatione* was written after the pendulum experiments that were probably done between August (or May) and November 1684 and thus dates from late 1684 or early $168^4/_5$, then a number of related items fall into place naturally. In the "augmented" *De motu*, given a probable date by Whiteside of December 1684, the relevant section of the tract is entitled "De motu corporum in mediis non resistentibus," as already noted. In the next version of *De motu*, the "initial revise" that Whiteside argues probably dates from the winter or early spring of $168^4/_5$, the same section revised is entitled "De motu corporum in spatijs non resistentibus."[69] The shift from "non-resisting media" to "non-resisting spaces" is a significant one. That change may well be an accurate reflection of Newton's realization in *De gravitatione* that the aethereal medium was mostly "empty spaces."

Then there is the matter of an item Herivel found to make its first appearance in *De gravitatione*: the "innate force" of bodies that is more or less the same as their inertia. Herivel thought *De gravitatione* probably to have been written in the late 1660s, leaving a gap after that time of at least fifteen years before the concept reappeared in the *De motu* papers.[70] With *De gravitatione* redated to late 1684 or early $168^4/_5$, however, the concept of "innate force" in that manuscript falls naturally into place in a developmental sequence of Newton's thinking on inertia that ends in the *Principia.*

It was in *De gravitatione* that Newton really broke with Descartes. One has seen that Newton had reasons for objecting to the Cartesian doctrine of the constancy of motion in the universe and that, following the lead of the Cambridge Platonists and perhaps also of Gassendi, he

69. Newton, *Mathematical Papers* (5, n. 15), vol. VI, 122.
70. Herivel, *Background to Newton's Principia* (4, n. 61), pp. 26–8.

had engaged in a search for active, nonmechanical principles with which to rectify the overly mechanized system of Descartes. But one has also seen that in the 1670s Newton never really discarded the Cartesian aether, seeing no difficulty in adding to it extrafine corpuscularian sources of activity. But in *De gravitatione* Newton was angry with Descartes; Descartes had misled him for a long time.

Newton dragged out his copy of Descartes's *Principia philosophiae* and cited chapter and verse in order "to dispose of [Descartes's] fictions."[71] First he disposed of relative motion, arguing that motion must be referred to immobile space rather than to surrounding bodies as Descartes had done; then just as thoroughly he disposed of Descartes's equation of matter with extension. Suddenly, it had become urgent for Newton to destroy those fundamental tenets of Descartes's philosophy because he had acquired experimental evidence against their viability from the new pendulum experiments. The retarding force of any corporeal aether, no matter how fine its particles, simply did not exist or was very small. Therefore, Newton said, body cannot be equated with extension. In order to remove the force of retardation from the great cosmic aethereal regions so that the comets and the planets can follow the laws they evidently follow without retardation, the aether must be conceived as incorporeal (or nearly so). Any residue of aethereal particles must be widely separated by extended empty spaces, whereas Descartes had insisted that extension did not exist apart from extended bodies. "However," Newton said,

> it is manifest that all this force [of retardation] can be removed from space only if space and body differ from one another; and thence that they can exist apart is not to be denied before it has been proved that they do not differ, lest a mistake be made by *petitio principii*.[72]

If the corporeal nature of the medium had to be rejected, then space and body had to differ. From the time of the writing of *De gravitatione* Newton's universe consisted mostly of space that was empty of body. The gravitational aether in the old sense of "the matter causing gravity" had been ruled out, and that was true both of the early kinetic circulatory aether and of the particulate tensional system of the 1678/9 letter to Boyle.

71. Newton, *Unpublished Papers* (2, n. 51), pp. 92 and 123; Harrison, *Library* (1, n. 6), item 509, p. 132, now Trinity College, Cambridge, NQ.9.116. The *Principia philosophiae* was included in Newton's copy of Descartes's *Opera philosophica* (3rd ed.; Amsterdam, 1656), incorrectly listed by Harrison as if the entire book were the *Principia philosophiae*: McGuire and Tamny, "Introduction," *Certain Philosophical Questions* (1, n. 22), p. 23, n. 24.
72. Newton, *Unpublished Papers* (2, n. 51), pp. 113 and 146–7, my brackets.

Newton was later to attempt to formulate a different sort of tensional aethereal system for gravity, as one may see in Chapter 7, but that one perforce incorporated much space that was empty of body. The older types of aether as a "cause" for gravity had departed forever from Newton's understanding of God's world. Even in 1686, in writing to Halley to dispute Hooke's claim to "the duplicate proportion" for gravity, when Newton resurrected his 1675 system momentarily in order to show Halley that he, Newton, had known the "cause" of gravity then, Newton hastened to add, "This [the gravitational aethereal system] was but an Hypothesis & so to be looked upon only as one of my guesses which I did not rely on...,"[73] thus disclaiming his past allegiance to it even while claiming it. Newton's gravity had become a "centripetal force" and so it remained even as Newton renewed his search for its "cause."

The vegetable spirit and gravity

In the past much confusion about Newton's aethers has arisen because investigators have tacitly assumed them to be unifunctional and usually have assumed that one function to be gravitational. But, as one can now see, there were other functions assigned in some of his speculative systems: primarily short-range and primarily mechanical but including one vegetative function that was quasi-cosmic as well as localized among the small particles of bodies.

Therefore it should come as no surprise to find that Newton's aether had not fully disappeared. The destruction of its gravitational function by no means destroyed its other uses. An aethereal explanation for short-range phenomena was not precluded by the removal of the corporeal particles previously responsible for gravitation. That was readily apparent to Newton, and so to a certain extent he set the problem of short-range phenomena to one side in order to complete his gravitational work on cosmic bodies, saying in the explication of Definition I of the *Principia*, "I have no regard in this place to a medium, if any such there is, that freely pervades the interstices between the parts of bodies."[74] Presumably

73. Newton, *Correspondence* (2, n. 14), vol. II, 435–41 (Newton to Halley, 20 June 1686), quotation from p. 440, my brackets.
74. Newton, *Principia* (Motte–Cajori: 2, n. 13), vol. I, 1; (Koyré–Cohen: 1, n. 9), vol. I, 1. Cohen has observed that this sentence was inserted only in the final draft sent to the printers, not being present in the immediately prior draft deposited in University Library, Cambridge, as the *De motu* lectures. Cf. Cohen, *Newtonian Revolution* (1, n. 8), p. 316, n. 19.

such a medium would have handled vegetative phenomena with its more active parts and mechanical phenomena with its less active ones. Perhaps Newton had in mind in this *Principia* passage some such concept as those he had considered in *De aere et aethere* for a "cause" for short-range repulsion: a "medium" not easily compressed, an "incorporeal nature" created by God that "seeks to repel," or a "sphere of most fluid and tenuous matter" surrounding the "hard and impenetrable" nuclei of particles. Whatever might be responsible for the short-range phenomena that Newton considered to be either vegetable or mechanical in nature, it was not relevant to the cosmic work in hand and so could be omitted.

But even though the effects of vegetation were primarily short-range, the activating spirit had traditionally been conceived in the alchemical literature as a cosmic agent. It was supposedly operative in organizing chaotic matter at the time of creation, and it was frequently illustrated as rays of light carrying divine power into the world of matter, as discussed in Chapters 2 and 3. In Newton's aethereal systems of the early to mid–1670s also, the vegetable spirit had a cosmic sweep, streaming toward the earth from the vast aethereal spaces beyond earth's atmosphere at first and then toward the sun and planets from that even vaster aethereal repository between the solar system and the fixed stars. Newton needed to maintain some residue of matter in space, and in fact some is in evidence in most of his statements on the cosmic aether. In *De gravitatione* he did not say that all space was empty of matter. Rather, he said there were "empty spaces" and that "by far the largest part of the aetherial space is void, scattered between the aetherial particles." Some particles remained in *De gravitatione*, and Newton continued to write about this residual cosmic aether in the published *Principia* as well.[75] Some of Newton's references to the aether in the *Principia* seem to be to an interparticulate variety designed to account for short-range phenomena, such as that of Definition I above, but others are clearly to a cosmic form with cosmic functions, as, for example, the cosmic aether in which the earth floated that was "nonresisting" or "free."[76]

But by the time of the writing of the *Principia* Newton had decided that his cosmic residue of aethereal particles must be composed of a matter inertially identical to that of other bodies because, were that not

75. Ibid., pp. 116–17 and 315–16, nn. 18–20. Cohen's discussion makes no distinction between Newton's older gravitational aether, his interparticulate aether for short-range phenomena, and the residual cosmic aether that is to be identified with the vegetable spirit.

76. Newton, *Principia* (Motte–Cajori: 2, n. 13), vol. I, 26; (Koyré–Cohen: 1, n. 9), vol. I, 71; Cohen, *Newtonian Revolution* (1, n. 8), p. 315, n. 18.

the case, the forms of bodies would affect their gravitational relationships in a way disproportionate to the quantity of matter they severally contained.

> If the ether, or any other body, were either altogether void of gravity, or were to gravitate less in proportion to its quantity of matter, then, because (according to *Aristotle, Descartes*, and others) there is no difference between that and other bodies but in *mere* form of matter, by a successive change from form to form, it might be changed at last into a body of the same condition with those which gravitate most in proportion to their quantity of matter; and, on the other hand, the heaviest bodies, acquiring the first form of that body, might by degrees quite lose their gravity. And therefore the weights would depend upon the forms of bodies, and with those forms, might be changed: contrary to what was proved in the preceding Corollary.[77]

With his customary reticence, Newton failed to mention that he himself was to be listed among the "others" who had thought with Aristotle and Descartes on this matter. But of course the condensation of aether to form all other bodies and the attenuation of all sorts of bodies to produce the air, which was in turn further attenuated into its "first principle" of aether, had been essential components of Newton's circulatory aethereal systems of the 1670s, and in the "Vegetation" manuscript he had explicitly stated that air, fumes, vapors, exhalations, even "whole clouds," would "lose their gravity" in their ascent to the aethereal regions. The discoveries of the mid–1680s had forced him to deny the possibility of such a loss of gravity as matter became aether. If one may trust the structure of the *Principia* itself to explain Newton's new insistence upon the ordinary materiality of the aether, it was the discovery that *all* ordinary bodies gravitate in proportion to "the quantities of matter which they severally contain" (Proposition VI, Theorem VI of Book III).[78] From that principle of universal gravitation there was no escape. And from it there followed as Corollary II the statement just quoted, that anything that did not gravitate, or gravitated less in proportion to its quantity of matter than did "body," could not become "body" simply by changing its form. If the vegetable spirit condenses into ordinary matter, then it too must be some form of material inertially homogeneous with ordinary matter even though its corpuscles would be unusually fine and active. The "body of light," or something closely related to visible light, probably

77. Newton, *Principia* (Motte–Cajori: 2, n. 13), vol. II, 413–14; (Koyré–Cohen: 1, n. 9), vol. II, 574–75. Cf. J. E. McGuire, "The origin of Newton's doctrine of essential qualities," *Centaurus 12* (1968), 233–60.
78. Newton, *Principia* (Motte–Cajori: 2, n. 13), vol. II, 411; (Koyré–Cohen: 1, n. 9), vol. II, 572.

continued to be Newton's paradigmatic exemplar of the sort of active material substance required. Certainly he continued to insist on the corpuscular nature of light, and in the *Principia* period began to press the "tails of comets" into service as a source of supply.[79] In Chapter 7 it will be necessary to consider the several uses Newton found for comets at greater length.

In the *Principia* Newton stated as a third corollary to universal gravitation that "All spaces are not equally full," and repeated the essential part of the argument on diminishing resistances from *De gravitatione*. But this time he carried the argument further, past the point of mere "tenuousness" for the aether, all the way to the point of an approximately zero quantity of matter: "And if the quantity of matter in a given space can, by any rarefaction, be diminished, what should hinder a diminution to infinity?"[80] So, having concluded that by far the largest part of the cosmic aether was empty spaces and so could not serve as a "cause" for gravity as he had once thought, Newton published in the *Principia* the "mathematical principles" of gravity only – "not defining in this treatise the species or physical qualities of forces, but investigating the quantities and mathematical proportions of them."

> I here use the word *attraction* in general for any endeavor whatever, made by bodies to approach to each other, whether that endeavor arise from the action of the bodies themselves, as tending to each other or agitating each other by spirits emitted; or whether it arises from the action of the ether or of the air, or of any medium whatever, whether corporeal or incorporeal, in any manner impelling bodies placed therein towards each other. In the same general sense I use the word impulse....[81]

Whether a medium "corporeal or incorporeal," whether bodily contact or the aether or the air or "spirits emitted," might be a "cause" of gravity or of vegetation, he would not then say. But in the post-*Principia* period Newton struggled on with each of the options still open to him. His older types of plenistic aether had been excluded by the new evidence and were never to return. But there remained as especially viable possibilities the "incorporeal medium" and the "spirits emitted," those quasi-material inhabitants of the gray area between the complete incorporeality of God and the full solidity of body. The search for a "cause" for gravity, as

79. Sara Schechner Genuth, "Comets, teleology, and the relationship of chemistry to cosmology in Newton's thought," *Annali dell' Instituto e Museo di Storia della Scienza di Firenze 10* (1985), 31–65. See also supra (2, n. 13).
80. Newton, *Principia* (Motte–Cajori: 2, n. 13), vol. II, 414; (Koyré–Cohen: 1, n. 9), vol. II, 575.
81. Ibid., (Motte–Cajori), vol. I, 192; (Koyré–Cohen), vol. I, 298.

well as a "cause" for vegetation, was not over. The details of it will constitute much of Chapters 6 and 7.

The fire at the heart of the world

One saw in earlier chapters that Newton made a close comparison of the vegetable spirit with the light from the sun and concluded that the vegetable spirit might be "the body of light." One also saw how his vegetable–gravitational aethereal system of 1675 came to be sun-centered. One found considerable evidence for Newton's study of Stoicism beginning in the 1670s and for his conversion to Arianism early in that decade. With Arianism he acquired the conviction that the Christ acted as the agent of a transcendent God, as the creative Logos that formed and transformed the world of matter according to the design of God's omnipotent will. Finally, Newton associated the vegetable spirit of alchemy so closely with the cosmic governance of the Arian Logos that Hermes, himself a personification of the operative alchemical spirit, became a pagan type of Christ.

In the unpublished treatise on the philosophical origins of gentile theology that he began to compose in the early 1680s, *Theologiae Gentilis Origines Philosophicae*, Newton brought all those themes, and others also, into a sharp focus that was informed by his belief in the *prisca sapientia*. The ancients had known the true religion before they corrupted themselves and became idolatrous, he said. Working his way backward through the several idolatrous traditions that had flourished among the Egyptians, the Chaldeans, and the Assyrians, Newton found behind the idolatry the uncorrupted religious truth that had once been universally recognized among them.[82] In addition, attributing to the ancients an understanding of Copernican heliocentrism, Newton said they had also known the true system of natural philosophy in which "the fixed stars stood immovable in the highest parts of the world; that under them the planets revolved about the sun," and that the earth, as one of the planets, "described an annual course about the sun."[83] That the ancients had known uncorrupted Truth in both religion and natural philosophy was fully consonant with Newton's general stance on primitive wisdom and

82. Westfall, "*Origines*" (5, n. 18); Manuel, *Isaac Newton, Historian* (3, n. 72), pp. 112–16.
83. University Library, Cambridge, Portsmouth Collection Add. MS 3990, f. 1, quoted in Westfall, "*Origines*" (5, n. 18), p. 15.

offers no surprise. But in the often reworked *Origines* manuscripts Newton carried his convictions somewhat further.

With their understanding of true natural philosophy, the ancients were enabled to create a form of religious structure and worship that adequately represented to human beings the structure of God's cosmos and suggested the study of nature as a means to satisfy human aspirations for knowledge of God – an extremely rational way of going about things, Newton thought.

> So then the first religion was the most rational of all others till the nations corrupted it. For there is no way ↗ (w^th^out revelation) ↙ to come to ye knowledge of a Deity but by the frame of nature.[84]

Newton spoke there as an early modern natural philosopher imbued still with Christian doctrines of long standing: nature being theologically transparent, a knowledge of "the frame of nature" led directly to a knowledge of the Deity and was indeed the preferred route to follow in the post-Reformation turmoil over valid interpretations of revelation.

The structure by which the ancients represented the world in the most ancient form of religion, Newton said, was "a fire for offering sacrifices [that] burned perpetually in the middle of a sacred place."[85] This arrangement, which Newton called a prytaneum, symbolized the cosmos, with the fire representing the sun at the center and the sanctified space around the central fire representing the entire world which was "ye true & real temple of God."

> The whole heavens they recconed to be ye true & real temple of God & therefore that a Prytaneum might deserve the name of his Temple they framed it so as in the fittest manner to represent the whole systeme of the heavens.[86]

The idea of the entire world as the true and real temple of God with the natural philosopher as priest had ancient roots and was often restated in the course of the scientific revolution. Copernicus placed the sun in the midst of "this most beautiful of temples," our world.[87] Kepler styled

84. Jewish National and University Library, Jerusalem, Yahuda MS Var 1, Newton MS 41, f. 7; also quoted in Westfall, "*Origines*" (5, n. 18), p. 25.
85. Jewish National and University Library, Jerusalem, Yahuda MS Var. 1, Newton MS 17.3, ff. 8–10, quoted in Westfall, "*Origines*" (5, n. 18), p. 24.
86. Yahuda MS Var. 1, Newton MS 41 (5, n. 84), f. 6, quoted in Westfall, "*Origines*" (5, n. 18), p. 24.
87. Nicolaus Copernicus, *De Revolutionibus Orbium Coelestium*, I, 10, in *Copernicus: On the Revolutions of the Heavenly Spheres. A New Translation from the Latin with an Introduction and Notes by A. M. Duncan* (Newton Abbot: David & Charles; New York: Barnes & Noble, 1976), p. 50.

himself a priest of God in the book of nature.[88] Some Hermeticists, including Ficino, adhered to the concept, as did some of the Cambridge Platonists, Sir Thomas Browne, and Robert Boyle.[89] Newton might have learned it from any of them, or from Philo.[90] It had considerable biblical justification,[91] and in addition Newton later cited Cicero, Macrobius, and Plutarch for similar expressions.[92]

Newton fitted the details of regional ritual into this concept as well: the movement of the priestly processions of the Egyptians showed that their theology was based on the science of the stars;[93] when the Jewish priests approached the altar, they circled the fire, lighting seven lamps to represent the planets moving around the sun.[94] Having been periodically recalled to the true religion by their prophets, the Jews never became so corrupted as others, but the rest of the nations, including Greece and Rome, ultimately adopted the Egyptian form of corruption in which the heavenly bodies and the natural elements were themselves falsely deified while the worship of the one true God fell into decay.[95]

Some form of the prytaneum probably was quite widespread in the ancient world. For ancient Greece, modern scholarship has documented its apparent ubiquity and its religious function, though it had other functions as well. The eternal flame symbolized the life of the city, and departing colonists took with them a spark of fire from the hearth of the mother city. In Greece, charity and rewards for public service were dispensed at the civic hearth, state dinners held, foreign ambassadors re-

88. Max Caspar, *Kepler*, tr. and ed. by C. Doris Hellman (London: Abelard-Schuman, 1959), pp. 374–6.
89. Harold Fisch, *Jerusalem and Albion. The Hebraic Factor in Seventeenth-Century Literature* (London: Routledge & Kegan Paul, 1964), pp. 200–21; idem, "The scientist as priest: a note on Robert Boyle's natural theology," *Isis* 44 (1953), 252–65.
90. Harry Austryn Wolfson, *Philo. Foundations of Religious Philosophy in Judaism, Christianity, and Islam* (Structure and Growth of Philosophic Systems from Plato to Spinoza II; 2 vols.; Cambridge, MA: Harvard University Press; London: Geoffrey Cumberlege, Oxford University Press, 1947), vol. II, 246–7.
91. I Kings 8: 27; Psalms 24: 1; Isaiah 66: 1.
92. Paolo Casini, "Newton: the Classical Scholia," *History of Science* 22 (1984), 1–58, esp. pp. 34 and 44–5 (nn. 50–1).
93. Westfall, "*Origines*" (5, n. 18), p. 19.
94. Ibid., pp. 24–5. As Westfall observed, this number of lamps is incorrect for a Copernican universe, but that constituted a problem Newton "did not pause to explicate."
95. Ibid., pp. 18–27.

ceived, a court of law convened, and the city archives housed; but the religious quality of the institution was pervasive, for it was the center of religious exercises and processions and the site of official sacrifices. The sacred space around the flame may even originally have been circular, as Newton seemed to think it was and as scholars continued to think until recently, though, if so, by the classical period Greek *prytaneia* were much enlarged over the original circle and had a generic rectangular form.[96]

But Newton was less interested in the sacred architecture of the Greeks than he was in that of the Jews. Presumably because he considered the Jewish religion to be less corrupted than any other, and consequently Jewish structures built to house the sacred flame to be truer cosmic representations than any other, Newton studied the details of the first Jewish temple, the Temple of Solomon, exhaustively. His conclusions on its exact dimensions and the functions of its several parts appeared after his death as Chapter V of his *Chronology of Ancient Kingdoms Amended.*[97] Whether the numerical proportions of the Temple, so carefully worked out by Newton in terms of the sacred cubit, entered in any way into his other cosmic calculations, or into his determinations of critical prophetic dates, or whether, conversely, proportions from other studies entered into his determination of Temple structure, one does not know, but passages suggestive of that possibility occur in some of his theological manuscripts.[98]

96. Stephen G. Miller, *The Prytaneion. Its Function and Architectural Form* (Berkeley: University of California Press, 1978).
97. Isaac Newton, *The Chronology of Ancient Kingdoms Amended. To which is Prefix'd, A Short Chronicle from the First Memory of Things in Europe, to the Conquest of Persia by Alexander the Great* (London: printed for J. Tonson in the Strand, and J. Osborn and T. Longman in Paternoster Row, 1728), pp. 332–46 and Plates I–III. Plate I has been reproduced in Helen Rosenau, *Vision of the Temple. The Image of the Temple of Jerusalem in Judaism and Christianity* (London: Oresko Books, 1979), p. 123. See also Newton's section of the Temple altar reproduced from a Babson manuscript in Joseph Rykwert, *On Adam's House in Paradise. The Idea of the Primitive Hut in Architectural History* (2nd ed.; Cambridge, MA: The MIT Press, 1981), p. 217. Although Rykwert does not give the Babson MS number, the reproduction is from the Sir Isaac Newton Collection, Babson College Archives, Babson Park, MA, Babson MS 434 (on the Temple of Solomon and written in the 1680s). Newton's ground plan of the Temple from Babson MS 434 has been reproduced in Manuel, *Isaac Newton, Historian* (3, n. 72), Plate 11 (facing p. 148).
98. See, for example, Jewish National and University Library, Jerusalem,

It is clear from the work of architectural historians, however, that Newton's perception of the Temple, or of the primitive prytaneum, as reflective of cosmic and divine reality was part of a long and distinguished tradition. God Himself had provided the "pattern" for the Temple in the account of David's presentation of the plans for the building to Solomon.[99] When it was constructed in the tenth century B.C., on a site with even more ancient associations with the holy, the First Temple became central to Jewish religious imagery, as it became central to Christian symbolism later. Restored around 500 B.C. by Zerubbabel (his building is called the Second Temple), destroyed, then rebuilt by Herod I not long before the birth of Jesus, the Third Temple was finally destroyed by the Romans in A.D. 70, and the Moslem Dome of the Rock now occupies its former place on Temple Mount, while the recently rediscovered Western Wall of Herod's Temple compound has become a prominent focus of popular Jewish piety.

Within Christendom the iconographic tradition of the First Temple stretches from antiquity to the Enlightenment, though with every age reading the few known facts in the light of its own preoccupations. Newton's work should be seen as a part of the post-Reformation effort of Christian humanists to reconstruct the original Temple accurately from the sources, mostly from Ezekiel's vision of the Temple, which, like the original "pattern," was thought to have been divinely inspired.[100] Newton probably hoped to move beyond his exact calculations, however, to a discovery of certain cosmic harmonic proportions, such as those he later thought the Greeks had encoded in their myths of Pan and Orpheus,[101] and in that expectation of the encapsulation of cosmic harmony in sacred architecture he probably owed something also to medieval cathedral builders and theologians, and to contemporary dialogue on Stonehenge as well.

Yahuda MS Var. 1, Newton MS 6, f. 18, published in Appendix B of Manuel, *The Religion of Isaac Newton* (3, n. 72), pp. 126–36, esp. pp. 134–5; Manuel's study of Babson MS 434 (5, n. 97) also led to the suggestion that Newton's extraordinary concern to establish the size of the sacred cubit stemmed from his belief that the Temple expressed important divine truths in physical terms: Manuel, *Isaac Newton, Historian* (3, n. 72), pp. 161–3.

99. I Chronicles 28: 11–19.

100. Rosenau, *Vision of the Temple* (5, n. 97); Ezekiel 40–3. For a generalized statement on the symbolism of the center, including fire altars and temples, as *imago mundi*, see also Mircea Eliade, *Images and Symbols. Studies in Religious Symbolism*, tr. by Philip Mairet (London: Harvill Press, 1961), pp. 37–56.

101. McGuire and Rattansi, "Newton and the 'Pipes of Pan' " (2, n. 2).

Christian imagination had found a strong source of stimulation in the prophecy of a New Jerusalem in the Book of Revelation, and a medieval church was conceived as an image of that celestial beatitude to come. John's vision of the heavenly city of the New Jerusalem was used in church dedication ritual, but so also was Ezekiel's vision of the Temple, since the latter was also thought to have prefigured the Celestial Jerusalem. But in addition to its function as an image of bliss to come into which the earthly faithful entered as into the heavenly sanctuary, the church edifice was constructed to represent the ordered beauty and meaning of God's structured cosmos. According to von Simson, medieval views on cathedral structure are to be traced to the influence of Augustine's *De musica* combined with the well-known dictum from the Wisdom of Solomon that God ordered everything by measure, weight, and number, and reinforced by Macrobius, Chalcidius, and Boethius. True beauty in their view could not be representational of confused sensory experience but was properly anchored in metaphysical reality and based on number: the visible harmonies of geometric construction and the audible harmonies of music are intimations of the ultimate harmony of God's design, of the sacred concord of God's eternal symphony. That the main dimensions of Solomon's Temple yield the proportions of the perfect consonances did not escape notice. Although Adam's fall had clouded man's understanding, the creation was thought of as a divine act of self-revelation, a theophany, and sometimes, from the thirteenth century onwards, God was represented in the iconography of creation using compasses to order the universe with geometric laws.[102] As late as the 1790s the geometric imagery of the creative compasses dominated William Blake's *The Ancient of Days* and his *Newton*, though with Blake they had become a negative symbol – not of creation but of restriction.[103]

The architects of the Gothic cathedrals had ordered their proportions by strictly geometric means, understanding that there were more-than-human values inhering in geometry. Von Simson argues that the devel-

102. Otto von Simson, *The Gothic Cathedral. Origins of Gothic Architecture and the Medieval Concept of Order* (2nd ed. rev.; Bollingen Series XLVIII; New York: Pantheon Books, 1962), pp. 8–38; for John's vision, Revelation 21; for God's mathematics, Wisdom of Solomon 11: 20b; for the proportions of Solomon's Temple, I Kings 6.

103. Anne Kostelanetz Mellor, *Blake's Human Form Divine* (Berkeley: University of California Press, 1974), pp. 136–40, 155–7, and Plates 34, 35, and 41; Anthony Blunt, "Blake's 'Ancient of Days.' The symbolism of the compasses," *Journal of the Warburg Institute* 2 (1938–9); 53–63.

opment of Gothic architecture marks a turning point from a mystical to a rational approach to truth, and that the design of a Gothic cathedral in a geometric attempt to reproduce the structure of the universe was an insistence upon the realization of the laws of truth, not unlike that of a modern scientific experiment.[104] Although the sort of truth sought was so different from the modern conception that von Simson's comparison of cathedral building to modern scientific experimentation may not be quite accurate, the medieval cathedral did faithfully reflect the medieval perception of the cosmos as one of ordered and spacious beauty, filled with love, light, and celestial harmonies.[105] The sacred space of Christendom allowed for a penetration of metaphysical realities into the mundane world just as the sacred spaces of antiquity had done.[106] And during the period of Newton's intense interest in the meaning of ancient temples, his contemporaries were debating whether ancient British monuments should be given a similar interpretation.[107]

For Newton's assimilation of the sun – the fire at the heart of the world – to the sacred flame on the altar and to the Deity, one must also reckon with venerable traditions. References to the central fire as cosmic "hearth" have been traced back to Philolaus; Plato's *Republic* stressed

104. von Simson, *Gothic Cathedral* (5, n. 102), pp. 34–9.
105. Cf. C. S. Lewis, *The Discarded Image. An Introduction to Medieval and Renaissance Literature* (reprint of the 1964 ed.; Cambridge University Press, 1979), esp. pp. 92–121.
106. Cf. the following essays in Truman G. Madsen, Ed., *The Temple in Antiquity. Ancient Records and Modern Perspectives*, with introd. essay by Truman G. Madsen (The Religious Studies Monograph Series, Vol. 9; Provo, UT: Brigham Young University Religious Studies Center, 1984): Hugh W. Nibley, "What is a temple?" pp. 19–37, esp. pp. 22–3; John M. Lundquist, "The common temple ideology of the ancient Near East," pp. 53–76; Frank Moore Cross, Jr., "The priestly tabernacle in the light of recent research," pp. 91–105; Richard J. Clifford, "The temple and the holy mountain," pp. 107–24; George MacRae, "The temple as a house of revelation in the Nag Hammadi texts," pp. 175–90.
107. The debate began in 1655 with the publication of Inigo Jones's ideas on Stonehenge and was carried on by Walter Charleton, Edmund Halley, William Stukeley, and others. See Inigo Jones, *The Most Notable Antiquity of Great Britain Vulgarly Called Stonehenge (1655)*, introd. note by Graham Parry (A Scolar Press Facsimile; reprint of the 1972 ed.; Menston, Yorkshire: The Scolar Press, 1973); Christopher Chippendale, *Stonehenge Complete* (Ithaca, NY: Cornell University Press, 1983), pp. 43–81. I am indebted to Barbara M. Stafford for calling my attention to the contemporary English context.

the importance of the sun as the cause of everything in the phenomenal world and the origin of birth and growth and nourishment.[108] Augustine reported that Zeno the Stoic thought that God Himself was fire; Stobaeus indicated Zeno's conviction that all the heavenly bodies, endowed with mind and wisdom, consisted of creative fire. Zeno seems to have held the general active principle of the universe to be fire, a notion accepted by later Stoics, and reported by Eusebius.[109]

The sun was the natural candidate for the most concentrated center of creative fire, and in antiquity was variously called the heart, eye, mind, and ruler of the world. Astral religion often tended toward sun worship; the Stoic Cleanthes was joined by the Neoplatonist Proclus in proclaiming the sun as master and guardian of all. Solar adoration occupied a prominent place in ancient hymnology, and "heliolatry" in its many cultic forms contended vigorously with early Christianity.[110]

Elements of Stoic and Neoplatonic thought associating the Deity with fire and light were absorbed by Christianity, where they were after all congruent with the God who first said "Fiat lux," who spoke to Moses out of a burning bush and led the Israelites with a pillar of smoke by day and a pillar of fire by night, and who sent the Pentecostal Spirit with tongues of flame. Of special importance in the transmission of Proclus's thought was the work of Pseudo-Dionysius (the Areopagite). Perhaps a late fifth-century Neoplatonic Christian but thought for many centuries to have been a disciple of St. Paul's and the first Bishop of Athens, Dionysius wrote several works that were to become important in the Middle Ages. Latin translations of them by John Scotus Eriugena stimulated the light aesthetic explicitly developed by the Abbot Suger of St.-Denis into the luminosity of Gothic style, while the later translations and commentary by Robert Grosseteste produced a major revival of ancient light metaphysics. "And what of the sun's rays?" Dionysius asked.

108. D. R. Dicks, *Early Greek Astronomy to Aristotle* (Aspects of Greek and Roman Life, General Ed. H. H. Scullard; reprint of the 1970 ed.; Cornell Paperbacks; Ithaca, NY: Cornell University Press, 1985), pp. 66, 99.

109. H. A. K. Hunt, *A Physical Interpretation of the Universe. The Doctrines of Zeno the Stoic* (Melbourne: Melbourne University Press, 1976), pp. 48–50. For a consideration of the possibility that these concepts may be traced back to Heraclitus, see A. A. Long, "Heraclitus and Stoicism," *Philosophia (Athens)* 5/6 (1975–6), 134–56.

110. S. Angus, *The Religious Quests of the Graeco-Roman World. A Study in the Historical Background of Early Christianity* (London: John Murray; New York: Charles Scribner's Sons, 1929), pp. 273–8.

> Light comes from the Good, and light is an image of this archetypal Good. Thus the Good is also praised by the name "Light", just as an archetype is revealed in its image. The goodness of the transcendent God... gives light to everything capable of receiving it, it creates them, keeps them alive, preserves and perfects them. Everything looks to it for measure, eternity, number, order. It is the power which embraces the universe. It is the Cause of the universe and its end.
>
> The great, shining, ever-lighting sun is the apparent image of the divine goodness, a distant echo of the Good. It illuminates whatever is capable of receiving its light and yet it never loses the utter fullness of its light. It sends its shining beams all around the visible world. ...It is responsible for the origins and life of perceptible bodies, nourishing them and causing them to grow, perfecting them, purifying them, and renewing them.[111]

Even though the classic metaphors for the sun as eye and heart of the world were never really lost in patristic writing, Grosseteste's restatement of them led to what McEvoy calls a "valuational heliocentrism," in which the sun occupied a central position in every sense except the physical.[112] Revived yet again with fresh infusions of Hermeticism, Neoplatonism, and Stoicism in the Renaissance, the association of the sun with the power, beauty, and majesty of divinity appeared in the writings of Ficino and Pico that formed the background to the work of Copernicus. Ficino said the sun signifies God "in the greatest degree" and that "God's eternal

111. Pseudo-Dionysius, *The Divine Names*, IV, 4; in *Dionysius the Areopagite, The Divine Names and The Mystical Theology*, tr. by C. E. Rolt, preface and conclusion by W. J. Sparrow-Simpson, intro. by C. E. Rolt (reprint of the 1920 ed.; Translations of Christian Literature; London: SPCK, 1979), pp. 91–2; quoted here from the new translation in Pseudo-Dionysius, *The Complete Works*, tr. by Colm Luibheid, foreword, notes, and translation collaboration by Paul Rorem, preface by Rene Roques, intros. by Jaroslav Pelikan, Jean Leclercq, and Karlfried Froehlich (The Classics of Western Spirituality, A Library of the Great Spiritual Masters; New York: Paulist Press, 1987), p. 74. See also von Simson, *Gothic Cathedral* (5, n. 102), pp. 50–8, 91–141; McEvoy, *Robert Grosseteste* (3, n. 39), pp. 69–146, and the following introductory essays in Pseudo-Dionysius, *Works*: Jaroslav Pelikan, "The odyssey of Dionysian spirituality," pp. 11–24, and Jean Leclercq, "Influence and noninfluence of Dionysius in the Western Middle Ages," pp. 25–32.

112. McEvoy, *Robert Grosseteste* (3, n. 39), pp. 200–3. On the centrality of the sun in medieval symbolism, see also H. Flanders Dunbar, *Symbolism in Medieval Thought and Its Consummation in the Divine Comedy* (New York: Russell & Russell, 1961).

power and divinity [are understood] through the sun."[113] Dionysius the Areopagite was a favorite of the Florentine Platonists, and Pico della Mirandola applied Dionysian metaphysics systematically in his *Heptaplus* to denote correspondences in the terrestrial, celestial, and supercelestial worlds and gain insight into the invisible world of the supercelestial by analogies with and differences from the visible:

> ... among us, fire is a physical element; the sun is fire in the sky, the celestial world; in the region above man, fire is the seraphic intellect. But see how they differ: the elemental fire burns, the celestial fire enlivens, the super-celestial fire loves.[114]

"Trismegistus called him [the sun] the visible God," Copernicus said, invoking that same tripartite Dionysian cosmos in which the sun was the visible representative in the celestial realm of unseen and unseeable reality in the supercelestial.[115]

Somewhat closer to Newton's own view, perhaps, is the related Hermetic doctrine of the sun as the channel by which creative activity passed from the supercelestial to the elemental or terrestrial realm.[116] The fe-

113. Marsilio Ficino, *De sole*, as translated and quoted in Paul Oskar Kristeller, *The Philosophy of Marsilius Ficino*, tr. by Virginia Conant (Gloucester, MA: Peter Smith, 1964), pp. 97–8. On the importance of the sun in Renaissance magic, see also Walker, *Spiritual and Demonic Magic* (2, n. 26).
114. Pico della Mirandola, *Heptaplus* (3, n. 61), p. 24; E. H. Gombrich, *Symbolic Images. Studies in the Art of the Renaissance, with 170 Illustrations. II.* (reprint of the 1972 ed.; Chicago: University of Chicago Press, 1985), pp. 152–3. The authenticity of Dionysius as a disciple of St. Paul was seriously challenged in the fifteenth century, especially by Lorenzo Valla; but although Pico and Ficino knew of Valla's position, they strongly defended the traditional identification. For the complicated story of the reception of the Pseudo-Dionysian corpus during the Renaissance and Reformation, see Karlfried Froehlich, "Pseudo-Dionysius and the reformation of the sixteenth century," in Pseudo-Dionysius, *Works* (5, n. 111), pp. 33–46.
115. Copernicus, *De revolutionibus*, I, 10, in *Copernicus: On the Revolutions* (5, n. 87), p. 50.
116. See especially Hermes Trismegistus, *Corpus Hermeticum* XVI, in *Corpvs Hermeticvm. Texte établi par A. D. Nock et traduit par A.-J. Festugière* (4 vols.; revised ed. of 1946–54; Collection des Universités de France publiée sous le patronage de l'Association Guillaume Budé; Paris: Société d'Édition ⟨⟨Les Belles Lettres⟩⟩, 1980–3), vol. II, 228–41; in Scott, Ed., *Hermetica. The Ancient Greek and Latin Writings which contain Religious or Philosophic Teachings ascribed to Hermes Trismegistus.* English tr. and

cundity of the sun, its action as demi-urge in generating life and in serving to form, conserve, and nourish all the infinite variety of things, as expressed in the Hermetic Corpus, was fully aligned with the original Platonic dictum on the sun as cause and origin of birth and growth and nourishment. This concept was adopted by the Copernicans also, and in Varenius's *Geographia generalis* it constituted the first argument for heliocentrism.[117] Newton's two editions of Varenius insured its presence in the Cambridge of the 1670s and 1680s as well.[118]

The idea also found many echoes in alchemical and Stoic literature. Divinity in its generative aspect, streaming from heaven in the light of the sun and stars (but especially in the light of the sun), might be conceived as the "fifth essence" or quintessence that was not strictly material as the four elements were, nor strictly spiritual in the incorporeal sense either – the link between heaven and earth, between spirit and matter, that the alchemists sometimes called a corporeal spirit and a spiritual body, and which constituted their active principle. It might also be conceived as the creative fire in the *pneuma*, or, as the author of *La lumière* had explained, "a corporeal spirit diffused through all nature, the principle of all vegetation, life, attraction, sympathy, and motion . . . the form informing all things . . . the lawful son of the sun and the true sun of nature."[119] Echoes of the same concept of course had already appeared in Newton's papers as the "fermental virtue" and in the notion that the "vegetable spirit" might be identified with "the body of light."

As the sun was the "visible God" in the *Corpus Hermeticum*, so was

notes by Walter Scott (1st ed., 1924; reprint of the 1982 ed.; 4 vols.; Boston: Shambhala, 1985), vol. I, 262–73, and vol. II, 428–57.

117. Varenius, *Geographia generalis* (5, n. 23), p. 59 of the 1650 ed., p. 56 of the 1671 ed.

118. Idem, *Geographia generalis*, ed. Newton 1672, (5, n. 24); 1681, (5, n. 25), p. 39 in both editions: "1. Sol non tantam fons lucis est, quae tanquam clarissima fax illuminat Tellurem, Lunam, Veneram & reliquos sine dubio planetas, sed etiam focus caloris & Vitalis Spiritus quo totum hoc universum foveri & sustentari videtur. Ideo medium locum omnium obtinere, & hos circa eum moveri probabile." In Newton's ed. of 1681 "foveri" becomes "fovere."

119. Taylor, "The idea of the quintessence" (4, n. 15); Atwood, *Hermetic Philosophy and Alchemy* (2, n. 26), pp. 75–97; for the quotation from *La lumière*, see Appendix C. See also Allen G. Debus, "The sun in the universe of Robert Fludd," in *Le soleil à la Renaissance* (3, n. 62), pp. 259–77 and following plates.

it also the "second God" and the "Son of God."[120] For Christian readers of the Hermetic materials, it was thus only a small and quite natural extrapolation to argue for the identification of Christ with the sun. Pico's typology identified the creation of the sun on the fourth day (the middle of the seven-day week) as a type of the birth of Christ in the midst of all the time allotted for history; Cardinal de Bérulle followed his lead so far as to predicate arguments for a Christocentric religion upon those for a heliocentric cosmos. Then there was the widespread identification of Christ with the philosopher's stone, the activating spirit of alchemy, the informing form of all matter, the quintessence that streamed from the heavenly bodies and that was "the lawful son of the sun and the true sun of nature."

It was apparently this complex of ideas that lay behind Newton's understanding that alchemy held the secrets of the first stage in the falling away of mankind from the original true religion.

> Now the corruption of this religion I take to have been after this manner: ffirst the frame of ye heavens consisting of Sun Moon & Stars being represented ⟨*illegible word, deleted*⟩ in the Prytanaea as ye real temple of the Deity men were led by degrees to pay a veneration to their sensible objects & began at length to worship them as the ↗ visible ↙ seats of ⟨*illegible word or symbol, deleted*⟩ divinity. ⟨ffor tis agreed that Idolatry began in ye worship of the heavenly bodies, *deleted*⟩ And because ye sacred fire was a type of ye Sun & all ye elements are parts of that universe wch is ye temple of God they soon began to have these also in veneration. For tis agreed that Idolatry began in ye worship of ye heavenly bodies & elements.[121]

Idolatry soon progressed beyond that stage, Newton thought, to the worship of dead men and even lower animals and statues,[122] but at first "the twelve Gods of the ancient Peoples were the seven Planets with the four elements and the quintessence of the Earth."[123]

It is his identification of the twelve gods of the ancient peoples with

120. Frances A. Yates, *Giordano Bruno and the Hermetic Tradition* (Chicago: The University of Chicago Press; London: Routledge & Kegan Paul; Toronto: The University of Toronto Press, 1964), pp. 7–9, 23, 36.
121. Yahuda MS Var. 1, Newton MS 41 (5, n. 84), f. 8r. I am indebted to Richard S. Westfall for having challenged me to find a connection between Newton's alchemy and his search for the original true religion.
122. Westfall, "*Origines*" (5, n. 18); Manuel, *Isaac Newton, Historian* (3, n. 72), pp. 103–21.
123. Jewish National and University Library, Jerusalem, Yahuda MS Var. 1, Newton MS 16.2, f. 1.

the seven ancient planets, the four traditional elements, and the earthly "quintessence" that yields insight into the linkage between alchemy and the search for the true primitive religion in Newton's work. The seven ancient planets included of course the sun and the moon, since those two heavenly bodies also wandered about the earth in the false natural philosophy of geocentrism, as well as the five bodies anciently known that still retained their planetary status in heliocentrism – Mercury, Venus, Mars, Jupiter, and Saturn. The four elements of Empedoclean–Aristotelian tradition were fire, air, water, and earth, whereas the alchemical quintessence carried Stoic, Neoplatonic, and Hermetic overtones of creative divinity, as noted above. These twelve gods of the earliest form of idolatry, that first dreadful corruption of true religion, were precisely those deities Newton identified with the materials of alchemical experimentation, as one may see in Plate 10.[124]

Of special interest is the inclusion of the quintessence in Newton's list. Variously identified, Newton said, by the various nations, as Isis, Juno, and Ceres, Newton himself called it "Quintessentia seu Elementorum Chaos i.e. Mundus" and noted that it was represented in alchemy as antimony or "Magnesia Gebri." "Magnesia is not fire nor air nor water nor earth," he said, "but all of them."

> It is fiery, aery, watery, earthy. It is heat and dryness, humidity and cold. It is watery fire and fiery water. It is a corporeal spirit and a spiritual body. It is the condensed spirit of the world.[125]

It is in short the quintessence of the alchemists that from below links earth with heaven, just as the heavenly quintessence from above links heaven with earth.[126]

The alchemical symbol for antimony (♁), be it noted, is based on the symbol of a redeemed earth utilized by medieval Christian kings, an orb or globe surmounted by a cross,[127] an emblem also carried by the Christ in his capacity as savior of the world, *Salvator Mundi.*[128] In

124. Babson MS 420 (1, n. 39), p. 2. See also the editorial note to Appendix E.
125. Babson MS 420 (1, n. 39), pp. 1–2: "Magnesia nec ignis est nec aer nec aqua nec terra, sed omnia. Est igneus, aereus, aqueus terreus. ↗ Est calidus, et siccus ⟨humid frigidus, *deleted*⟩ humidus et frigidus. ↙ Est ignis aquosus et aqua ignea ⟨*illegible words, deleted*⟩ Est spiritus corporalis et corpus spirituale. Est condensatus spiritus mundi. . . ."
126. Cf. Taylor, "The idea of the quintessence" (4, n. 15), esp. Fig. 1, p. 257.
127. J. M. Bak, "Medieval symbology of the state: Percy E. Schramm's contribution," *Viator* 4 (1973), 33–63, esp. 35–6.
128. George Ferguson, *Signs & Symbols in Christian Art with Illustrations from Paintings of the Renaissance* (A Galaxy Book; New York: Oxford

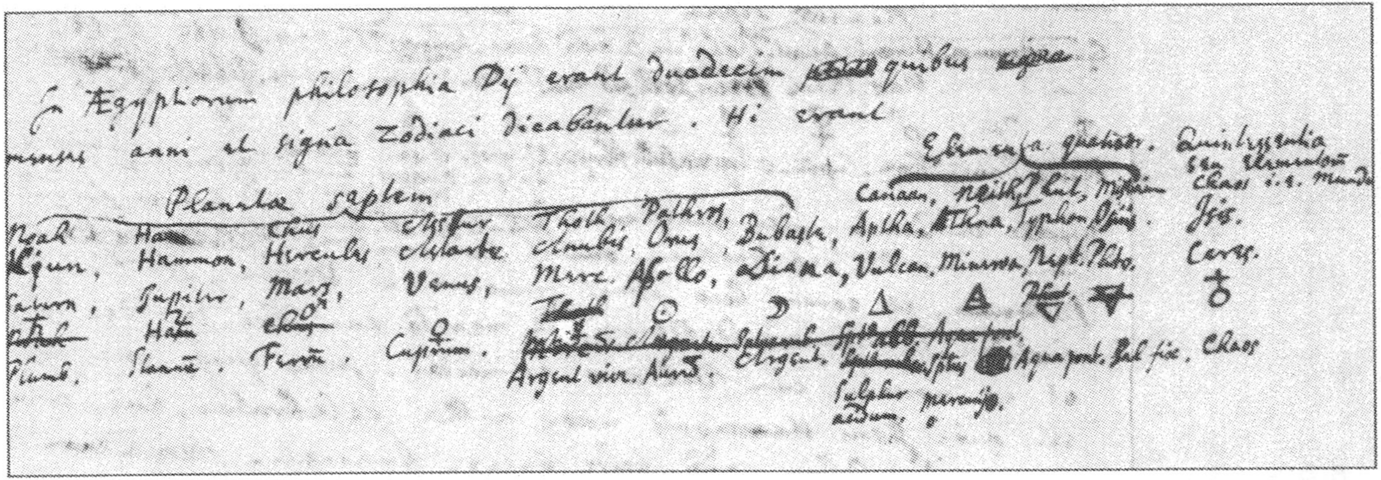

Plate 10. Babson MS 420, p. 2, The Sir Isaac Newton Collection, Babson College Archives, Babson Park, MA. Reproduced by permission of the Babson College Archives.

Newton's correlation of alchemical materials with "the twelve Gods of the ancient Peoples," antimony (or the magnesia of Geber) is the only substance carrying access to the divine active principle, recalling of course, in this late alchemical manuscript, Newton's identification of magnesia as the only species that revivifies in his alchemical "Propositions" paper of the late 1660s. The seven traditional metals in the late list (lead, tin, iron, copper, mercury, gold, and silver) are identified with the seven traditional planetary deities (Saturn, Jupiter, Mars, Venus, Mercury, Apollo, Diana, or their more ancient counterparts), while the more or less ordinary chemicals of Newton's final list – "Sulphur acidum," "Sp^{tus} mercurij," "Aqua pont.," and "Sal fix." – are identified with the traditional four elements of fire, air, water, and earth, respectively, and with Vulcan, Minerva, Neptune, and Pluto (or their more ancient counterparts). Only antimony or "Magnesia Gebri," the quintessence that contains within it the form of everything, is associated with the concept from *La lumière* of "the lawful son of the sun and the true sun of nature."

What one finds, then, in this linkage between Newton's work on the original religion and his work on alchemy, is yet another level of meaning in the symbolism of the ancient prytaneum conceived as cosmic image. The "fire at the heart of the world" was also to Newton the creative fire at the heart of matter, the informing form of all ordinary substances, hidden but active in the elemental or terrestrial realm. As the fire at the center of the prytaneum was "a type of ye Sun," so likewise was it a type of the alchemical quintessence, the fermental virtue, the vegetable spirit, the creative fire in the Stoic *pneuma*, the active principle of fermentation that gave the many and various forms to formless matter in accordance with "the ideas and will of a Being necessarily existing," the Arian Logos still active in the ongoing creation of the world.

Nevertheless, since Newton identified the basic materials of alchemical experimentation with the twelve *idolatrous* deities of early corruption, it is clear that he did not view received alchemical doctrine as entirely uncorrupted. Although alchemy might hold in its ancient codes some of the *least* corrupted aspects of religious and natural knowledge, it represented nonetheless a declension from original Truth. The worship of the twelve idolatrous deities had been a corruption of true ancient religion, just as geocentrism had been a corruption of true ancient natural philosophy, both deviating from the correct central focus on "the fire at the heart of the world." Exactly so had the alchemy inherited by the

University Press, 1966), s.v. "Globe," p. 175; Hall, *Dictionary of Subjects and Symbols in Art* (2, n. 67), s.v. "Salvator Mundi," p. 271.

seventeenth century deviated from its correct focus on "the fire at the heart of the world" in micromatter, its encoded secrets having been transmitted in garbled form and its original meaning and purpose having been lost.

Its true purpose was the demonstration of divine providence in the daily interactions of the small particles of matter. The purpose of the study of alchemy was not to gratify curiosity about natural transformations nor to become skilled in producing the spectacular phenomena of natural magic nor to enrich one's self with alchemical gold. The only purpose of learning to make gold or to perform the other unusual operations reported in the alchemical literature was to obtain experimental verification of one's decoding of the enigmatic texts, to match laboratory success with one's interpretations of the secretive language of alchemy and thus know that one's solution to the puzzle was correct. In this area Newton seems to have approached the problem in precisely the same way he approached the interpretation of prophecy. The purpose of attempting to interpret prophecy was *not* to foretell "times and things," Newton said, nor "to gratify men's curiosities" about the future. God had given the prophecies so that human beings might match them with their fulfillments and so produce "a convincing argument that the world is governed by providence." Just as historical facts might enable the interpreter of prophecy to choose between possible interpretations of the mysterious prophetic words, so laboratory results might enable the philosophical alchemist to choose between possible interpretations of the occult knowledge buried in alchemical texts. In either case, for Newton, it was only the firm correspondence of fact with interpretation that would enable him to do what he most wanted to do: provide an irrefutable demonstration of God's providential action in the world.[129]

The idea of "the fire at the heart of the world" thus held significance for Newton at many different levels. In the first instance, he attached historical significance to it, as it was the eternal flame in the center of sacred space in the true religion of the ancients. But the entire world was represented there, the world itself being the true and real temple of God, and the antique ordering of religious structure and worship around the prytaneum had been for the purpose of representing God's cosmos to the people, a rational way of providing a knowledge of God by the frame of nature. In the second instance then, Newton found deep religious meaning: in his efforts to restore true religion he, like Kepler, could be a priest of God in the book of nature and could provide a rational way

129. Cf. supra, 3, pp. 84–6 and n. 79.

to the knowledge of God by the restoration of the true natural philosophy of heliocentrism. The cosmos represented by the prytaneum was of course sun-centered, and so ancient sacred architecture might be presumed to carry encoded echoes of cosmic harmonies applicable to heliocentric planetary motions, and in that concept one may detect a third level of meaning: for Newton, as for Pythagoras, Plato, the author of the Wisdom of Solomon, Augustine, and so many more, the metaphysical and the physical met in number, weight, and measure. The focus of worship in the round prytaneum was naturally the eternal flame, the "type" or emblem of the sun, itself an emblem in the celestial world of the beneficent Deity in the supercelestial world and of the creative hidden activity of the Deity in the elemental world. The sun was from antiquity the emblem of divine power and governance, and as Newton wrote the *Principia* in the 1680s and pursued his alchemical work, the fourth and fifth levels of meaning of "the fire at the heart of the world" unfolded for him: the sun was the center of the penetrating gravitational attraction that held his heliocentric world together as it was also the source and celestial counterpart of his fermental virtue that animated, shaped, and governed the realm of passive matter.

Had Newton successfully completed his work in this area, then we would really have known something.

Conclusion

The 1680s marked a critical turning point in Newton's developing understanding of modes of divine action in the world. Before 1680 he had considered gravity to operate mechanically through the impulsion of fine material particles, or, latterly, through the density gradients of a material tensional system. Although he had partially fused the mechanical gravitational aether with the active vegetative spirit for a time in the 1670s, when he composed his letter to Boyle in 167$^{8}/_{9}$ that union had been dissolved. But in any case Newton had never thought in the 1660s and 1670s that the action of gravity was direct evidence of God's *potentia absoluta*, and so he had never really focused his attention upon it. Gravity was, in its regular mechanical operation, only evidence of God's *potentia ordinata*. Newton entered the decade of the 1680s, therefore, resolved to find immediate evidence of divinely *willed* activity in his alchemical and theological work, where nonmechanical principles of design were to be found operating, producing prophecy and its fulfillment and a "variety of things" that "could arise from nothing but the ideas and will of a Being necessarily existing."

Recalled to celestial dynamics in 1684 by Halley's visit, Newton discovered that no mechanical cause could be used to explain gravity; no material aether in either a kinetic or a tensional form would serve. First, two lines of his many investigations came together and reinforced each other: the mathematical and the observational. What he could deduce mathematically matched the empirical reality of the Keplerian area law rather precisely – much more closely than it should have done in the presence of a corporeal fluid aether. Newton was thus led for the first time to question the existence of a Cartesian-type aether. He had never before doubted a corporeal aether to be the "cause" of gravity, as the continuity of aethereal speculations in his manuscripts demonstrates. Then, when he did begin to question the existence of the mechanical aether, he devised a new set of pendulum experiments in a subtle test for its putative presence on earth. The terrestrial experimental evidence being negative, he extrapolated it to the supposed aethereal reservoir of the heavens – to conclude that no dense corporeal fluid existed there, "that by far the largest part of the aetherial space is void, scattered between the aetherial particles." A line of experimental investigation had reinforced the lines of mathematics and observation, which was enough to force a break with Descartes.

It was also enough to reopen, or perhaps open for the first time for Newton, the question of the place of gravity in God's scheme of things. The confluence of three lines of investigation provided a certainty about the *operation* of gravity that neither Newton nor anyone else had ever had before, but it raised a serious problem about the *cause* of gravity. If gravity had no mechanical cause, must it not then be assigned a non-mechanical cause just as fermentation had been? Was there then at work in gravity a designing, organizing principle – a formal or a final cause in the Aristotelean sense? Just such a principle was embedded in the vegetable spirit, carrying divinity into the world, putting the divine will into effect. But the vegetable spirit provided a *vehicle*, a mode of transport between pure, active spirit and passive matter. With its intermediate, quasi-material ontological status, it had done just that in the alchemical literature, and had served as the link and bond between heaven and earth in more senses than one. Just so had the prophetic spirit operated, bringing messages from God to man; just so did the Christ mediate. Where, then, was the vehicle for the operation of gravity? If the heavens – the great aethereal spaces, as Newton still called them – were in reality mostly empty spaces "scattered between the aetherial particles," what was left to subsume the operation of gravity?

Newton was not prepared to answer that question in 1687; hence the equivocation in the *Principia* and related papers. But there were certain sources of knowledge still open to him for exploration: the wisdom of the ancients, both sacred and natural; the ideas of more recent thinkers; and, finally, new experimental evidence. His detailed struggles with those options will be the focus of the next two chapters.

6

Modes of divine activity in the world: after the Principia, *1687–1713*

Introduction

After the *Principia* was published in 1687, Newton continued his search for evidence of divine activity in the world in several different ways. He apparently took seriously the program of restoring true natural philosophy that he had himself set forth in the *Origines* papers, as the best way of restoring true religion, for his work in the post-*Principia* period focused much more on natural philosophy than on theology.

In natural philosophy three principal areas of concentration may be detected. First and foremost came Newton's continued pursuit of alchemy. Next came an intense concern to find the cause of gravity. He had of course just demonstrated that gravity could not be accounted for by mechanical principles; by his own well established dichotomy of the 1670s, if gravity was not mechanical then it must be nonmechanical, active, and divine. Finally Newton made an immense effort properly to incorporate comets into his new system of the world. The significance of these diverse activities for Newton's goal of restoring true religion will become apparent in the extended discussions of each one in this chapter and the next.

According to Westfall, relatively few of Newton's surviving theological papers date from the period immediately following the first edition of the *Principia*, and those that do introduce no new themes. Westfall argues that Newton's beliefs were well fixed by the end of the 1680s, especially his Arianism, and showed little variation in subsequent decades; when Newton renewed his work on this set of papers after about 1710, he reworked old ideas only, sometimes however toning down their radicalism, and certainly setting the date of the expected apocalypse further and further in the future.[1]

My own survey of that set of papers has produced a conclusion in essential agreement with Westfall's, though I might place a few more of

1. Westfall, "Newton's theological manuscripts" (5, n. 17).

the papers in the late 1680s and the 1690s. It is also possible to argue that Newton's theological papers do reflect the introduction of one radically new idea during this period, as will appear in Chapter 7, where that new idea is relevant to Newton's second solution to the problem of a cause for gravity. Newton's emphasis on idolatry continued, those beliefs and practices that seemed to him to represent mankind's declination from the true worship of the one true God. Nothing seemed to shake Newton's voluntarism, and, though his first solution to the problem of a cause for gravity may have shaken his Arianism briefly, the theological papers convincingly document his reaffirmation of that particular heresy.

One should not assume, however, that Newton's interest in theological issues had in any way declined even though he shifted the focus of his study for the most part during the last period of his life from the written records of Scripture, church fathers, sacred ritual, church history, and so forth. He had after all explored all of that exhaustively already. Perhaps because of the remarkable success of the *Principia* itself in restoring true natural philosophy, Newton shifted his focus to *more* study of natural philosophy as the *best* way to restore true religion. He sought the border where natural and divine principles met and fused, and both theological and religious issues present themselves vigorously in every episode described in this chapter and the next.

Newton, Nicolas Fatio de Duillier, and alchemy

A renewal of his alchemical work absorbed much of Newton's time and energy between 1687 and 1696. Indeed, even before publication of the *Principia* in the summer of 1687, Newton picked up his alchemical experimentation just where he had abandoned it after Halley's visit in 1684. Sending Book I of the *Principia* to London, where it was received by the Royal Society on 28 April 1686,[2] Newton returned to the laboratory even while the manuscript was in transit, on 26 April. "Experim^{ts} April 26^{t}. 1686," he recorded.

> Apr. 26 Wednesday I sublimed sp^{t} of spelt^{r} (wch two years before had been dissolved in distilled sp^{t} of ♁ & sublimed wth ✳ (two parts w th three) then wth its weight of 🜅 y^{e} ✳ destroyed, the salt filtred & now sublimed.)[3]

2. Idem, *Never at Rest* (1, n. 4), pp. 444–5.
3. MS Add. 3973 (1, n. 38), f. 19a, and MS Add. 3975 (5, n. 16), f. 75v. ♁ = antimony, ✳ = sal ammoniac, 🜅 = aqua fortis.

Laboratory notes written between 26 April 1686 and the day of Newton's departure for London and the Mint in 1696 comprise close to 55,000 words, his other alchemical papers of the same period an estimated 175,000 words.[4]

One may begin to appreciate the extent, content, and significance of Newton's alchemical enterprise during the post-*Principia* period through a documentary reconstruction of his alchemical relationship with Nicolas Fatio de Duillier (1664–1753), the young Swiss mathematician who flashed like a brilliant meteor into Newton's life shortly after the *Principia* was published. Newton and Fatio developed an unusual intimacy and even spoke of living in close proximity,[5] and the last three of the eight manuscripts chosen to document this study are all demonstrably associated with their friendship, as is a part of Newton's work on Hermes Trismegistus in Keynes MS 28. There are as well many, many more of Newton's books and papers that may be so associated.

In "Praxis," written about the middle of the 1690s, Newton cited a letter from Fatio just as he would any other alchemical reference: "This powder amalgams with mercury and purges out its feces if shaken together in a glass [Epist. N. Fatij]."[6] That seems perfectly straightforward: Fatio had sent Newton some information that Newton accepted as fact

4. These are my estimates of the laboratory records in MSS Add. 3973 (1, n. 38) and 3975 (5, n. 16) and of the late alchemical papers listed in Dobbs, *Foundations* (1, n. 1) that are now scattered from Glasgow to Jerusalem and from Cambridge to Stanford.
5. Newton, *Correspondence* (2, n. 14): vol. III, 241 (Newton to Fatio, 24 Jan. $169\frac{2}{3}$); vol. III, 242–3 (Fatio to Newton, 30 Jan. $169\frac{2}{3}$); vol. III, 262–3 (Fatio to Newton, 9 March $169\frac{2}{3}$, and Newton to Fatio, 14 March $169\frac{2}{3}$). See also: Frank E. Manuel, *A Portrait of Isaac Newton* (Cambridge, MA: The Belknap Press of Harvard University Press, 1968); Charles A. Domson, "Nicolas Fatio de Duillier and the Prophets of London: An Essay in the Historical Interaction of Natural Philosophy and Millennial Belief in the Age of Newton" (Yale University: Ph.D. dissertation, 1972); Margaret C. Jacob, *The Newtonians and the English Revolution 1689–1720* (Hassocks, Sussex: The Harvester Press, 1976); Hillel Schwartz, *Knaves, Fools, Madmen, and That Subtile Effluvium. A Study of the Opposition to the French Prophets in England, 1706–1710* (University of Florida Monographs, Social Science Number 62; A University of Florida Book; Gainesville: The University Presses of Florida, 1978); idem, *The French Prophets. The History of a Millenarian Group in Eighteenth-Century England* (Berkeley: University of California Press, 1980); Westfall, *Never at Rest* (1, n. 4).
6. Babson MS 420 (1, n. 39), p. 13. For Newton's original English, see Appendix E.

and incorporated into his treatise. But one must of course recognize that the black powder in question was no ordinary chemical. It played a central role in the alchemical process Newton described in the "Praxis" and required an involuted and, one may safely assume, idiosyncratic preparation to give it its special properties. For Fatio to have been able to experiment with the powder, Newton must either have guided him in its preparation or have provided him with a prepared sample.[7] At the very least Newton's simple reference to Fatio's letter implies an extensive mutual exchange of information.

A similar exchange apparently was in progress in the first half of 1693. In his experiments and observations dated from December 1692 through June 1693, Newton was primarily concerned with metallic fermentation. First setting out common knowledge regarding fermentation in beer making, he then attempted to find analogous reactions in the mineral kingdom, patiently trying out systematically varied mineral combinations. The red precipitate of iron, for example, did not work, but iron ore did, causing the material to swell to four or five times its original size and to be everywhere full of little bubbles. "So then the ferment lies only in the ore of iron and not in the red precipitate."[8] One may infer that Newton relayed his interest in metallic fermentation to Fatio during the late winter of 169²⁄₃ or the spring of 1693 (if not before), perhaps in a letter or letters now lost, for in May Fatio responded to that interest: "If You be curious Sir of a metallick putrefaction and fermentation. . . . ," he said to introduce the subject. Then Fatio described the mineral "trees" he had just seen produced by a new friend. His friend had a specially prepared mercury to which he had added some filings of pure gold.

> These matters being put in a sealed egg in a sand heat do presently swell, and puff up, and grow black and in a matter of seven days go through the coulours of the Philosophers. After which time there grows a heap of trees out of the matter. . . . there is plainly a life and a ferment in that composition.[9]

The recipient of Fatio's letter was back in his laboratory the next month, noting and recording the following facts: the two serpents ferment well with salt of lead, tin, and copper, better with salt of lead and copper,

7. Ibid., pp. 12–13. The black powder, called "our Pluto" and other esoteric names, is made of the "rod & ye male & female serpents" (proportions: 3, 1, 2), fermented and digested together for 65–70 days until the material becomes "rotten." Cf. Appendix E.
8. MS Add. 3973.8 (1, n. 38), ff. 2v–3r. Cf. Appendix D.
9. Newton, *Correspondence* (2, n. 14), vol. III, 265–7 (Fatio to Newton, 4 May 1693).

and best with salt of lead alone: the addition of mercury enhanced the process and had even started the fermentation up again when it had seemed to be over, so that "the matter swelled much with a vehement fermentation."[10] Beside Newton's systematic experimentation, Fatio's report looks simplistic, yet one cannot doubt the identity of their interests in mineral fermentation during that period.[11]

Were more Newton–Fatio letters extant, one could expect them to reveal other areas of common interest, but very little survives of what may have been an extensive correspondence. Fatio's letter of 4 May 1693 gives a hint of deliberate destruction: he asked Newton to burn the letter when he had finished with it. One must turn therefore to quite a different way of relating Newton's alchemy in the post-*Principia* period to his friendship with Nicolas Fatio. Another of the exemplar manuscripts for this study shows how that may be done: through the sudden appearance among Newton's alchemical papers of materials from the French.

All of Newton's pre-*Principia* alchemical papers are in English or Latin and show no residue of the activity of translation. He seems to have used English and Latin with equal ease, and in taking notes simply followed the language of his source.[12] As Keynes long ago noticed in the papers he had collected, however, some do contain translations, apparently done by Newton himself.[13] A number of peculiarities suggest the translator's

10. MS Add. 3973.8 (1, n. 38), f. 4r (entry for June 1693). Cf. Appendix D.
11. The reaction Fatio reported on 4 May 1693 was actually quite similar to the one Newton had recorded in Keynes MS 18 some fifteen years earlier, to the one Robert Boyle had described, and to the one Newman has recently discovered in the prototype of Keynes MS 18 found among the Boyle papers. Cf. Dobbs, *Foundations* (1, n. 1), pp. 175–87, 251–5; Newman, "Newton's 'Clavis' " (1, n. 33). President Kenneth Rogers of The Stevens Institute of Technology has kindly suggested to me that these reported treelike structures are probably to be explained as fractals. See, for example, Leonard M. Sander, "Fractal growth," *Scientific American 256*, No. 1 (January 1987), pp. 94–100, esp. the photographs of natural fractals on p. 100 that resemble the branches, roots, and leaves of trees; also the discussion of computer simulation of fractal "trees" and other ramified and branching structures in Mort La Brecque, "Fractal applications," *Mosaic 17*, No. 4 (Winter 1986–7), 34–48, esp. 39–40; also James Gleick, *Chaos. Making a New Science* (New York: Viking Penguin, 1987), pp. 307–14.
12. Cf. Dobbs, *Foundations* (1, n. 1), pp. 22–3.
13. A. N. L. Munby, "The Keynes collection of the works of Sir Isaac Newton at King's College, Cambridge," *Notes and Records of the Royal Society of London 10* (1952), 40–50.

struggles to render the French (and they all are translations from the French) into a language more comfortable to him.[14] In places Newton entered a French word or phrase above the line as if not sure of his translation. Elsewhere he made corrections that can best be interpreted as refinements of the translation. Sometimes little wavy lines appear in the text; probably they fill overly large blanks Newton had left for consultation with his dictionary, his grammar, or his friend. All of the manuscripts in which these peculiarities occur are written in the bold, expansive handwriting of the late 1680s and early 1690s, the period when Newton's friendship with Fatio was at its height, and "Out of La Lumiere sortant des Tenebres" is one of them.

Newton's source for "Out of La Lumiere" is readily identified as *La lumière sortant par soy méme des tenebres*, published in Paris in 1687, as noted in Chapter 1.[15] The book was in that portion of Newton's library presented by the Pilgrim Trust to Trinity College, Cambridge, in 1943, and it is still extant there.[16] There are no notes in the book although many pages have been dog-eared in the characteristic manner that indicates Newton's concentrated interest. But not content with dog-earing the book and then compiling the translated abstract of it in "Out of La Lumiere," Newton came back to the same material and abstracted his own abstract in a document he entitled "Lumina de tenebris."[17] "Lumina de tenebris" is written in a much later hand, showing that Newton's interest in the material carried over well into the eighteenth century.

The cosmogonic dimension of alchemy is an important theme in *La lumière*, and the tack taken in the book of relating the alchemical process to the first chapter of Genesis may well have been the aspect that attracted Newton.

> As the world was formed of a chaos or materia prima which was void and without form . . . : so in our work the stone is made of a chaos or materia prima which by putrefaction is void without form and black or dark. . . .[18]

14. The *locus classicus* regarding Newton's difficulty with French comes from Fatio: Newton, *Correspondence* (2, n. 14), vol. III, 68–9 (Fatio to Huygens, 24 Feb. 16⁸⁹/₉₀): "Il a quelque peine à entendre le François mais il s'en tire pourtant avec un Dictionaire."
15. *La lumière* (1, n. 37). Cf. also Appendix C.
16. Harrison, *Library* (1, n. 6), p. 184, item no. 1003, now Trinity College, Cambridge, NQ.16.117.
17. Sotheby lot no. 41; The Sir Isaac Newton Collection, Babson College Archives, Babson Park, MA, Babson MS 414 A.
18. Yahuda MS Var. 1, Newton MS 30 (1, n. 36), f. 1v. Cf. Appendix C.

In this Newton continued to study an idea that had become important to him at least by the early 1680s, sometime before he met Fatio, but he never lost interest in it, and there is reason to believe he discussed it with Fatio. An echo of it appears among the Fatio papers in Geneva: "there are diverse Alchemists who believe that in this Chapter Moses also had in view some of their most considerable operations."[19] Domson has argued that Fatio composed that manuscript during or after 1692, as he absorbed Newton's method of scriptural exegesis, and it would indeed be unreasonable to suppose that Newton and Fatio did not discuss the alchemical interpretation of Genesis together during that time. *La lumière* remained high in Newton's favor, ranking sixth in his eighteenth-century list of the best authors.[20]

Three of Newton's translations derive from the two volume *Bibliothèque de philosophes [chymiques]* (Paris: 1672–8) that had once been Fatio's.[21] Fatio left the books with Newton sometime before 14 February 169²⁄₃,[22] and Newton subsequently bought them from him.[23] From the first volume Newton translated "Le Livre de Nicolas Flamel" and "La

19. Nicolas Fatio de Duillier, "Hypotheses," Bibliothèque Publique et Universitaire de Genève, MS. français 605, as quoted and translated in Domson, "Nicolas Fatio de Duillier" (6, n. 5), pp. 62–3.
20. Sotheby lot no. 2; King's College, Cambridge, Keynes MS 13, f. 4r. Newton's list is somewhat repetitious and duplicates have been omitted in calculating the ranks given here and in succeeding notes.
21. *Bibliotheque de philosophes [chymiques,] ou recueil des oeuvres des auteurs les plus approuvez qui ont ecrit de la pierre philosophale. Tome premier. Contenant sept Traitez qui sont énoncey dans la page suivante. Avec un Discourse, servant de Preface, sur la verité de la Science, & touchant les Auteurs qui sont dans ce Volume. Et une Liste des Termes de l'Art, & des Mots anciens qui se trouvent dans ces Traitez, avec leur explication. Par le Sieur S. D. E. M. Tome second. Qui contient cinq Traitez énoncey dans l'autre page, & nouvellement traduits. Avec des Remarques & les diverses Leçons. Une Lettre Latine sur le Livre intitulé Icon Philosophiae Occultae. Vne Preface sur l'obscurité des Philosophes, & sur les Traittez de ce Tome, & leurs Auteurs. Et une Table des Matieres. Par le Sieur S. Docteur en Medecine.* (2 vols.; Paris: Chez Charles Angot, ruë Saint Jacques, au Lyon d'or, 1672–8). The first volume survives as Trinity College, Cambridge, NQ.16.94, but the second volume has disappeared. See Harrison, *Library* (1, n. 6), p. 104, item no. 221.
22. Newton, *Correspondence* (2, n. 14), 245–6 (Newton to Fatio, 14 Feb. 169²⁄₃).
23. Ibid., vol. III, 260–3 (Newton to Fatio, 7 March 169²⁄₃; Fatio to Newton, 8 March 169²⁄₃; Fatio to Newton, 9 March 169²⁄₃; Newton to Fatio, 14 March 169²⁄₃).

Table d'Emeraude" of Hermes Trismegistus, and from the second volume he translated Hermes's "Les Sept Chapitres."

Entitled by Newton "The Book of Nicholas Flamel conteining The explication of the Hieroglyphical Figures wch he caused to be put on the Church of the SS. Innocents at Paris," the first of these is now at the Massachusetts Institute of Technology.[24] It shows all the characteristic signs of translation, and it constitutes a complete copy of Flamel's work even though Newton had been familiar with the work – he had indeed used its ideas extensively – during the preceding twenty years.

To emphasize Newton's prior familiarity with Flamel, one may consider the following facts. Flamel's work on the hieroglyphic figures was first published in Paris in 1612.[25] There is no evidence that Newton ever saw a copy of that rare first edition, but one knows that he temporarily had the first English edition[26] in his possession, probably in the early

24. Sotheby lot no. 25; Massachusetts Institute of Technology, Cambridge, MA, 246/N56/17--/ Rare Book Collection, Institute Archives and Special Collections Department, MIT Libraries. I am indebted to Helen Slotkin of the MIT Libraries for calling my attention to this manuscript and providing me with a xerox copy of it.

25. *Trois Traictez de la Philosophie natvrelle, non encore imprimez. Scavoir, le secret livre dv tres-ancien Philosophe Artephivs, traictant de l'Art occulte & transmutation Metallique, Latin François. Plus les figvres hieroglyphiqves de Nicolas Flamel, ainsi qu'il les a mises en la quatriesme arche qu'il a bastie au Cimetiere des Innocens à Paris, entrant par la grande porte de la rue S. Denys, & prenant la main droite; auec l'explication d'icelles par iceluy Flamel. Ensemble Le vray Liure du docte Synesivs Abbé Grec, tiré de la Bibliotheque de l'Empereur sur le mesme sujet, le tout traduict par P. Arnavld, sieur de la Cheuallerie Poicteuin* (Paris: Chez la vefue M. Gvillemot & S. Thibovst, au Palais, en la galerie des prisonniers, 1612). There is a facsimile reprint, with commentary and other material: Claude Gagnon, *Description du Livre des Figures Hiéroglyphiques attribué à Nicolas Flamel, suivie d'une réimpression de l'édition originale et d'une reproduction des sept talismans du Livre d'Abraham, auxquels on a joint le Testament authentique dudit Flamel* (Montréal: Les Editions de l'Aurore, 1977). Cf. also Ferguson, *Bibliotheca Chemica* (3, n. 43), vol. I, 47–8. Flamel's illustrations as well as some of the controversies surrounding Flamel and his work may be found in J. van Lennep, *Art & Alchimie. Étude de l'iconographie hermétique et de ses influences*, préface de Serge Hutin (Collection Art et Savoir; Bruxelles: Éditions Meddens avec le concours de la Fondation Universitaire de Belgique, 1966), pp. 168–83 and Plate 175.

26. Nicolas Flamel, *Nicholas Flammel, His Exposition of the Hieroglyphicall*

1670s, for he made transcripts of the tracts by Artephius and Pontanus that were printed with the English Flamel of 1624.[27] A man of Newton's diligence who had already copied out half a book in longhand might reasonably be expected to copy the rest of it.[28] But such was not the case: Newton's first reading of Flamel resulted only in extensive abstracts, a carefully drawn pen-and-ink reproduction of the hieroglyphic figures, and a few tentative interpretations of Newton's own.[29] However, there can be no doubt that Newton knew this particular alchemical text well. An early alchemical anthology of his contains some Flamel,[30] and ideas from Flamel appear early and late in the regimen papers.[31] One is forced to the conclusion that in the MIT manuscript Newton was translating a work he already knew intimately.[32] Presumably one sees here a forceful example of Newton's scholarly habits. As in biblical studies, so also in

Figures which he caused to be painted upon an Arch in St. Innocents Church-yard, in Paris. Together with The secret Books of Artephivs, And the Epistle of Iohn Pontanus: Concerning both the Theoricke and the Practicke of the Philosophers Stone. Faithfully, and (as the Maiesty of the thing requireth) religiously done into English out of the French and Latine Copies. By Eirenaevs Orandvs, qui est, Vera veris enodans (London: imprinted at London by T. S. for Thomas Walkley, and are to bee solde at his Shop, at the Eagle and Childe in Britans Bursse, 1624).

27. Sotheby lot no. 7; King's College, Cambridge, Keynes MS 14. Cf. Dobbs, *Foundations* (1, n. 1), pp. 130–1.
28. On the basis of the Sotheby description, the present writer suggested that Sotheby lot no. 25 (the MIT manuscript) might actually be such a transcription. Cf. ibid., p. 131, n. 5.
29. Sotheby lot no. 35 (part); Jewish National and University Library, Jerusalem, Yahuda MS Var. 259.3.
30. Sotheby lot no. 26; King's College, Cambridge, Keynes MS 25. Cf. Dobbs, *Foundations* (1, n. 1), p. 132.
31. Dobbs, "Newton's copy" (1, n. 5).
32. Newton's translation corresponds quite well with the version in *Bibliothèque* (6, n. 21), vol. I, 49–97 except for a minor omission. Another French version was in his library: *Philosophie naturelle de trois anciens philosophes renommez Artephius, Flamel, & Synesius. Traitant de l'Art occulte, & de la Transmutation metallique. Augmentée d'un petit Traité du Mercure, & de la Pierre de Philosophes de G. Ripleus, nouvellement traduit en François. Derniere edition* (Paris: Chez Laurent d'Houry, sur le Quay des Augustins, à l'Image Saint Joan, 1682); Harrison, *Library* (1, n. 6), p. 217, item no. 1309, now Trinity College, Cambridge, NQ.16.93. This book, however, is based on the first (Arnauld) edition of 1612, and the Flamel text is not organized quite as Newton's is, nor is it quite as extensive.

alchemy: he studied all available variants of his texts. A fresh turn of phrase or even a word inadvertantly omitted elsewhere might serve to unlock the secrets of the great work.

In translating the two Hermetic tracts from the *Bibliothèque* Newton similarly worked with texts long available to him in other forms,[33] and the new versions he prepared were only part of his long-term study of Hermes Trismegistus.[34] His English translation of the *Emerald Tablet* rests now with a Latin version of the text he had copied earlier and with his own *Commentarium*; it is reproduced in full in Appendix B.[35] The *Emerald Tablet* is so short and had been published in so many places that one might question the assertion that it is Newton's own translation from the French of the *Bibliothèque*. But Newton's English is not exactly like any other contemporary English or Latin version that I have been able to discover, and it reflects several peculiarities of the French.[36] Furthermore, Newton interlineated an alternative translation of "le moyen"

33. Both were, for example, in his *Theatrum chemicum* (3, n. 44). An English version of the *Emerald Tablet* was also in another book that one knows Newton used because of his occasional references to John Sawtre: *Five Treatesis of the Philosophers Stone. Two of Alphonso King of Portugall, as it was written with his own hand, and taken out of his Closet: Translated out of the Portuguez into English. One of John Sawtre a Monke, translated into English. Another written by Florianus Raudorff, a German Philosopher, and translated out of the same Language, into English. Also a Treatise of the names of the Philosophers Stone, by William Gratacolle, translated into English. To which is added the Smaragdine Table. By the Paines and Care of H. P.* (London: printed by Thomas Harper, and are to be sold by John Collins, in Little Brittain, near the Church door, 1652). Newton once owned this book, but it is now missing: Harrison, *Library* (1, n. 6), p. 144, item no. 621. For some of the references to Sawtre, see Dobbs, "Newton's copy" (1, n. 5).
34. Insofar as possible I have reconstructed Newton's study of Hermes in the editorial note to Appendix B.
35. Keynes MS 28 (1, n. 35). Cf. Appendix B.
36. One may take, for example, this sentence from Newton's English version (ibid., f. 2r): "Its force or power is entire if it be converted into earth." That sentence in the Latin version of the same manuscript (ibid., f. 6r) is: "Vis ejus est integra si versa fuerit in terram." Another Latin version [*Theatrum chemicum* (3, n. 44), vol. VI, 715] reads: "Potentia ejus perfecta est si mutatur in terram." The English of 1652 [*Five Treatesis* (6, n. 33), sig. A4v] has the following: "...his force and power is perfect, if it be turned into Earth." On the other hand, the French [*Bibliothèque* (6, n. 21), vol. I, 1] would yield Newton's sentence rather precisely: "Sa force ou puissance est entiere, si elle est convertie en terre."

in one spot,[37] and at the end of his version he said, "See yᵉ ffrench Bibliotheque."[38] It is equally clear that *The Seven Chapters* is Newton's own translation from the same source.[39]

The appeal for Newton of the Hermetic texts lay in their supposed antiquity. A strong believer in the doctrine of *prisca sapientia*, Newton thought the more ancient the document the less extensive its corruption from original pure knowledge. Hermes was the oldest and so was and remained at the top of Newton's list of "Authores optimi."[40]

A mantle of antiquity also validated the work of Flamel, who wrote in the first person a circumstantial tale of acquiring a mysterious book, the cover of which was of copper engraved with strange figures and the letters of an ancient language, and the enigmas of which could be interpreted to him only by one who recognized it as a lost treasure of antique Jewish wisdom. Flamel furthermore gave an explication of "hieroglyphick" figures he himself had had placed in a Parisian church in the fourteenth century to commemorate and encode his own successful procedure for making the philosopher's stone – just the sort of veiled information Newton always set himself to decode.

Especially when the material hinted at access to the truly ancient in alchemy were Newton's efforts intensified, for access to the Truth in alchemy would open access to the Truth of original religion, as one saw in Chapter 5, alchemy being the earliest form of the corruption that had ultimately obscured the primitive true religion. Newton's technique for deciphering ancient alchemy was so similar to his technique for understanding the "hieroglyphick" language of the prophets – which he was teaching to Fatio at just this time – that one must suppose he shared the alchemical with Fatio as well. Indeed the procedures for alchemical and biblical exegesis could hardly be kept distinct sometimes, as one may see with the interpretation of Genesis.

Studies on another "hieroglyphick" work complete the present list of

37. Keynes MS 28 (1, n. 35), f. 2r. Cf. Appendix B.
38. Ibid., f. 2v. He then listed all the relevant citations in *Theatrum chemicum* (3, n. 44) as well, but none of them could really have yielded his version.
39. *Bibliothèque* (6, n. 21), vol. II, 1–5, 14–20, 27–29, 32–5, 55–7, 63–4, 72–4 (each chapter being followed by extensive remarks by the editor that Newton did not translate): Sotheby lot no. 30, King's College, Cambridge, Keynes MS 27. A misleading transcript of this manuscript has been published: Churchill, "The Seven Chapters" (1, n. 42). Although Churchill does not indicate it, the bracketed material in the original is partly Newtonian and interpretative, partly translations of alternative readings supplied by the editor. In places Newton compressed the text and made minor omissions.
40. Keynes MS 13 (6, n. 20), f. 4r.

translations. The work in question was *Le triomphe hermetique* by A. T. Limojon, Sieur de St. Didier.[41] The book itself, surviving at Trinity College, Cambridge, contains notes in Newton's hand and has many pages turned down in his characteristic manner. But still not entirely comfortable with his French apparently, Newton made a full translation of the last section of the book, the *Lettre Aux vrays Disciples d'Hermes, Contenant six principales clefs de la Philosophie Secrete*,[42] and of the interpretation of the "hieroglyphick" frontispiece. The translation was into Latin: "Epistola ad veros Hermetis discipulos continens claves sex principales Philosophiae secretae."[43] No Latin version of this work seems ever to have been published; Newton's Latin follows the French closely except for an occasional error of omission; the usual evidence of translation appears; one cannot doubt it is Newton's own translation.

But Newton was not through with Didier, as he always called Limojon. Soon Newton composed a document entitled "The method of ye work."[44] About twelve of its thirty-five pages are devoted to abstracts – in English this time – of Didier's keys and emblematic frontispiece. Didier's method provided the framework for the entire composition, and Newton went on to compare it with the methods of d'Espagnet, Sendivogius, Basilius Valentinus, Eirenaeus Philalethes, the authors of *Manna* and *Thesaurus thesaurorum*, Pearce the black monk, Aristotle, Hermes, Theodorus Mundanus, Elucidarius (Christopher the Parisian), the figures of Abraham the Jew, John de Monte Snyders, and the author of *Instructio de arbore solari*. Newton had two other tracts by Didier as well,[45] but from his position of centrality in "The method of ye work," where he filled 34 percent of the

41. *Le triomphe hermetique* (2, n. 70); the "hieroglyphick" frontispiece is reproduced as Plate 3, in Chapter 2.
42. Ibid., pp. 123–53.
43. Sotheby lot no. 20; King's College, Cambridge, Keynes MS 23.
44. Sotheby lot no. 17; King's College, Cambridge, Keynes MS 21.
45. *Le tres-ancien duel des Chevaliers*, in *Divers traitez de la philosophie naturelle. Sçavoir, la turbe des philosophes, ou le code de verité en l'art. La parole delaissée de Bernard Trevisan. Les deux traitez de Corneille Drebel Flaman. Avec le tres-ancien duel des Chevaliers. Nouvellement traduit en François, par un Docteur en Medicine* (Paris: Chez Jean d'Houry a l'Image S. Jean, au bout du Pont-neuf, sur le Quay des Augustins, 1672); also, *Lettre d'un philosophe sur le secret du grand oeuvre, ecrite Au sujet de ce qu'Aristée a laissé par écrit à son Fils, touchant le Magistere philosophique. Le Nom d l'Autheur est en Latin dans cett' Anagramme. Dives Sicut Ardens, S.* (La Haye: Chez Adrian Moetjens, 1686): Harrison, *Library* (1, n. 6), pp. 134 and 179, items nos. 531 and 950, now, respectively, Trinity College, Cambridge, NQ.16.118 and NQ.16.95[4].

space, Didier began to slip downhill. In a later manuscript, "Of ye first Gate,"[46] abstracts from Didier comprise only 22 percent of Newton's compilation, and in the eighteenth-century list of the best authors Didier had sunk to seventeenth place.[47] Yet the fascination he held for Newton, with his keys to the secret and his "hieroglyphick" emblem, had certainly been intense for a period of time.

It has been tacitly assumed in this section that the efflorescence of translation from the French one finds in Newton's alchemical papers during the late 1680s and early 1690s is closely connected to his friendship with Nicolas Fatio. The reader should be aware that this is an assumption. Fatio did not correct Newton's translations; nothing in Fatio's hand appears in these papers, and all the corrections are in Newton's hand. Yet three of the works translated came from books Fatio had owned and sold to Newton; two others came from books published on the continent while the relationship was in its first white heat. Fatio was knowledgeable about Newton's difficulty with French, and he knew Newton might want help reading it: regarding Huygens's *Traité de la lumière* Fatio said to Newton, "It beeing writ in French you may perhaps choose to read it here with me."[48] It thus seems a natural assumption that Fatio provided Newton with some assistance, or at least moral support, in his translations of French alchemical texts. It also seems very likely that the content of the translated works, which in every case centered on the problem of the interpretation of ancient hidden knowledge, was of considerable mutual interest, especially as they did in fact discuss just that sort of thing with regard to prophecy.

In what is to follow one's assurance of mutual interest and contact between Newton and Fatio in their alchemical work diminishes exponentially. But other suggestive evidence there is, and it is presented here in the order of the diminishing conviction it carries. Even if it does not speak to an alchemical partnership between Newton and Fatio, it continues to illuminate the exceedingly high level of alchemical activity Newton himself maintained in the post-*Principia* period.

First, there is one manuscript among Newton's papers that is actually in French, "Le Procede Universelle pour faire la Pierre Philosophale laquelle l'auteur dit davoir faict quatre fois."[49] It is written in that same bold handwriting Newton employed in the late 1680s and early 1690s.

46. Sotheby lot no. 93; King's College, Cambridge, Keynes MS 53.
47. Keynes MS 13 (6, n. 20), f. 4r.
48. Newton, *Correspondence* (2, n. 14), vol. III, 390–1 (Fatio to Newton, 24 Feb. 16$^{89}/_{90}$).
49. Sotheby lot no. 35 (part), Jewish National and University Library, Jerusalem, Yahuda MS Var. 259.1.

The nineteenth-century syndicate that first described Newton's alchemical papers, and also Sotheby and Company in 1936, suggested Jodochus a Rhe as the author,[50] but I have been unable to substantiate that suggestion. Although Newton did study Rhe's works rather extensively,[51] he did not attribute this particular tract to him. Furthermore, no French edition of Rhe's works seems ever to have been published, nor are there known manuscript remains of his with this title. Since Newton was so conscientious about entering exact citations in his notes on printed material, their absence in "Le Procede Universelle" probably indicates that the material came from an anonymous circulating manuscript, possibly one Fatio had acquired from a continental contact.

Next, there are quite a number of alchemical–chemical medical books in French in the surviving remnant of Newton's library in addition to those already mentioned.[52] There were almost certainly others at one

50. *Catalogue of the Portsmouth Collection* (1, n. 27), p. 14; *Catalogue of the Newton Papers* (1, n. 27), p. 5.
51. Newton had Rhe's major works: Joannus Rhenanus, *Opera chymiatrica, quae hactenus in lucem prodierunt omnia à plurimus, quae in prioribus editionibus irrepserant, mendis vindicata & selectissimis Medicamentis aucta, ing; unum fasciculum collecta, Quorum Catalogum versa indicabit pagina* (Francofurti: Apud Jacobum Gothofredum Seylerum, 1668): Harrison, *Library* (1, n. 6), p. 227, item no. 1397, now Trinity College, Cambridge, NQ.8.90. He also had tracts edited by Rhe: *Harmoniae chymico-philosophicae, sive philosophorvm antiquorum consentientium, hactenus quidem plurrimùm desideratorum, sed nondum in lucam publicam emissorum, Decas II. Collecta studio & industria Ioannis Rhenani M. D. Cum elencho singulorum huius Decadis* (Francofurti: Apud Conradvm Eifridvm, 1625): Harrison, *Library* (1, n. 6), p. 157, item no. 740, now Trinity College, Cambridge, NQ.16.122². He made notes on Rhe's *Opera*: Sotheby lot no 89; now King's College, Cambridge, Keynes MS 50. He set out to make comparisons of "Basilius Valentinus & Jodachus a Rhe" in Sotheby lot no. 110, now British Library, London, Add. MS 44,888, though the surviving material covers only Basilius. He twice mentioned "Jodocus & Basil" in tandem in a memorandum on a conversation with an unnamed London alchemist: Sotheby lot no. 45, now King's College, Cambridge, Keynes MS 26, published in D. Geoghegan, "Some indications of Newton's attitude towards alchemy," *Ambix* 6 (1957–58), 102–6, also published in Newton, *Correspondence* (2, n. 14), vol. IV, 196–9. Newton mentioned Rhe (but not Basilius) in another version of the memorandum: Sotheby lot no. 46, now Department of Special Collections, The University of Chicago Library, Chicago, IL, The Joseph Halle Schaffner Collection of Scientific Manuscripts, Box 1, Folder 47. Rhe was not to make it to the short list of "Authores optimi," however: Keynes MS 13 (6, n. 20), f. 4r.
52. The following, in order of publication, are still extant in Trinity College,

time. A list among the Ekins Papers entitled "Books for Mr Newton" mentions *La physique des anciens* (Paris: 1701), which is now missing,[53] and other chemical–alchemical tracts are thought not to have survived.[54] Although it is not known when Newton purchased all of his French books, the Ekins list dates the purchase of ten of his French alchemical books to a time between 1702 (last date of publication: 1701) and mid-1705 (when Newton was knighted and after which he would have been called Sir Isaac).[55] The same list records additional purchases of the collected works of Paracelsus and Rhe in Latin.[56] Most writers on the Newton–Fatio relationship have assumed that the friendship had cooled by 1702, but the Ekins list hints that they may have resumed the study

Cambridge: (1) René de la Chastre, *Le prototype... de l'art Chimicq;...* (Paris, 1620) (now NQ. 16.111); (2) Joseph Du Chesne, *Traicté de la matiere* ... (Paris, 1626) (now NQ.16.132); (3) D. l'Agneau, *Harmonie mystiqve* ... (Paris, 1636) (now NQ.16.129); (4) Joseph Du Chesne, *Recveil des plvs cvievx et rares secrets...* (Paris, 1648) (now NQ.16.100); (5) [Jean d'Espagnet], *La philosophie natvrelle...* (Paris, 1651) (NQ.10.81); (6) Basilius Valentinus, *Azoth...* (Paris, 1659) (now NQ.16.124[1]); (7) Bernard Trevisan, *Traicté...de l'oevf des Philosophes...* (Paris, 1659) (now NQ.16.124[2]); (8) Basilius Valentinus, *Les dovze clefs de philosophie...* (Paris, 1660) (now NQ.16.124[3]); (9) [P. M. de Respour], *Rare experiences svr l'esprit mineral...* (Paris, 1668) (now NQ.16.57); (10) [H. d'Atremont], *Le tombeau de la pauvreté...* (Paris, 1673) (now NQ.16.109); (11) Luigi de' Conti, *Discours philosophiques...* (Paris, 1678) (now NQ.16.110); (12) Guilliaume Lamy, *Dissertation sur l'antimoine* (Paris, 1682) (now NQ.16.107); (13) *Deux traitez nouveaux sur la philosophie naturelle...* (Paris, 1689) (now NQ.16.30); (14) *Le texte d'alchymie et le songe-verd* (Paris, 1695) (now NQ.16.147; (15) [Gasto Claveus?], *Le filet d'Ariadne* ... (Paris, 1695) (now NQ.8.60); (16) *Le Parnasse assiegé...* (Lyon, 1697) (now NQ.16.5); (17) Nicolas Venette, *Traité des pierres...* (Amsterdam, 1701) (now NQ.8.58); (18) Nicolas Lemery, *Traité de l'antimoine...* (Paris, 1707) (now NQ.16.128). See Harrison, *Library* (1, n. 6), items nos. 897 (p. 173), 540 (p. 135), 901 (p. 174), 539 (p. 135), 1311 (p. 217), 127 (p. 95), 169 (p. 99), 130 (p. 95), 1372 (p. 224), 1619 (p. 250), 437 (p. 124), 908 (p. 175), 511 (p. 132), 1607 (p. 249), 619 (p. 144), 1263 (p. 212), 1675 (p. 256), and 939 (p. 178) (in order of publication).

53. Bodleian Library, Oxford, MS New College 361 (Ekins Papers), Vol. II, f. 78, cited in ibid., p. 9.
54. Ibid., pp. 28–57.
55. In (6, n. 52), nos. (2), (4), (9), (13), (14), (15), (16), and (17); also *Divers traitez* (6, n. 45) and the missing *La physique des anciens*.
56. Paracelsus, *Opera omnia medico chemico-chirurgica...* (Genevae, 1658): Harrison, *Library* (1, n. 6), p. 210, item no. 1242, now missing; Rhe, *Opera* (6, n. 51).

of alchemy together sometime between 1702 and 1705, after Fatio's return from three years in Switzerland.

Finally, Fatio himself left some alchemical manuscripts.[57] They contain transcripts (all in French) of major and minor works attributed to Raymond Lull; minor works of Eirenaeus Philalethes, Michael Maier, Hortulain, and Flamel; also a few anonymous pieces. The collection is not extensive; the transcripts of Lull comprise the bulk of it. Although Newton studied these alchemists, it would be difficult to find alchemists he had not studied, and there is little to link Fatio's transcripts directly with any Newtonian material. But Fatio indicated that he had translated some of his material from the Latin of the six-volume *Theatrum chemicum* and from the *Musaeum Hermeticum* of 1678. Newton owned both of those works,[58] and it is not impossible that Fatio found his texts in Newton's volumes. But there is no positive indication that he did, nor does one know when Fatio's alchemical papers were written. In any case, when compared with Newton's heroic struggles with the literature of alchemy, Fatio's papers reflect at most a passing and superficial interest in the field.

Nonetheless, the evidence makes it moderately certain that Newton and Fatio shared a variety of alchemical activities, literary and experimental. There is reason to believe that one focal point lay in the interpretation of ancient occult knowledge, another in signs of life in the mineral kingdom as attested by fermentation. Those foci were closely related aspects of Newton's search for evidence of God's action in the world, but it is not so clear that Fatio ever fully understood or participated in that search. Most of the evidence for their joint alchemical activities has come of necessity from Newton's papers, for Fatio did not leave much in this area; their relative degree of involvement in alchemy is probably accurately reflected by that disproportion. In the interpretation of ancient prophecy, on the other hand, also an attempt to comprehend God's action in the world, the two men were closer together for a time. Their mutual involvement was most intense in both alchemy and prophecy in the late 1680s and the 1690s. Their alchemical work may have resumed at a lesser level between 1702 and 1705, if one accepts the hint given by the Ekins book list.

But Fatio's involvement with prophecy became ever more consuming. Leaving the dry bones of academic interpretation behind, Fatio heard the voice of God directly from the French Prophets who arrived in London

57. Bibliothèque Publique et Universitaire de Genève, MSS français 605/5, 605/8, 605/9.
58. *Theatrum chemicum* (3, n. 44); *Musaeum Hermeticum* (3, n. 40).

in 1706. Newton, reportedly restrained by other friends from attending the prophetic sessions, discussed the meaning of the new prophecies with Fatio as late as January 1706/7.[59] After Fatio became in effect a recording secretary for the inspired group, however, he probably no longer had time for alchemy, whereas Newton continued to labor over his alchemical papers occasionally[60] and to incorporate alchemical ideas into other aspects of his work, as one may see in Chapter 7.

A cause for gravity?

Newton found himself in a difficult situation in the *Principia* and post-*Principia* years with respect to a causal principle for gravitation. A brief recapitulation of his position will serve to emphasize his difficulties and to prepare the stage for his continuing search.

During his undergraduate years, probably about 1664, Newton accepted the Cartesian postulate of a dense aether, acting mechanically by impact, as the cause of terrestrial gravitation. Manuscripts written during the next twenty years reflect his continuous commitment to some form of mechanical aether as the principle of gravitation. Even though in those papers Newton sometimes associated the mechanical gravitational aether with the quasi-spiritual "active principle" of alchemy that was responsible for certain nonmechanical natural processes, by the end of the decade of the 1670s his gravitational aether was again entirely mechanical in its operation.

However, in 1684 or early 1684/5, when he was engaged in writing the *Principia*, Newton rejected all forms of a dense mechanical aether as a cause of gravitation, apparently because he had recognized their incompatibility with celestial motions. In 1684, when he derived mathematically the general area law for bodies revolving about a center of force (Theorem 1 in the various versions of *De motu* and then Proposition I, Theorem I of Book I of the *Principia*), he found that his mathematical demonstration

59. Margaret C. Jacob, "Newton and the French prophets: new evidence," *History of Science* 16 (1978), 134–42.
60. Sotheby lot nos. 24, 33, 40, and 41 – now, respectively: Sir Isaac Newton Collection, Babson College Archives, Babson Park, MA, Babson MS 421; King's College, Cambridge, Keynes MS 30; Jewish National and University Library, Jerusalem, Yahuda MS Var. 1, Newton MS 30; and Sir Isaac Newton Collection, Babson College Archives, Babson Park, MA, Babson MS 414 – were probably all written after 1707, at least in part, and many others appear to have been reworked and modified.

matched Kepler's empirically justified area law quite precisely. Within the context of the mechanical philosophy a problem immediately arose: the observationally verified area law of Kepler should *not* fit so closely with the exact area law derived mathematically by Newton if the heavens were filled with the aethereal medium postulated by the mechanical philosophers. Unless the medium is somehow disposed to move with exactly the same variable speed that the planetary body exhibits, the planet should encounter enough resistance from the medium to cause an observed deviation from the mathematical prediction, just as projectiles in the terrestrial atmosphere are observed to deviate from mathematical prediction. Consequently, Newton made a dramatic break with the orthodox mechanical philosophy of his day, the philosophy that was generally understood among advanced thinkers at the time to be the most promising method of approaching the study of the natural world. He did not reject the entire system of mechanical thought, but he did reject one of its most basic assumptions: that force could be transferred only by the impact of one material body with another.

The focus must now turn to Newton's search for a different sort of causal principle for gravitation during the ensuing twenty-five years. It seems one must recognize that Newton was dismayed by the fact that a combination of mathematics, observation, and experimentation had forced him to abandon the mechanical causal principle he had accepted for so long. It has often been assumed that he was content to be an early version of a philosophical positivist and to let his mathematical principles stand alone, but a close scrutiny of the surviving evidence from Newton's papers will not sustain that interpretation. Quite the contrary. Newton had always accepted the Renaissance view of history as a declination from an original golden age, a time in which there had existed an original pure knowledge of things both natural and supernatural, an ancient wisdom subsequently lost or garbled through human sin and error and through temporal decay. By the time of the writing of the *Principia* in the mid-1680s, Newton had already been engaged for a long time in attempts to restore the original truths once known to mankind by decoding obscure alchemical texts and by searching ancient records for the original pure religion. So when the (for him, modern) mechanical explanation of gravity failed, it was only natural that he should turn to ancient sources in an attempt to recapture the truer explanation of gravity once known to the wise ancients. What he found there in antiquity he then incorporated into his cosmic system.

One may now approach the historical evidence that sustains this interpretation. In "A Treatise of the System of the World" probably written in 1685, when he still thought the *Principia* would consist of only two books and of which this treatise would be the second, Newton began with material from his work on the origins of gentile theology.[61]

> It was the ancient opinion of not a few in the earliest ages of philosophy, That the fixed Stars stood immoveable in the highest parts of the world; that under the Fixed Stars the Planets were carried about the Sun; that the Earth as one of the Planets, described an annual course about the Sun, while by a diurnal motion it was in the mean time revolved about its own axe; and that the Sun, as the common fire which served to warm the whole, was fixed in the center of the Universe.
>
> This was the philosophy taught of old by *Philolaus, Aristarchus* of *Samos, Plato* in his riper years, and the whole sect of the *Pythagoreans*. And this was the judgment of *Anaximander*, more ancient than any of them, and of that wise king of the *Romans, Numa Pompilius*; who, as a symbol of the figure of the world with the Sun in the center, erected a temple in honour of *Vesta*, of a round form, and ordained perpetual fire to be kept in the middle of it.[62]

Not, on this occasion, attempting to define the original religion, Newton nevertheless used his material from antiquity to validate heliocentrism, an astronomical system in which "the fire at the heart of the world" had been symbolized by the ancient fire altars in the original pure religion.

But what had the ancients said about gravity? How were the motions of the celestial bodies performed as they moved about the central fire?

He had searched for but had as yet found no satisfactory answer. The most ancient had agreed that the motions "were performed in spaces altogether free, and void of resistance" (just as his own new system required) until the "whim of solid orbs" was introduced by Eudoxus, Callipus, and Aristotle "when the ancient philosophy began to decline." The motion of comets being incompatible with solid orbs, those bodies had been "thrust down below the Moon," but recent observations had restored them to their rightful place "among the celestial bodies" where the Chaldeans had correctly placed them before the Greek "fictions" prevailed. But, he regretfully noted, "we do not know how the ancients explained" the fact that the planets are "retained within any certain

61. Isaac Newton, *A Treatise of the System of the World Translated into English*, intro. by I. Bernard Cohen (facsimile reprint of the 2nd London ed. of 1731; London: Dawsons of Pall Mall, 1969), "Introduction," pp. vii–viii; Westfall, "Newton's theological manuscripts" (5, n. 17), esp. p. 137. For a similar version of the same material, see Casini, "Newton: the Classical Scholia," (5, n. 92), quotation on pp. 1–2. For Newton's work on the origins of gentile theology, see Westfall "*Origines*" (5, n. 18).
62. Newton, *System* (6, n. 61), pp. 1–2.

bounds in their free spaces" and "drawn off from the rectilinear courses, which, left to themselves, they should have pursued, into regular revolutions in curvilinear orbits." The ideas of later philosophers who "pretend to account for it" are all to be rejected, so he will treat the action of the force responsible for planetary restraint only "in a mathematical way" and "call it by the general name of a centripetal force, as it is a force which is directed towards some center."[63]

Statements such as the above, and the famed "hypotheses non fingo" of the *Principia*, are often taken to exemplify Newton's positivistic approach, but that is rather to misread Newton.[64] By positivistic one means here simply that the mathematical law *describing* gravity is sufficient and that no explanation of the *cause* of gravity is needed. One can hardly say Newton was uninterested in finding a causal explanation for gravity. He had for years thought he had had one, a material one, and, when that one failed, he wanted another. In the context of his statement in the "System," it is quite apparent that his "mathematical way" was a less than satisfactory alternative for him, put forward only because he could not determine what the ancients had thought about it and had not himself been able to formulate a good answer either.

There is much more evidence that Newton really did want a causal explanation for gravity; indeed there is a record of his continued search for one. He was even willing for a short time to give some credence to a new hypothesis of mechanical causation propounded by his friend Nicolas Fatio de Duillier. Fatio's basic supposition was,

> That the Aether, or such like subtle matter, had a very violent agitation; so as to pass every-way, in great-Circles, about the earth at the rate as to surround the Globe in 1 h. 25'. By which he shewed that a Body driven by the impulse of this motion, would be carried towards the center.[65]

Much given to collecting the endorsements of famous men for his idea, Fatio later reported Newton's approval in these strong terms:

> Sir Isaac Newton's Testimony is of the greatest weight of any. It is contained in some Additions written by himself at the End of his

63. Ibid., pp. 3–5.
64. Cf. I. Bernard Cohen, "Hypotheses in Newton's philosophy," *Physis 8* (1966), 163–84.
65. From the Journal of the Royal Society for 4 July 1688, quoted in Bernard Gagnabin, "De la cause de la pesanteur. Mémoire de Nicolas Fatio de Duillier, Présenté à la Royal Society le 26 février 1690. Reconstitué et publié avec une introduction par Bernard Gagnebin, Conservateur des manuscrits à la Bibliothèque publique et universitaire de Genève," *Notes and Records of the Royal Society of London 6* (1949), 105–60, on p. 115.

> own printed Copy of the first Edition of his Principles, while he was preparing it for a second Edition, And he gave me leave to transcribe that Testimony. There he did not scruple to say *That there is but one possible Mechanical cause of Gravity, to wit that which I had found out*: Thô he would often seem to incline to think that Gravity had its Foundation only in the arbitrary will of God....[66]

Although that copy of the *Principia* has never been found, there is reason to think that it did once exist and that Fatio quoted Newton accurately, for in a draft addition to the *Principia*, never used but probably written about 1690, Newton attempted to set the "necessary conditions" for a mechanical explanation of gravity, adding that the "unique hypothesis" by which gravity might be explained that way "was first devised by the most ingenious geometer Mr. N. Fatio."[67]

Gregory reported late in 1691, however, that "Mr Newton and Mr Hally laugh at Mr Fatios manner of explaining gravity."[68] For Newton, Fatio's hypothesis had probably foundered on the problem of the lack of retardation of celestial bodies moving in the subtle fluid. In the fragmentary and tentative addition to the *Principia* in which he praised Fatio, Newton tried to organize the particles of gross bodies into "networks" with pores large enough to allow "the gravitating cause" to operate just as much on the interior particles as on the exterior ones. Otherwise, "gravity would cease to be proportional to the [quantity of] matter." But even though he could create hypothetical "net-like figures" of any desired rarity to allow for the passage of "the gravitating cause" to the interior, still he knew "that in a heaven more filled with matter" solid gross bodies, without pores, "would lose a larger part of their motion... and even more if they were not solid bodies."[69] The "net-like figures" were required for the free passage of "the gravitating cause" to the interior of bodies. But porous bodies with that sort of structure would experience even more retardation than solid ones, whereas in actuality there was no evidence of retardation at all. The "necessary conditions" for Fatio's mechanical explanation of gravity thus were incompatible with each other. So Newton was left thinking that perhaps "Gravity had its Foundation only in

66. Bibliothèque publique et universitaire de Genève, Papiers Fatio No. 3 (Inv. 526), f. 66, quoted in Gagnebin, "De la cause" (6, n. 65), p. 117.
67. Cohen, *Introduction* (1, n. 9), pp. 184–5; Newton, *Unpublished Papers* (2, n. 51), pp. 312–17, quotations from pp. 313 and 315.
68. Memorandum by David Gregory, 28 Dec. 1691 (from the original in the Library of the Royal Society of London), in Newton, *Correspondence* (2, n. 14), vol. III, 191–2.
69. Newton, *Unpublished Papers* (2, n. 51), pp. 312–17.

the arbitrary will of God." Could he find support elsewhere for that radical proposition?

If there were no material aether in the heavens, then what was there? God, of course. "It is true," Newton had written already in the 1660s in his student notebook, "God is as far as vacuum extends, but he, being a spirit and penetrating all matter, can be no obstacle to the motion of matter; no more than if nothing were in its way."[70] God, being spirit, constituted no obstacle to motion; in addition He penetrated all things.

The question of "penetration" had concerned Newton again about 1680–4 in his alchemical *Commentarium* on the *Emerald Tablet* of Hermes Trismegistus, the handwriting of which is indistinguishable from that of various papers associated with the early stages of the writing of the *Principia*. There, in his commentary, Newton made some revealing corrections in his text, especially revealing if the *Commentarium* was indeed written about the time that he was being forced to make a break with corporeal causality for gravity. Hermes had referred to the penetration of the alchemical active entity (it "penetrates every solid thing" in Newton's translation; *omnemque solidam penetrabit* in his Latin text), and in commenting on that Newton revealed the direction his thoughts were taking. Beginning to write *ob subti* – probably "on account of its subtility" – he canceled those words (which of course would apply to a subtle corporeal fluid), and, after several attempts at a new formulation, ended by saying *vim penetrantem spiritus* – "the penetrating force of spirit."[71] Only spirit could penetrate and act in the way the evidence seemed to require. And so it was to be for the next three decades that Newton gave preferential consideration to a spiritual cause for gravity. As he summed up the problem of penetration in 1713 in the General Scholium to the *Principia*, "the power of gravity"

> must proceed from a cause that penetrates to the very centres of the sun and planets, without suffering the least diminution of its force; that operates not according to the quantity of the surfaces of the particles upon which it acts (as mechanical causes used to do), but according to the quantity of the solid matter which they contain. . . .[72]

The cause must penetrate, and he had tried to devise a netlike structure

70. McGuire and Tamny, *Certain Philosophical Questions* (1, n. 22), pp. 408–9.
71. Keynes MS 28 (1, n. 35), f. 6v. Cf. Appendix B for a reconstruction of Newton's several attempts at the formulation of this phrase.
72. Newton, *Principia* (Motte–Cajori: 2, n. 13), vol. II, 546; (Koyré–Cohen: 1, n. 9), vol. II, 764.

for matter in the 1690s to allow for the penetration of Fatio's mechanical aether, but had been forced to conclude that only a spiritual cause could penetrate adequately without constituting a frictional drag on the motions of the planets and comets and producing a slowing down of their motions that was contrary to the phenomena.

Two tentative solutions to the problem of a cause for gravity emerged in Newton's later years. One was, that the omnipresent supreme Deity subsumed gravity directly. The other was, that an intermediate agent existed that could account for gravity (as the vegetable spirit did for vegetation) yet not constitute a drag on the motion of the heavenly bodies. One may consider each of these possibilities in turn, the first in the following section, the second, and later one, in Chapter 7.

Gravity and ancient wisdom: the first solution

The first tentative solution to the problem of a cause for gravity to emerge in Newton's later years, then, was that the omnipresent supreme Deity subsumed gravity directly. There can be no doubt that Newton thought God to be literally omnipresent. As Gregory recorded on 21 December 1705,

> The plain truth is, that he [Newton] believes God to be omnipresent in the literal sense; And that as we are sensible of Objects when their Images are brought home within the brain, so God must be sensible of every thing, being intimately present with every thing: for he supposes that as God is present in space where there is no body, he is present in space where a body is also present.[73]

Henry More and other contemporaries were already treating the Deity as an incorporeal yet three-dimensional Being whose immensity constituted infinite three-dimensional space, and it is possible to see Newton's ideas as the "fruition of a long tradition" extending from Aristotle through Newton, a tradition in which Aristotle's finite plenum was slowly and by painful steps converted into the void, infinite, three-dimensional framework of the physical world required by classical physics. Newton's God-filled space was the penultimate development in the process by which concepts of space were developed by attributing to space properties derived from the Deity; after Newton's time, the properties remained with the space while the Deity disappeared from consideration.[74]

73. Hiscock, Ed., *Gregory's Memoranda* (2, n. 16), p. 30.
74. Grant, *Much Ado about Nothing* (4, n. 5), esp. pp. 182–264.

But there were theological complexities of which Gregory seemed unaware when he reported Newton's conviction of God's omnipresence. Simply to describe God as "present" wherever there is no body and also wherever there is body is to give a priority to body and the void that is theologically unacceptable. The Creator must be conceived as prior to His creation, and Newton's conception surely owed something to the theological construct of God "as the ground of all being" in which God is the place of the world but the world is not His place. Newton sometimes used the Hebrew word *māqôm* (place) as an expression for God's omnipresence, and in the General Scholium cited many of the Old Testament texts upon which Jewish theologies of space were based.[75] "In him are all things contained and moved," Newton said, citing the passages from Kings, Psalms, Job, and Jeremiah that the rabbis and Jewish philosophers had pondered.[76] In addition he cited texts from "Moses" in Deuteronomy, from John, and from Acts, thus bringing in the "sacred writers" from the most ancient Hebrew authority through Christ himself and earliest Christian antiquity.[77]

The force of ancient wisdom drawn from the sacred writers thus helped form and also helped validate Newton's conception of God's omnipresence, as the force of ancient wisdom drawn from the most ancient natural philosophers helped validate his heliocentrism. But the old knowledge had been fragmented, corrupted, and partially lost: Newton concluded his validating footnote in the General Scholium with a comment on corruption that in effect explained that *all* of the ancients were not to be trusted. "The Idolaters supposed the sun, moon, and stars, the souls of men, and other parts of the world, to be parts of the Supreme God, and therefore to be worshipped; but erroneously."[78] This caution, this qualification, this selectivity with respect to ancient wisdom, accords with and reinforces the impression scholars have long had of a sane and cautious balance in Newton's general methodology, as well as with the

75. Brian P. Copenhaver, "Jewish theologies of space in the scientific revolution: Henry More, Joseph Raphson, Isaac Newton and their predecessors," *Annals of Science* 37 (1980), 489–548.
76. Newton, *Principia* (Motte–Cajori: 2, n. 13), vol. II, 545; (Koyré–Cohen: 1, n. 9), vol. II, 762; I Kings 8: 27; Psalms 139: 7–9; Job 22: 12–14; Jeremiah 23: 23–4. For the references to the prior Jewish tradition, see Copenhaver, "Jewish theologies" (6, n. 75), p. 546, n. 220.
77. Newton, *Principia* (Motte–Cajori: 2, n. 13), vol. II, 545; (Koyré–Cohen: 1, n. 9), vol. II, 762; Deuteronomy 4: 39; Deuteronomy 10: 14; John 14: 2; Acts 17: 27–8.
78. Newton, *Principia* (Motte–Cajori: 2, n. 13), vol. II, 545; (Koyré–Cohen: 1, n. 9), vol. II, 762.

interpretation of Newton's methodology offered in Chapter 1: it was only when the various lines of his many investigations converged that he was convinced of the relative validity of his findings. After all, great matters were at stake.

Although undoubtedly Newton's conception of God's omnipresence was richer than that required for a conception of physical space, enriched as it was by a profound appreciation of the religious awe of God's immediacy, ubiquity, and plenitude that the ancients had expressed, the sacred writers had given Newton no assistance in the matter of gravity. It was one thing to let them speak for him and reinforce his own sense of God's presence everywhere. It was quite another thing to ask them *how* God's omnipresence subsumed the mathematical laws of universal gravity he had found. On that point the sacred writers were silent.

But perhaps the ancient natural philosophers were not entirely so. Newton had been talking with Gregory about God's omnipresence in 1705 with respect to possible publication of his ideas in the *Optice* of 1706. "But if this way of proposing this his notion be too bold," Gregory recorded immediately after the passage on omnipresence, "he [Newton] thinks of doing it thus."

> *What Cause did the Ancients assign of Gravity.* He believes that they reckoned God the Cause of it, nothing els, that is no body being the cause; since every body is heavy.[79]

What did Newton mean when he told Gregory that the ancients thought God was the cause of gravity? By late 1705, when he spoke to Gregory about it, he had been studying the issue for over twenty years, for already early in 1685 he had indicated in "A Treatise of the System of the World" that "we do not know how the ancients explained" it. What had he found between 1685 and 1705?

Much of Newton's work on the problem was done in connection with the abortive second edition of the *Principia* that was to have been edited by Nicolas Fatio de Duillier or David Gregory in the 1690s.[80] Fatio first mentioned Newton's efforts in a letter to Huygens in 169½[81] and Gregory noted them in 1694.[82] The principal fruit of Newton's work, the so-called Classical Scholia, were never published by Newton himself; they were, however, designed as scholia for Propositions IV–IX of Book III of the 1687 *Principia*. The manuscripts, in Newton's hand, passed to David

79. Hiscock, Ed., *Gregory's Memoranda* (2, n. 16), p. 30.
80. Cohen, *Introduction* (1, n. 9), pp. 177–99.
81. Newton, *Correspondence* (2, n. 14), vol. III, 193–5, (Fatio de Duillier to Huygens, 5 Feb. 169½.
82. Ibid., vol. III, 335–6, 338 (Gregory's Memoranda, 5, 6, 7 May 1694).

Gregory (who used them freely in composing the Preface to his own *Astronomiae physicae & geometricae elementa* of 1702), and they remain now with the Gregory manuscripts in the Library of the Royal Society.[83] Of primary interest here is the scholium for Proposition IX.

"So far I have expounded the properties of gravity," Newton said, referring to the propositions and scholia prior to Proposition IX. "Its cause I by no means recount." The latter statement seems to indicate that in the early 1690s Newton was by no means convinced that the explanation he had by then extracted from the ancients was correct. "Yet I shall say what the ancients thought about this subject," he continued.

> Thales regarded all bodies as animate, deducing that from magnetic and electrical attractions. And by the same argument he ought to have referred the attraction of gravity to the soul of matter. Hence he taught that all things are full of Gods, understanding by Gods animate bodies. And in the same sense Pythagoras, on account of its immense force of attraction, said that the sun was the prison of Zeus.... And to the mystical philosophers Pan was the supreme divinity inspiring this world with harmonic ratio like a musical instrument and handling it with modulation, according to that saying of Orpheus 'striking the harmony of the world in playful song'. Thence they named harmony God and soul of the world composed of harmonic numbers.... But the souls of the sun and of all the Planets the more ancient philosophers held for one and the same divinity exercising its powers in all bodies whatsoever....[84]

As with his work on the origins of gentile theology, Newton's search for ancient opinions on gravity led him to make eclectic use of authorities and to mold them into a coherent whole that spoke to his own purposes. Behind the ancient play on harmony was the Pythagorean–Platonic metaphysics of a recalcitrant world of matter, a chaos, that was divinely ordered and maintained by a "soul of the world composed of harmonic numbers." But the ancient doctrine had passed through many hands, often Christian or Stoic ones, on its way to Newton; thus, the supervisory harmonic function of the pagan demi-urge could be read by Newton as "the providence and preservation of God," as Newton's contemporary Thomas Stanley called it.[85] Stanley was actually quoting the pseudo-

83. McGuire and Rattansi, "Newton and the 'Pipes of Pan' " (2, n. 2); Casini, "Newton: the Classical Scholia" (5, n. 92); Gregory MS 247, Library of the Royal Society of London. Drafts of several items exist in the Portsmouth Collection, University Library, Cambridge.
84. Isaac Newton, draft scholium to proposition IX, as translated in McGuire and Rattansi, "Newton and the 'Pipes of Pan' " (2, n. 2), p. 119; original Latin in Casini, "Newton: the Classical Scholia" (5, n. 92), p. 33.
85. S. K. Heninger, Jr., *Touches of Sweet Harmony. Pythagorean Cosmology*

Plutarchian *De placitis philosophorum*, where the statement was ascribed to Pythagoras himself,[86] and Newton probably read it in the *De placitis philosophorum*, from which he drew material for some of the Classical Scholia.[87]

Since Newton was later to cite Pythagoras in his validating footnote in the General Scholium, it is of some interest to trace the Pythagorean tradition as Newton received it. Justin Martyr had already moved in the direction of a Christianization of Pythagoras in the second century A.D. in a text that became firmly established in patristic literature and continued to be repeated through the eighteenth century. "God is one," Justin had Pythagoras say in what purported to be a direct quotation.

> And He is not, as some think, outside the world, but in it, for He is entirely in the whole circle looking over all generations. He is the blending agent of all ages; the executor of His own powers and deeds; the first cause of all things; the light in heaven; the Father of all; the mind and animating force of the universe; the motivating factor of all the heavenly bodies.[88]

In addition to having the works of Justin himself,[89] Newton had several other works that repeated Justin on Pythagoras.[90] Although one does not

and Renaissance Poetics (San Marino, CA: The Huntington Library, 1974), pp. 201–14, quotation from Stanley's *History of Philosophy* (2 vols.; London, 1656–60) on p. 204. Newton once owned Stanley's *Historia philosophiae Orientalis* (Amstelodami, 1690) and the third edition of his *History of Philosophy* (London, 1701): Harrison, *Library* (1, 6), p. 242, item nos. 1551–2. Both items are, however, missing from that portion of Newton's library that survives at Trinity College, Cambridge.

86. Heninger, *Touches of Sweet Harmony* (6, n. 85), p. 230, n. 18.
87. Casini, "Newton: the Classical Scholia" (5, n. 92), e.g., pp. 39–40, nn. 4–5, 17–18.
88. Justin Martyr in his *Exhortation to the Greeks*, as quoted in Heninger, *Touches of Sweet Harmony* (6, n. 85), p. 202; for a selection of later authorities repeating Justin's rendition of Pythagorean thought, see ibid., pp. 202–3 and 229, n. 9.
89. Justin Martyr, *Opera*... (Coloniae, 1686): Harrison, *Library* (1, n. 6), p. 170, item no. 868, surviving as Trinity College, Cambridge, NQ.18.25[1], with some pages dog-eared.
90. For example, Clement of Alexandria, *Opera*... (Lutetiae Parisiorum, 1641); Cyril, *Opera*... (Lutetiae Parisiorum, 1631); Hierocles, *Commentarius in aurea Pythagoreorum carmina*... (Parisiis, 1583); Stanley, *History of Philosophy* (6, n. 85): Harrison, *Library* (1, n. 6), respectively, item nos. 398 (p. 120), 476 (p. 128), 762 (p. 159), and 1552 (p. 242). The *Opera* of Clement of Alexandria is now missing, as is the work by Stanley, but Cyril's *Opera* and Hierocles' *Commentarius* are extant as, respectively, Trinity College, Cambridge, NQ.11.48 and NQ.9.78, both with some signs

know Newton's exact source(s) for his conception of the Pythagorean philosophy, Justin's interpretation may well have informed Newton's views. As Heninger has observed, two conceptions of the Deity are implicit in Justin's words: one, the Deity as all-pervasive spirit infusing the universe; the other, the Deity as workman building and ordering the cosmos.[91] Both were important to Newton.

In conjunction with the Pythagorean–Platonic emphasis on number and world harmony, the vision of God as cosmic worker led to the ancient concept of "the music of the spheres" so fully assimilated later to the dictum from the Wisdom of Solomon that God had ordered everything by number, weight, and measure, and so thoroughly absorbed into Christian aesthetics through Augustine's *De musica*. It was the concept that undergirded Newton's conviction that ancient prytanea and temples were true geometric representations of God's cosmos; it was the concept that in the 1690s came to support his tentative belief that the ancients had recognized the mathematical laws of gravity and had hidden their knowledge in musical myths or "figures."[92]

of dog-earing. One should also note in Newton's library the presence of several other works relevant to Pythagorean–Platonic thought and ancient music: Iamblichus, *De mysterius liber*... (Oxonii, 1678) [Harrison, *Library* (1, n. 6), p. 166, item no. 827, now NQ.11.26], idem, *De vita Pythagorica liber*... (Amstelodami, 1707) [Harrison, *Library* (1, n. 6), p. 166, item no. 828, now NQ.8.75, with dog-earing]; Marcus Meibomius, *Antiquae musicae auctores septem*... (2 vols.; Amstelodami, 1652) [Harrison, *Library* (1, n. 6), p. 190, item no. 1060, now NQ.8.36 & 37, including the musical works of Aristides Quintilianus, Nicomachus Gerasenus, and Martianus Capella, among others]; Ficino's edition of Plato, *Opera*... (Francofurti, 1602) [Harrison, *Library* (1, n. 6), p. 218, item no. 1325, now missing]; Claudius Ptolemy, *Harmonicorum libri III*... (Oxonii, 1682) [Harrison, *Library* (1, n. 6), p. 222, item no. 1357, now NQ.10.36]. See also Penelope Gouk, "Newton and music: from the microcosm to the macrocosm," *International Studies in the Philosophy of Science 1* (1986), 36–59, and idem, "The harmonic roots of Newtonian science," in *Let Newton Be!*, ed. by John Fauvel, Raymond Flood, Michael Shortland, and Robin Wilson (Oxford: Oxford University Press, 1988), pp. 100–25.

91. Heninger, *Touches of Sweet Harmony* (6, n. 85), pp. 209–10.
92. The brief exchange of letters between young John Harington and Newton drew together ideas on musical and mathematical harmonies, sacred architecture, and macrocosmic harmony: Newton, *Correspondence* (2, n. 14), vol. IV, 272–5 (Harington to Newton, 22 May [1698?]; Newton to Harington, 30 May 1698). I am indebted to Marion S. Trousdale for calling my attention to this exchange of letters. See also R. J. van Pelt, *Tempel van de Wereld, de Kosmische symboliek van de tempel van Salomo* (HES Uit-

By the early 1700s, in an unpublished draft associated with Query 23 of the *Optice* of 1706, Newton asked his question about the cause of gravity again, but this time gave a somewhat firmer answer.

> By what means do bodies act on one another at a distance? The ancient philosophers who held Atoms and Vacuum attributed gravity to atoms without telling us the means unless in figures: as by calling God harmony representing him & matter by the God Pan and his Pipes, or by calling the Sun the prison of Jupiter because he keeps the Planets in their Orbs. Whence it seems to have been an ancient opinion that matter depends upon a Deity for its laws of motion as well as for its existence.[93]

The cause of gravity was God, acting by harmonious mathematical laws to order and maintain cosmos; Newton drew there upon a many-textured tradition of thought that stemmed from remotest antiquity and was an integral part of Christendom.

Newton later cited Pythagoras in his validating footnote to the General Scholium, however, and there he was in fact pointing to the other side of Justin's interpretation of the Pythagorean Deity – the Deity as all-pervasive spirit (or soul or mind) infusing the universe. "In him are all things contained and moved," Newton said.

> This was the opinion of the Ancients. So *Pythagoras*, in *Cicer. de Nat. Deor.* lib. i. *Thales, Anaxagoras, Virgil,* Georg. lib. iv. ver. 220; and Aeneid, lib. vi. ver 721. *Philo Allegor*, at the beginning of lib. i. *Aratus*, in his Phaenom. at the beginning. So also the sacred writers. . . .[94]

One has already reflected upon the concepts Newton found in the "sacred writers" regarding the omnipresence of God; one has seen that they gave him no insight into the question of a cause for gravity. One has also found via the draft Classical Scholium to Proposition IX of Book III of the *Principia* and the draft Query for the *Optice* of 1706 that Newton came to think it probable that one line of ancient thought (associated with Pythagoras) had encoded its knowledge of gravity in musical myths and in talk about divine and celestial harmonies. In their own way, Newton assumed, the mathematical and musical ancients had stated "that matter depends upon a Deity for its laws of motion." But that line of

gevers/Utrecht: H & S, 1984), esp. pp. 323–9 and (English abstract) pp. 397–8.

93. Isaac Newton, draft Query 23, University Library, Cambridge, Portsmouth Collection MS Add. 3970, f. 619r, as quoted in McGuire and Rattansi, "Newton and the 'Pipes of Pan' " (2, n. 2), p. 118.

94. Newton, *Principia* (Motte–Cajori: 2, n. 13), vol. II, 545; (Koyré–Cohen: 1, n. 9), vol. II, 762.

ancient thought had still not explained to Newton exactly *how* the Deity put the laws of motion into effect. There is reason to believe that Newton was not fully satisfied with the musical–mathematical analogy, and when he cited Pythagoras in the General Scholium (as described in Cicero's *De natura deorum*), Newton was pointing in a different direction.

In Book I of Cicero's dialogue *On the Nature of the Gods*, the speaker is Velleius, the proponent of Epicurean theology, and the beliefs of Thales, Anaxagoras, and Pythagoras are explicated by Velleius simply in order that he may refute them. Newton, however, apparently ignored the Epicurean critiques of the older views and appropriated instead the reports of the ancient theologies that served his own purposes.

> Thales of Miletus, who was the first person to investigate these matters, said that water was the first principle of things, but that god was the mind that moulded all things out of water.... Then there is Anaxagoras, the successor of Anaximenes; he was the first thinker to hold that the orderly disposition of the universe is designed and perfected by the rational power of an infinite mind.... As for Pythagoras, [he] believed that the entire substance of the universe is penetrated and pervaded by a soul of which our souls are fragments. ...[95]

The idea of the Deity as an infinite mind, soul, or spirit penetrating and pervading all things was evidently what attracted Newton in those Ciceronian passages, for, as one may see in what follows, that idea was also what he drew from Virgil.

In Virgil, in both the *Georgics* and the *Aeneid*, Newton's references point to divine mind, spirit, rational intent, life – all permeating the cosmos. In the *Georgics* Virgil had this to say:

> Observing these patterns, their intricacy, their justness,
> Some men conclude that bees, as they drink the air,
> Draw in with it a trace of divine intention.
> They say that a spirit fills earth and sky and sea;
> That man and his fellow-creatures, the wild and the
> tame,
> Take from it the delicate stuff of their life.
> There is no dying, they believe, when the body fails;
> But, released, the life leaps lightly away to its place
> In the depth of heaven, a brightness pointing the stars.[96]

95. Cicero, *De natura deorum*, I, x–xi, in *Cicero in Twenty-Eight Volumes. XIX. De nature deorum, Academica*, tr. by H. Rackham (The Loeb Classical Library, ed. by E. H. Warmington; Cambridge, MA: Harvard University Press; London, William Heinemann, 1972).
96. Virgil, *Georgics*, IV, 219–27, in Virgil, *The Georgics*, tr. by Robert Wells (Manchester: Carcanet New Press, 1982).

In the *Aeneid* the following:

> "First, know, a soul within sustains the heaven
> and earth, the plains of water, and the gleaming
> globe of the moon, the Titan sun, the stars;
> and mind, that pours through every member, mingles
> with that great body."[97]

In both of these references to Virgil, one finds that the Deity is everywhere and mingled with everything.

It is of considerable interest for the present line of analysis that Virgil is known to have come under Stoic influences,[98] for in a draft intended for the unimplemented edition of the *Principia* in the 1690s Newton himself called attention to "the Stoics" as having had the right ideas.

> ... [T]hose ancients who more rightly held unimpaired the mystical philosophy as Thales and the Stoics, taught that a certain infinite spirit pervades all space *into infinity*, and contains and vivifies the entire world. And this spirit was their supreme divinity, according to the Poet cited by the Apostle. In him we live and move and have our being.[99]

Stoicism had a systematic physics of course, and of fundamental importance for a fresh understanding of Newton's cosmos in the post-*Principia* period is his new use of the Stoic concept of *pneuma* or aether, for that proved to be the essential factor, the factor, for Newton, linking the divine will with the world of matter. In contrast to his earlier use of a material *pneuma* or aether, Newton came in the post-*Principia* period to utilize a spiritual *pneuma*. As he indicated in this draft, the Stoics taught him about an "infinite spirit" pervading everything, a spirit that "contains and vivifies the entire world."

One Stoic influence on Virgil, as on Newton himself, was the Stoic poet Aratus, whose didactic poem on the heavens opened with a hymn to Zeus as the all-pervasive, providential Deity.

> The sky is our song
> and we begin with Zeus; for men cannot speak

97. Idem, *Aeneid*, VI, 724–7, in Virgil, *The Aeneid of Virgil, A Verse Translation by Allen Mandelbaum. With Thirteen Drawings by Barry Moser* (Berkeley: University of California Press, 1981).
98. Cf. *Virgil: Georgics I & IV*, ed. by H. H. Huxley (London: Methuen & Co., 1963), pp. 162–3; Jean Martin, *Histoire du texte des Phénomènes d'Aratos* (Etudes et Commentaires XXII; Paris: Librairie C. Klincksieck, 1956).
99. Isaac Newton, draft scholium, University Library, Cambridge, Portsmouth Collection MS 3965.12, f. 269, as quoted in McGuire and Rattansi, "Newton and the 'Pipes of Pan' " (2, n. 2), p. 120.

> without giving Him names: the streets are detailed
> with the presence of Zeus, the forums are filled,
> and the sea and its harbors are flooded with Zeus,
> and in Him we move and have all our being.
> For we are His children, and He blesses our race
> with beneficent signs, and wakes man to his work,
> directing his mind to the means of his life.[100]

Aratus studied under Zeno the Stoic in Athens around 227 B.C., and, though this prologue to his astronomical work is thought to be his own, it is similar in some ways to the Stoic Cleanthes's "Hymn to Zeus."[101] The famous Pauline statement on God as the ground of all being in Acts 17: 27–28 was probably drawn from the prologue of Aratus, for the poem was immensely popular in Hellenistic and Roman circles, and Newton seemed to indicate his knowledge of that filiation in his draft comment of the 1690s when he mentioned "the Poet cited by the Apostle." In the *Principia*'s General Scholium, of course, Newton cited the prologue of Aratus directly: "*Aratus*, in his Phaenom. at the beginning."

But even more fundamental to Newton's new perspective was the work of Philo and his successors. Newton's reference in the *Principia* was to Philo's *Allegorical Interpretation* of the Mosaic account of creation. At the beginning of Book I, Philo treated the words of Moses in Genesis 2: 1 – "And the heaven and the earth and all their world were completed." Again, Newton selected a passage that focused on mind, for Philo argued that Moses called the mind "heaven." Moses "had already told of the creation of mind and sense-perception," Philo said:

> he [Moses] now fully sets forth the consummation of both. He does not say that either the individual mind or the particular sense-perception have reached completion, but that the originals have done so, that of mind and sense-perception. For using symbolical language he calls the mind heaven, since heaven is the abode of natures dis-

100. Aratus, *Phaenomena*, ll. 1–9 (tr. by Stanley Frank Lombardo, 1976); cf. Stanley Frank Lombardo, "Aratus' *Phaenomena*: An Introduction and Translation" (The University of Texas at Austin: Ph.D. dissertation, 1976), esp. pp. vii, 1–3, 30–2, 54–7, 102–3, quotation from p. 102; *Sky Signs: Aratus' Phaenomena*, intro. and tr. by Stanley Lombardo, with illust. by Anita Volder Frederick (Berkeley, CA: North Atlantic Books, n.d.), unpaginated "Introduction," p. 1, and unpaginated n. 7.
101. Lombardo, "Aratus' *Phaenomena*" (6, n. 100), pp. 1–3 and 102; Cleanthês, *Hymn to Zeus*, in T. F. Higham and C. M. Bowra, Eds., *The Oxford Book of Greek Verse in Translation* (Oxford: Clarendon Press, 1938), pp. 533–5, no. 483 (tr. by Michael Balkwill).

> cerned only by mind, but sense-perception he calls earth, because sense-perception possesses a composition of a more earthly and body-like sort. "World," in the case of mind, means all incorporeal things, things discerned by mind alone: in the case of sense-perception it denotes things in bodily form and generally whatever sense perceives.[102]

Not only did Philo identify "mind" with "heaven," but also he indicated that "world" (κόσμος, cosmos) "in the case of mind" meant "all incorporeal things," and those are almost certainly the passages that attracted Newton's interest, for what one finds in Philo is a Platonizing Stoicism in which the traditionally material Stoic Deity has been dematerialized – made "incorporeal."

The more orthodox of the ancient Stoics had certainly held everything to be material, including the *pneuma* or aether as Deity, probably in conscious reaction to the philosophical inadequacies of Aristotle's Unmoved Mover.[103] The Stoics maintained the four traditional elements of earth, water, air, and fire, supposed to constitute the entire universe. These elements were ranked in order of ascending tonicity or tensional activity and divinity, and the cosmic *pneuma* that filled the heavens and permeated the entire universe was thought to be composed of the hottest possible combination of fire and air, the most tense, active, and divine substance in the world – in short, the Stoic Deity. But the cosmic *pneuma*, as active principle, permeated and blended completely with body (conceived as passive principle), the tensional powers of the *pneuma* holding the body together.

Stoic insistence upon the corporeality of the cosmic divine *pneuma* had always made the Stoic theory of total blending somewhat problematic. Are corporeal *pneuma* and corporeal body to fuse completely and so become something that is neither God nor matter? Are two corporeal bodies to occupy the same space at the same time? In their origins Stoic arguments on blending more or less avoided those issues by treating the *pneuma* in its motion through matter as a shaping, cohesive force,[104] but another solution to the difficulty was to create a dualistic system of matter

102. Philo, *Allegorical Interpretation of Genesis II., III (Legum Allegoria)*, I, 1, in *Philo in Ten Volumes with an English Translation by F. H. Colson and G. H. Whitaker* (The Loeb Classical Library; London: William Heinemann; New York: G. P. Putnam's Sons, 1929), vol. I, 140–473.
103. Hahm, *Origins of Stoic Cosmology* (4, n. 14), pp. 3–28, esp. pp. 14–15.
104. Hunt, *Physical Interpretation* (5, n. 109), pp. 47, 50; Robert M. Todd, *Alexander of Aphrodisias on Stoic Physics. A Study of the De mixtione with Preliminary Essays, Texts, Translation and Commentary* (Leiden: E. J. Brill, 1976); Hager, "Chrysippus' theory of pneuma" (4, n. 14).

and spirit, a Platonizing Stoicism in which the corporeal (but active and divine) Stoic *pneuma* was made spiritual and incorporeal but still mingled and blended with every body. That was the solution chosen by Philo, by many of the Church Fathers, and by the prominent sixteenth-century Renaissance Stoic Justus Lipsius. It was also the solution Newton needed.

Philo Judaeus of Alexandria (c. 25–20 B.C. – c. A.D. 50 was in some ways probably quite central to Newton's understanding of ancient wisdom. Influenced by both Platonism and Stoicism, Philo provided a synthesis of Hellenistic and Judaic thought by writing commentaries on many Old Testament books and characters (e.g., *On the Account of the World's Creation Given by Moses; On Abraham; On Noah's Work as a Planter*) and on various theological topics (e.g., *On Providence; On the Unchangeableness of God*), and also by allegorical exegesis of Scripture (e.g., *Allegorical Interpretation of Genesis II*, which Newton cited). Modern scholarly opinions on the originality and philosophic importance of Philo have been sharply divergent and contradictory, but there can be no doubt of his historical significance since his writings became a prime source for patristic exegesis and he influenced, among others, Justin Martyr, Clement of Alexandria, Origen, Eusebius, Ambrose, St. John Damascene, Basil the Great, and St. Augustine.[105] Newton owned Philo's *Opera*,[106] and, when he read the works in that volume, he would have been enabled to find Plato and the Stoics in remotest Jewish antiquity. For, although modern scholars understand Philo to have read Platonic and Stoic ideas *into* Scripture, Philo himself and also his Christian followers understood matters quite differently, insisting that the Platonism and Stoicism came *out of* Scripture itself.[107] Newton, believing in an original pure wisdom, through his reading of Philo thus projected the

105. David Winston, *Logos and Mystical Theology in Philo of Alexandria* (Cincinnati: Hebrew Union College Press, 1985), pp. 9–12, provides a recent survey of modern opinion; see also Samuel Sandmel, *Philo of Alexandria. An Introduction* (New York: Oxford University Press, 1979); Max Pulver, "The experience of the pneuma in Philo," tr. by Ralph Manheim, in *Spirit and Nature. Papers from the Eranos Yearbooks*, ed. by Joseph Campbell (Bollingen Series XXX.1; Princeton: Princeton University Press, 1982), pp. 107–21, originally published in German in *Eranos-Jahrbuch 13* (1945); Erwin R. Goodenough, *An Introduction to Philo Judaeus* (2nd. ed.; Oxford: Basil Blackwell, 1962).
106. Philo, *Omnia quae extant opera*... (Lutetiae Parisiorum, 1640): Harrison, *Library* (1, n. 6), p. 216, item no. 1300, now missing.
107. Sandmel, *Philo* (6, n. 105), p. 28; Goodenough, *Introduction* (6, n. 105), pp. 94–7.

knowledge of an incorporeal heavenly *pneuma* and divine mind back onto Moses himself.

There were almost certainly other influences in Newton's mental world pushing him toward a Philonic interpretation of Stoicism. Except for Tertullian, all of the early patristic writers had tended to spiritualize the material Stoic Deity (as Supreme God) even while evincing a certain tendency toward materialism with respect to other spiritual beings (such as angels and demons) and allowing for the mingling and penetration of spirit and matter in a way that indicated no sharp dualism in their thought between the spiritual and material realms.[108] Their ambiguity was similar to that of the seventeenth-century alchemists and physicians. But a sharper dualism was present in Neoplatonic and Hermetic writings, and, in the revival of Stoic thought in the Renaissance, Stoic materialism was curtailed as it was Christianized. Most of the interest in Stoic thought in the Renaissance was focused on ethics and moral philosophy, but with Justus Lipsius there was also a full-scale effort to Christianize Stoic physics and metaphysics, Lipsius even using some of Philo's Christian followers as authorities.

The Stoicism of Lipsius in the sixteenth century was similar to that of Philo in the first century, for Lipsius argued that the supposed materialistic monism of the Stoics was simply a disguised dualism. Drawing upon a distinguished line of Stoic reasoning regarding the two universal principles – active and passive – he made his point quite effectively with quotations from Diogenes Laertius, Seneca, and Cicero. The two principles are variously called, he said: *Efficiens* and *Patiens*, Reason and Matter, Cause and Matter, God and Matter, Force and Matter. But only the active principle should be conceived as eternal. The Stoics had made an error, Lipsius thought, in conferring eternity upon matter itself, and, their material Deity being of necessity joined with passive matter in all of nature, they had thus been led into a materialistic pantheism. Lipsius proposed to resolve these problems by de-materializing the Deity: God as active principle was to be conceived as present to the world as an eternal spirit, vivifying, creative, providential, as the eternal Cause and the ground of all being. Lipsius identified the Cause with the seminal reasons or spermatic logoi of the Stoics, that act to inform the passive, sluggish material principle, that from the latter generate the four formed elements of earth, air, fire, and water. Following the *Corpus Hermeticum* and one of the Christian writers influenced by Philo, St. John Damascene, Lipsius argued that God, the Divine Fire, *contains* the forms of all things,

108. M. Spanneut, *Le Stoïcisme des pères de l'Eglise: de Clément de Rome à Clément d'Alexandrie* (Paris: Seuil, 1957), pp. 288–91, 394–6, 426.

the Stoic forms thus becoming very similar to the Platonic Eternal Ideas. Lipsius further interpreted the Stoics who had denied form to the Fiery Breath to have meant only that God has no *visible* form, and, again basing his argument on the *Corpus Hermeticum*, concluded that God reveals no form to human vision because He is incorporeal. God reveals Himself, however, in the forms of all corporeals. That is what is meant by Christians, Lipsius went on, when they say no one sees God but yet God appears in all things. That is also what the Stoics meant, he thought, when they identified God with nature or with the world. The pantheistic interpretations of Stoic intentions in equating God with the world being unacceptable to Lipsius, he argued away Stoic materialism, and, relying on Augustine to criticize Tertullian's materialism, insisted that God, being incorruptible and incorporeal, cannot be mixed with matter in any real sense (as Stoic theory on total blending had required). The all-pervasiveness of God was to be understood as forming and animating all things, but not as the form or soul itself: as ruling principle of all but blended with nothing. For Lipsius, God is in the world but more than the world. Conversely, the world is not by its own nature God, although the world communicates with and participates in the Divine Intelligence. God is Spirit. Lipsius thought there was little or no difference between the meaning of Spirit and the Stoic terms Reason, Soul, Mind, Intellect – if, that is, the Stoic terms were correctly interpreted – and he expressed his Christianized Stoic convictions in a passage reminiscent of Aratus and the Apostle Paul: *quia nos & omnia vivimus, et movemur, et sumus in Deo.*[109]

109. Léontine Zanta, *La renaissance du stoïcisme au XVIe siècle* (Bibliothèque littéraire de la renaissance, nouvelle série, tome V; Paris: Librairie ancienne Honoré Champion, Edouard Champion, 1914), esp. pp. 225–30; Saunders, *Justus Lipsius* (4, n. 26), pp. 121–37; Justus Lipsius, *Physiologiae stoicorum libri tres: L. Annaeo Senecae alliisq. scriptoribus illustrandis*, in Justus Lipsius, *Justi Lipsi V. C. Opera omnia, postremum ab ipso aucta et recensita: nunc primum copioso rerum indice illustrata* (4 vols.; Vesaliae: Apud Andraeam ab Hoogenhuysen et Societatem, 1675), vol. IV, 823–1006, esp. *Physiologiae stoicorum*, I, iii–ix, in *Opera*, vol. IV, 837–53, quotation from vol. IV, 851. Newton is not known to have owned Lipsius's *Physiologiae stoicorum* even though two other of Lipsius's works were in his library: *Epistolarum selectarum chilias*... (Genevae, 1611) and *Roma illustrata, sive Antiquitatum Romanarum breviarium*... *Ex nova recensione A. Thysii*... *Postrema ed.* (Amstelodami, 1657): Harrison, *Library* (1, n. 6), p. 180, item nos. 959–60, now, respectively, Trinity College, Cambridge, NQ.8.135, and missing.

Or, in King James's English, "In him we live and move and have our being" – the very point for which Newton cited Paul and "the Stoics" in the 1690s and Paul and the Stoic Aratus in the General Scholium.

On more than one specific point Newton seems to follow Lipsius. Some Stoics placed the center of the divine creative fire in the sun itself, "the fire at the heart of the world." Later Stoics often placed the divine creative fire in the *pneuma* or aether, however, and in the sixteenth century Lipsius said that the distinction between sun and aether as the seat of the divine active principle was not a significant one.[110] Since the *pneuma* was conceived as penetrating all things in any case, the Lipsian point of view had much to recommend it, and Newton seemed to echo it in the draft Scholium to Proposition IX when he attributed to "the more ancient philosophers" the idea that "the souls of the sun and of all the Planets" were "one and the same divinity exercising its powers in all bodies whatsoever."

The Philonic and Lipsian points of view evidently recommended themselves to Newton in general, and Newton's own views on the divine *pneuma* in the post-*Principia* period resemble their approaches much more than they do those of most of the ancient Stoics and much more than the new views resemble Newton's own views of the 1670s. One cannot be certain that Newton read Lipsius's *Physiologiae stoicorum*, but he did read Philo. And since the work of Lipsius depended at least in part on Philonian influences, perhaps a direct reading of Lipsius would have been supererogatory. Newton would in any case have preferred the more ancient authority as being closer to the fount of Truth. The tenor of Philonic Stoicism was to shift from a materialistic monism toward a Platonic dualism, within which there existed a distinction between spirit and matter that was quite foreign to most of the ancient Stoics. Under the dispensation of a Platonizing Stoicism with its incorporeal Deity as the active principle, Newton was able to use the Stoic concept of the Deity as a tensional force, binding the parts of the cosmos together, penetrating and mingling with all bodies, without the inconveniences attendant upon the corporeal *pneuma* or aether.

In the matter of gravity, Newton had recognized that no material, mechanical cause would serve and had been forced to make a break with corporeal causality. The evidence he had in hand denied the presence of the corporeal aether. Yet gravity acted, and it seemed to act as if it penetrated to the very centers of bodies. Only spirit could penetrate in

110. Lipsius, *Physiologiae stoicorum* (6, n. 109), II, xiii, in *Opera* (6, n. 109), vol. IV, 927–30.

that way without constituting a frictional drag by acting on the surfaces of bodies and/or on the surfaces of their internal parts. In the context of Stoic thought, as interpreted by Philo and Lipsius, the all-pervasive spirit was of course the active principle, acting everywhere to penetrate and bind the passive principle of matter. The conceptualization of gravity as active principle, as subsumed by the literal omnipresence of God, as a spiritual force binding all together, was to serve Newton for many years. Stoic the idea was certainly, but Newton used it in its Platonizing version, in which the Deity was wholly immaterial, noncorporeal, yet all-pervasive.

With the important exceptions of Sambursky and Kubrin,[111] historians of science have largely ignored the affinities between Newton's doctrines and Stoic thought, but classicists have not been so negligent, as one may see from the brief examination of a Ciceronian text. Cicero made explicit the concept of the Deity as tensional force at the heart of the world and of all the matter in the world as he described Stoic philosophy in Book II of his *De natura deorum*, a work Newton certainly read.

> For all its [the world's] parts in every direction gravitate with a uniform pressure towards the centre. Moreover all bodies conjoined maintain their union most permanently when they have some bond encompassing them to bind them together; and this function is fulfilled by that rational and intelligent substance which pervades the whole world as the efficient cause of all things and which draws and collects the outermost particles towards the centre.[112]

In the modern Loeb Classical edition, from which this translation is drawn, the concept is considered to be so explicit, in fact, that the word Newton coined for his central attractive force, "centripetal," is used to gloss Cicero's words. Indeed, Newton's word has frequently been used to gloss the Ciceronian text at least since the middle of the nineteenth century.[113]

The concept, if not the word "centripetal," undoubtedly had a lengthy pre-Newtonian history also. Hunt has argued that Zeno, the founder of Stoicism, probably explained the coherence of the cosmos by "centripetal force" (Hunt using Newton's word anachronistically)

111. Sambursky, *Physics of the Stoics* (2, n. 25); Kubrin, "Newton and the cyclical cosmos" (2, n. 33); idem, "Providence and the mechanical philosophy" (2, n. 33).
112. Cicero, *De natura deorum* (6, n. 95), II, xlv, 115–16.
113. I have traced it as far back as 1850: Cicero, *M. Tullii Ciceronis, De natura deorum libri tres*, erklaert von G. F. Schoemann (Leipzig: Weidmannsche Buchhandlung, 1850), p. 157.

and that Stobaeus recorded a similar idea.[114] In his commentary on Lucretius, *De rerum natura*, I, 1052–82, Bailey attributed similar views to Parmenides and Plato as well as to the Stoics (including Chrysippus and Cicero in the passage quoted above). Lucretius had undertaken to refute the idea, since his own infinite universe could have no center, and Bailey somewhat sadly observed, "This doctrine contained of course the germ of the theory of gravitation, and it is rather pathetic to see Lucr. rejecting an idea which is now universally accepted...."[115] There is, in addition, considerable evidence that the concept continued to be a living force in European intellectual life from the classical period until Newton's own time.[116] Thus, in some ways, ironically, Newton was entirely correct in his sense of affinity with antiquity in this matter, though one may still be allowed to doubt that the wise ancients had really done the mathematics that Newton attributed to them.

In 1706, in what was then Query 23 of the *Optice* and what became Query 31 of the 1717/18 *Opticks*, Newton openly designated gravity as an active principle in his most general statement on the forces associated with matter, the particles of which he described as "solid, massy, hard, impenetrable, [and] moveable."

> It seems to me farther, that these Particles have not only a *Vis inertiae*, accompanied with such passive Laws of Motion as naturally result from that Force, but also that they are moved by certain active Principles, such as is that of Gravity, and that which causes Fermentation, and the Cohesion of Bodies.[117]

Within the context of Stoic doctrines, the divinity of Newton's active principles now seems perfectly obvious. Defined against the passive principle of matter, they are the eternal Cause, *Efficiens*, Mind, Reason, Force.

Although Westfall and I have both argued that Newton came to view gravity as an active principle by analogy with the active alchemical agent, that story no longer seems quite so straightforward. Active and divine

114. Hunt, *Physical Interpretation* (5, n. 109), pp. 57–8.
115. Lucretius, *Titi Lvcreti Cari De Rervm Natvra Libri Sex*, ed. with Prolegomena, Critical Apparatus, Translation, and Commentary by Cyril Bailey (3 vols.; Oxford: Clarendon Press, 1947), vol. II, 782–3.
116. Michael Lapidge, "A Stoic metaphor in late Latin poetry: the binding of the cosmos," *Latomus* 39 (1980), 817–37; James Hutton, " Spenser's 'Adamantine Chains': a cosmological metaphor," in Luitpold Wallach, Ed., *The Classical Tradition. Literary and Historical Studies in Honor of Harry Caplan* (Ithaca, NY: Cornell University Press, 1966), pp. 572–94, esp. pp. 583–8.
117. Newton, *Opticks* (2, n. 15), pp. 400–1.

the alchemical agent was certainly; in the *Emerald Tablet* Hermes Trismegistus even spoke of its quality of "penetration." Alchemical thought carried much Stoicism into the seventeenth century and focused a great deal on the active principle. Newton's original dichotomy between vegetable and mechanical was certainly stated in an alchemical context early in the 1670s, and that original dichotomy is restated in the *Opticks* passage in the terminology of active and passive.

But the alchemical agent was focused on micromatter. It was "that which causes Fermentation, and the Cohesion of Bodies," Newton said in the *Opticks*, and during most of the post-*Principia* period he apparently did not think he could make a direct analogy between it and the cosmic gravitational cause because he conceived the alchemical vegetable spirit as at least quasi-material and so subject to some of the same disabilities as his early mechanical versions of a material aether so far as gravity was concerned. He was later to have doubts on this point, and to put forward suggestions for a cosmic gravitational cause that *was* analogous to the vegetable spirit, but not until after new electrical experimentation suggested a new interpretation to him about 1710, as one may see in the next chapter.

Between 1684 and 1710 Newton's conceptualization of the "cause" of gravity was probably more directly mediated by ancient cosmic thought and especially by the Platonizing version of ancient Stoicism than by alchemical doctrines of an active principle at work in the realm of micromatter. The analogies at work in the evolution of Newton's thought on the cause of gravity would seem to have been cosmic ones based on the mathematical and musical harmonies of ancient Pythagorean–Platonic metaphysics and on the Divine Mind of Presocratic and Stoic thought.

When Newton spoke to Gregory about these matters late in 1705, Gregory recorded one more analogy that had become central to Newton's thinking. Presumably because he had immersed himself in ancient ideas about the cosmic Divine Mind, Newton made an effort to understand that infinite Divine Mind by analogy with finite human minds. He said to Gregory, "as we are sensible of Objects when their Images are brought home within the brain, so God must be sensible of every thing, being intimately present with every thing." As Newton paraphrased the ancient ideas, God is omnipresent and in him all creatures live and move and have their being. Newton's Deity can be aligned easily enough with the God of Judeo–Christian tradition, and indeed must be in many ways, but His intimacy, the fact that He is mingled or mixed with everything – as in the Virgilian and Aratean passages Newton cited – is really not traditional and establishes that Newton's God had affinities also with the divine rational principle of the Presocratics, Philo, and the Stoics: the active principle as Mind. It was that aspect of Newton's Deity that at last seemed to him to offer, by analogy, an explanation of the *how* of gravitation.

As we sense objects *indirectly* when their images are brought home to our little sensories, the centers of perception in our brains, so God perceives objects *directly*, "being intimately present with every thing." Then in the *Opticks* Newton established the correlative analogy of willed activity, all space being as it were God's "boundless uniform Sensorium." If we will to move objects, we can do so only indirectly, sending messages from brain through nerves to bodily parts, which are thereby enabled to put our wills into effect. But God is "void of Organs, Members or Parts" and so acts directly. He is

> a powerful ever-living Agent, who being in all Places, is more able by his Will to move the Bodies within his boundless uniform Sensorium, and thereby to form and reform the Parts of the Universe, than we are by our Will to move the Parts of our own Bodies.[118]

Thus, by analogy with the indirect willed activity of the human mind, the Divine Mind wills directly "to form and reform the Parts of the Universe," and that is what we call gravitation.

When Newton then referred to his "active principle" of gravitation and compared it to the "active principle" of fermentation, the latter concept was quite clearly based on alchemical ideas. But it is doubtful that Newton came to treat gravity as an "active principle" simply by analogy with the "active principle" of alchemy. Though he undoubtedly had been thinking about the alchemical active principle for many years, his new conceptualization of the gravitational principle was almost certainly mediated by the rigorous distinctions he had found in Philo and in Renaissance Stoicism between active and passive principles, between Divine Mind and Matter. And it was of course the conceptualization of the active gravitational principle as Divine Mind that led Newton into a consideration of space as the *sensorium dei* – the sensory of God – and into considerable trouble with Leibniz after the publication of the *Optice* of 1706.[119]

118. Ibid., p. 403.
119. Appearing first in *Optice* (1706), Newton's initial statement on the *sensorium dei* seemed to indicate full identification between space and God's sensory; the insertion of a *tanquam* (as it were) turned identification into analogy in most copies, but Leibniz probably read one of the originals: Alexandre Koyré and I. Bernard Cohen, "The case of the missing *Tanquam*. Leibniz, Newton, and Clarke," *Isis* 52 (1961), 555–66. Leibniz

Conclusion

It is in the two and one-half decades after the initial publication of the *Principia* that the Janus faces of Isaac Newton become most fully visible to the historian; it is also the period in which Newton's commitment to the restoration of Truth becomes most apparent. Recognizing, as did almost everyone else who could understand it, that his *Principia* depicted a cosmic system rather more in line with phenomena than Descartes's, Newton nevertheless saw his achievement not as something new under the sun but rather as a step toward the restoration of the true natural philosophy of the ancients.

Furthermore, universal gravity offered him an unexpected opportunity to demonstrate direct divine activity in the created world. Until he had been forced by his own discoveries to doubt and then to reject mechanical causation for gravity, Newton seems never to have perceived the possibility that gravitational phenomena might carry evidence of God's direct and immediate relationship with the cosmos, for he generally classified mechanical laws as the sort of secondary causes that God had originally ordained to inhere in nature and which God utilized indirectly to effect His will. Somewhat reluctantly publishing his mathematical principles of gravity only, Newton continued his search for the cause of gravity, looking for hints in ancient theologians and natural philosophers, exploring the necessary conditions for Fatio's distinctive mechanical explanation. Finding that Fatio's system foundered on the lack of retardation in the

later insisted, to Newton's indignation, that Newton had given God an "organ": H. G. Alexander, Ed., *The Leibniz–Clarke Correspondence, Together with Extracts from Newton's Principia and Opticks*, with intro. and notes by H. G. Alexander (Philosophical Classics, General Ed.: Peter G. Lucas; Manchester: Manchester University Press, 1956); Alexandre Koyré and I. Bernard Cohen, "Newton & the Leibniz–Clarke correspondence with notes on Newton, Conti, & Des Maizeaux," *Archives internationales d'histoire des sciences* 15 (1962), 63–126. Leibniz's reactions are explored more fully in Howard R. Bernstein, "Leibniz and the *Sensorium Dei*," *Journal of the History of Philosophy* 15 (1977), 171–82. It has been claimed that Newton's term (*sensorium dei*) is symbolic of a metaphysics particularly fruitful for science, utilized by Aristotle, Newton, Einstein, and de Broglie: Jean Zafiropulo and Catherine Monod, *Sensorium dei dans l'hermetisme et la science* (Collection d'études anciennes publiée sous le patronage de l'Association Guillaume Budé; Paris: Société d'Édition «Les Belles Lettres», 1976). For Newton's early ideas of the human sensorium, see McGuire and Tamny, "Physiology and Hobbesian epistemology," in *Certain Philosophical Questions* (1, n. 22) pp. 216–40.

motion of heavenly bodies, just as all other mechanical hypotheses did, Newton concluded that only spirit could penetrate to the centers of bodies without causing retardation. Finding evidence that the wise ancients had recognized a universal Divine Mind that penetrated and organized all matter and that constituted the ground of all being for the cosmos, Newton came to believe that the force of gravitation rested solely in "the arbitrary will of God." Gravity was thus evidence of the *potentia dei absoluta*, and Newton wrestled with some mode of making such breathtaking divine activity comprehensible to limited human minds by utilizing the analogies of cosmic music and of infinite space as the divine sensory.

He also renewed his immense efforts to demonstrate divine activity in micromatter. If gravitational phenomena demonstrated macrocosmic divine activity, and if his own mathematical principles of natural philosophy helped restore true cosmic natural philosophy and thus led humanity closer to the restoration of true religion, then to complete his system he must also restore solid and irrefutable evidence of divine activity in the organization of the microcosm, where, to him, all the glorious variety of natural forms could only come from the ideas and will of the Divine Mind. Plunging again into the exhaustive alchemical experimentation that had been interrupted by the writing of the *Principia*, attempting again and again to decode and explicate alchemical texts – the more ancient the better – Newton struggled to find the true ancient alchemy behind the corrupted and erroneous versions that had come down to him. The divine alchemical principle, the vegetable spirit, the vital agent of life in all three kingdoms of nature – the lost knowledge of that occult principle of life could be restored also if he could but unravel the phenomena of mineral fermentation.

Some previous misconstruals of Newton's work disappear from the perspective of the present chapter. Newton's writing of the Classical Scholia during this period of his life has sometimes been considered anomalous. Why should a person who had just made the greatest *new* discoveries of the ages claim the ancients had known them all already? Clearly, Newton did not think his discoveries were really new; they were restorations of what had been lost. Similarly, Newton's publication of his mathematical principles without an accompanying statement on the *cause* of gravity has often led to anachronistic arguments for Newton's positivism. Assertions of Newton's positivism can only be made in the future by disregarding the manuscript record of his continuing search for a cause for gravity, a search that lasted the rest of his life.

In conclusion, as one reviews the first twenty-five years of Isaac Newton's post-*Principia* activities, one has the overwhelming impression of the backward-looking focus of his gaze. The moderns having failed him in the matter of a cause for gravity, he found it incumbent upon himself to master the remnants of ancient knowledge upon that subject; more recent alchemical knowledge also proving flawed, he focused more and more upon the decoding of the cryptic ancient texts. But the answers he found, or thought he found, in the ancient literature during these years were not wholly satisfactory, for they did not fully coalesce with conclusions he had drawn from other areas of study, and after 1710 he made some rather radical revisions in his views.

7

Modes of divine activity in the world: after the Principia, *1713–1727*

Introduction

The regular operation of the laws of nature was usually considered evidence of *potentia dei ordinata* in voluntarist theology. Predicated upon secondary causes that God had instituted in the beginning and that He maintained in orderly fashion because of His beneficence, such laws illuminated the ordinary concourse of nature, whereas cases in which God acted "without means" pointed to miraculously willed acts outside the ordinary concourse of nature. But nothing in that theological stance really precluded the possibility that God might will *directly* some *regular* law of nature in which He acted "without means." Indeed, in the Augustinian view of law and miracle that Newton later stated explicitly in the context of the Leibniz–Clarke debate, the only real difference between law and miracle lay in human perceptions. The "miraculous" appeared to be so to human beings only because such events "happen seldom," as will be developed later in this chapter.

If gravity was due solely to "the arbitrary will of God," as Newton had suggested to Fatio, and if the supreme and omnipresent Deity willed directly the regular operations of gravity, as Newton seems really to have concluded by the time of the publication of the *Optice* of 1706, that direct activity of God in a regular law of nature might still have been perceived by Newton as orthodox voluntarist theology. Certainly that solution to the problem of a cause for gravity constituted a resounding answer to those atheistical tendencies inherent in the mechanical philosophy detected so many years before by the Cambridge Platonists, and, under their tutelage, by Newton himself, for it caused God to be most intimately involved with all events in the natural world.

However, for an Arian theologian that solution would have generated considerable unease: the Supreme God should not be directly responsible for the moment-to-moment movements of all the particles of matter in the universe. The Supreme God does nothing by Himself that He can do by others, Newton said repeatedly. As Gwatkin pointed out with orthodox revulsion in the nineteenth century, fourth-century Arianism was strongly influenced by philosophical attitudes and methods, to the extent that, starting from an acceptable belief in the unity of God, the Arians were led unacceptably into postulates of God's absolute simplicity and absolute isolation from the created world, making of the Supreme God a Being hidden from the knowledge of man and requiring an intermediary for intercourse with the world.[1] While Newton's position may never have been as extreme as that Gwatkin delineated, Newton did repeat over and over again in his theological papers the statement that the Supreme God is He whom no eye has seen nor can see.

Invisibility would not of course preclude activity, but there is nevertheless some evidence in his theological papers that Newton searched for a possible cosmic mediator between God and the world. Whether Newton conceived this mediator as an agent for gravitation or for the alchemical activation of matter, that set of papers gives no indication, however: Newton's tendency toward a single-minded pursuit of the partial knowledge available in each of his many fields of study insured that he adhered to the terminology of the field, and so he wrote theology like a theologian and not like a natural philosopher.

The problem, as it had always been for gravity since 1684, was the issue of corporeality. No dense aether made up of ordinary matter would do since celestial bodies showed no retardation. For the alchemical agent it was possible to retain a few scattered particles of matter in the vast aethereal spaces but such scattered particles were quite incapable of yielding a material cause for gravity. Could there, on the other hand, be some other substance that would constitute a nonretarding aether? A substance intermediary between the full corporeality of body and the full incorporeality of the Supreme Deity? The obvious candidate for such a substance was the cosmic Arian Christ, and in some places Newton does indeed appear to have been searching Scripture for information on the nature of Christ's body.

At first there is evidence of Newton's frustration. In a fragmentary passage probably written in the 1690s or early 1700s and now preserved in Jerusalem, Newton had focused on the nature of Christ's body before and after his incarnation.

> ...body as he had after his resurrection such a body he had before his incarnation. And therefore as his ⟨natural, *deleted*⟩ ↗ mortal ↙

1. Henry Melvill Gwatkin, *Studies of Arianism, Chiefly Referring to the Character and Chronology of the Reaction Which Followed the Council of Nicaea* (2nd ed.; Cambridge: Deighton Bell and Co.; London: George Bell and Sons, 1900), pp. 20–8.

> body by the resurrection became an immortal body, so his immortal body by the incarnation became a mortal one. And it is as easy to beleive the one as the other.[2]

But, rallying to continue the search, Newton quoted from the beginning of the Gospel according to John, observed that "John thought it no absurdity to speak of the Word as a being visible & tangible," then recorded passages in which "Christ himself" represented "the state wch he was in," and concluded with a resounding passage that emphatically distinguished Father and Son in the matter of "incorporation":

> But the father is a pure spirit ⟨intangible, *deleted*⟩ invisible intangible & immovable, being alike in all places & incapable of incorporation. ffor he is the invisible God whom no eye hath seen nor can see.[3]

But by the last period of his life, probably in the 1710s or 1720s from the appearance of the handwriting, Newton had been able to reconcile himself to the concept of a "spiritual body," for he recorded the following similar but subtly different passage.

> And he who by his resurrection has changed his mortal flesh into an immortal spirituall body might by his incarnation change his immortal spiritual body into a body of flesh. ⟨ffor, *deleted*⟩ ffor whereas the father is the invisible God whom no eye hath seen nor can see & therefore is totally incorporeal, the son ↗ ⟨Word, *deleted*⟩ ↙ before his incarnation & the Holy Ghost have appeared in visible shapes upon several occasions & therefore had spiritual bodies.[4]

After a passage on Christ's various appearances, Newton concluded, "For nothing is tangible but body."[5]

However, as to what exactly might constitute a "spiritual body," Scripture gave no clue. It is conceivable that Newton's interest was strictly theological: such an interest is surely consonant with his Arian Christology. Christ is not of the same substance with the Father, yet, as first-created, Christ occupies a unique niche intermediate between God and the world and between God and humanity, all of which implies that Christ's substance is intermediary also between the incorporeal and the corporeal. Nonetheless, it does not seem to have been a question that concerned Newton until the post-*Principia* period, when he needed to find a physical substrate for "forces" that he had earlier explained by corporeal aethers of either a mechanical or an active sort.

Now with respect to the vegetable spirit, Newton had already decided

2. Yahuda MS Var. 1, Newton MS 15.3 (2, n. 50), f. 66v. The arrows up and down indicate Newton's interlineations.
3. Ibid.
4. Yahuda MS Var. 1, Newton MS 15.5 (2, n. 50), f. 96r.
5. Ibid., f. 96v.

in the 1680s that it should be inertially homogeneous with ordinary matter, though active, and should be condensable into ordinary matter. The demise of his cosmic gravitational aether had of course forced a reconsideration of aethereal explanations for short-range phenomena, both mechanical and vegetable, and in *De aere et aethere* he had considered several possibilities as a "cause" for short-range repulsion: a "medium" not easily compressible, "an incorporeal nature" created by God, and a "sphere of most fluid and tenuous matter" surrounding the "hard and impenetrable" nuclei of particles.

Perhaps during the *Principia* period Newton also thought seriously about the void of ancient atomism. His copy of the *De rerum natura* of Lucretius, albeit showing signs of concentrated study, could not have been purchased before 1686, its date of publication.[6] Newton was later, however, to pin his argument for interstitial spaces empty of ordinary matter to certain experimental facts: that light can be transmitted through "pellucid" bodies without being stifled and lost, as it surely would be if it hit a truly solid particle, and also that magnetic and gravitational forces could pass through even very dense and opaque bodies "without any diminution."[7] Newton called his interstitial spaces "pores," but they were not really void spaces since they were permeated by "forces," and in the end all he seems to have accepted from ancient atomism was the limited divisibility of matter. The primitive particles never wear out or break, and their permanence guarantees that nature will be "lasting," with the "Changes of corporeal Things . . . placed only in the various Separations and new Associations and Motions of these permanent Particles. . . ."[8]

The result was the matter theory of the *Opticks*, Newton's well-known hierarchical system of parts and pores arranged in three-dimensional netlike patterns, "until the Progression end in the biggest Particles on which the Operations in Chymistry, and the Colours of natural Bodies depend, and which by cohering compose Bodies of a sensible Magnitude."[9] Only at the lowest level of the hierarchies do the unbreakable "primitive" particles exist; at intermediate levels there are relatively complex subunits that carry the properties of contemporary chemical elements or principles (watery, oily, saline, earthy, etc.); at the highest and most complex level of the hierarchies there are the units of ordinary chemical substances.[10] The force holding the particles together in stable arrange-

6. Harrison, *Library* (1, n. 6), p. 183, item no. 990.
7. Newton, *Opticks* (2, n. 15), pp. 266–9.
8. Ibid., p. 400.
9. Ibid., p. 394.
10. Dobbs, *Foundations* (1, n. 1), pp. 219–21; idem, "Newton's alchemy and

ments was the "active principle" of cohesion; the force working to change particulate relationships was the "active principle" of fermentation. The "pores" were filled with these "active principles."

The question was, what sort of physical substrate might subsume these "active principles" or "forces." Cohesion and fermentation had of course originally been nonmechanical functions of the vitalistic vegetable spirit appearing in Newton's earlier papers entangled in the gravitational aether. The particles of the alchemical spirit had in the 1670s differed from those of the gravitational aether primarily in their smaller size and greater activity; all of Newton's speculative aethers then had been composed of the one universal catholic matter of mechanical philosophy that comprised all gross bodies as well. Indeed, Newton already thought then that aethereal substances and ordinary bodies were interconvertible. Even after he eliminated the gravitational aether in the 1680s, there was still some sort of residual cosmic aether in Newton's system though it was no longer dense. Those widely scattered residual particles, though they undoubtedly had a cosmic function, one to be considered later, did not seem likely candidates for the "active principles" of cohesion and fermentation that operated interstitially. What then were Newton's remaining options for the "active principles" in micromatter?

The remaining options for the *cosmic* "active principle" of gravitation were, as Newton had delineated them, an "incorporeal medium" or "spirits emitted." For gravity he had chosen the "incorporeal medium" of the omnipresent Deity and had striven mightily to sustain that choice with the wisdom of antiquity, but there is reason to suppose that, for the theological reason just noted, he was not fully satisfied with that choice even for gravity. In the case of the "active principles" of micromatter, there is little evidence that Newton seriously considered the option of an "incorporeal medium," though he did of course toy with the idea of an "incorporeal nature" for repulsion in *De aere et aethere*.

It was rather the remaining option of "spirits emitted" that accorded rather well both with alchemical literature and with the generalized medical and chemical doctrines of the seventeenth century that were permeated with the tenuous shaping spirits of the Stoic spermatic logoi. As one saw in Chapter 5, Newton summarized the alchemical position in the 1690s, calling the active agent both "Magnesia" and "quintessence": the agent is a "corporeal spirit," a "spiritual body," and "the condensed spirit of the world." Not only had the alchemical literature suggested a concept to him, but the alchemical concept must also have seemed con-

his theory of matter" (1, n. 5); idem, "Conceptual problems in Newton's early chemistry" (1, n. 20) and the literature cited there.

gruent to him with the notions of Christ's "spiritual body" that he had gleaned from Scripture. One must also recall that in the 1670s Newton had thought that the vegetable spirit might be the "body of light," and that behind both Christian and alchemical doctrines on light there stretched a lengthy metaphysical tradition of light as both symbol and agent of divinity.

Gravity, alchemy, and electricity: the second solution

With this complex of ideas in his mind, and no doubt driven by subtle metaphysical considerations of light as a divine agent, Newton participated in and observed experimentation on light and electricity before the Royal Society between 1703 and 1708.[11] The interactions of light and matter were the focus of the queries at the end of the first edition of the *Opticks* in 1704; such interactions were also the focus of Newton's experiments with concave mirrors in 1704, experiments that led eventually to revised and new queries on the subject. Remaining convinced that light particles were intrinsically active, partly at least because of their very small size, and that they contributed activity to bodies, Newton demonstrated to his own satisfaction that bodies and light are mutually convertible: "The changing of Bodies into Light, and Light into Bodies, is very conformable to the Course of Nature, which seems delighted with Transmutations."[12] The great cyclical interchange between light and the grosser forms of matter envisioned by Newton in his final Query 30 echoed his earlier idea that the alchemical agent might be the "body of light." Light could be condensed into body: he had demonstrated it, he thought, by driving light particles into various materials by the use of his seven concave mirrors with coincident foci. Conversely, all bodies could be caused to emit light by the application of sufficient heat. These were hardly the gentle transmutations of true vegetation, which required a much milder warmth and illumination, but the "changing of Bodies

11. J. L. Heilbron, *Electricity in the 17th and 18th Centuries. A Study of Early Modern Physics* (Berkeley: University of California Press, 1979), pp. 229–36; idem, *Physics at the Royal Society during Newton's Presidency* (Los Angeles: William Andrews Clark Memorial Library, University of California, Los Angeles, 1983), pp. 52–66.
12. Newton, *Opticks* (2, n. 15), pp. 374–5; Heilbron, *Physics at the Royal Society* (7, n. 11), pp. 52–8; McMullin, *Newton on Matter and Activity* (2, n. 61), pp. 84–94.

into Light, and Light into Bodies" was there for the Royal Society and all the world to see.

> For all fix'd Bodies being heated emit Light so long as they continue sufficiently hot, and Light mutually stops in Bodies as often as its Rays strike upon their Parts, as we shew'd above.[13]

What did Newton think he had demonstrated, if anything, beyond the obvious? Since light is the most active of all bodies known to us, and enters into the composition of other bodies, why may light not be "the chief principle of activity" in them, he asked in a suppressed draft for the *Optice* of 1706, and in the published *Optice* asked whether bodies might not receive their active power from the light that entered into their composition. But in the *Opticks* of 1717/18 he equivocated with the suggestion that bodies receive "much" of their activity from particles of light.[14] The "body of light" might constitute a divine agent, and be intimately involved in the activation of matter, but perhaps even so it did not subsume all the operations of the vegetable spirit. Perhaps an aethereal medium – perhaps even an active one – was required to mediate between the particles of light and the grosser particles of ordinary matter. Evidently what caused Newton to revise his formulations on light as the sole or chief source of activity in matter, beginning with Hauksbee's work on the cold light from "phosphors" in 1703, was the connection made between light and electricity.

Newton himself had done some electrostatic experiments years before, reported to the Royal Society in his "Hypothesis" of 1675. The "electric effluvia seem to instruct us," he then said, "that there is something of an aethereall Nature condens'd in bodies." Rarified by friction into "an aethereall wind," the electric effluvium could move light objects about in "odd motions" and by condensing again into the body "cause electricall attraction." He conceived this substance as a vibrating medium,

13. Newton, *Opticks* (2, n. 15), p. 374.
14. McMullin, *Newton on Matter and Activity* (2, n. 61), pp. 85–6; Isaac Newton, *Optice: sive de Reflexionibus, Refractionibus, Inflexionibus & Coloribus Lucis libri tres. Authore Isaaco Newton, Equite Aurato. Latine reddidit Samuel Clarke, A. M. Reverendo admodum Patri ac D^no^ Joanni Moore Episcopo Nowicensi a Sacris Domesticis. Accedunt Tractatus duo ejusdem Authoris de Speciebus & Magnitudine Figurarum Curvilinearum, Latine scripti* (Londini: Impensis Sam. Smith & Benj. Walford, Regiae Societatis Typograph. ad Insignia Principis in Coemeterio D. Pauli, 1706), p. 319: "& Annon fieri potest, ut Corpora vim suam actuosam accipiant a particulis Luminis, quae in eis componendis insunt?"; Newton, *Opticks* (2, n. 15), p. 374.

the vibrations of which were responsible for the reflection and refraction of light and

> the cheif meanes, by wch the parts of fermenting or putrifying Substances, fluid Liquors, or melted burning or other hott bodyes continue in motion, are shaken asunder like a Ship by waves, & dissipated into vapours, exhalations, or Smoake, & Light loosed or excited in those bodyes....[15]

A speculative connection between light and electricity was clearly already present in Newton's mind in 1675, but lacking then was experimental evidence of the connection, for the principal electrical phenomena Newton reported to Oldenburg were those of the "odd motions" of light bodies, nothing at all in 1675 about the generation of light by electrical activity. Presumably those early ideas on electrical effluvia had been shaken by the events of the 1680s in any case.

Newton refused public speculation about an interstitial aethereal medium, electrical or otherwise, in 1704 and 1706. Queries 1–16 in the first two editions of his optical work treat the interaction of light and matter sui generis, without the mediation of any other substance that might be hidden in the pores of matter.[16] He even *seemed* to be suggesting that body acts upon light at a distance.

> Query 1. Do not Bodies act upon Light at a distance, and by their action bend its rays, and is not this action (*caeteris paribus*) strongest at the least distance?[17]

Similarly, in Query 4, where he asked whether the rays of light falling on bodies do not begin to bend "before they arrive at the Bodies."[18] All

15. Newton, *Correspondence* (2, n. 14), vol. I, 364–6 (Newton to Oldenburg, 7 Dec. 1675). For the immediate contemporary context of Newton's association of the emission of light with fermentation and putrefaction, see E. Newton Harvey, *A History of Luminescence from the Earliest Times until 1900* (Memoirs of the American Philosophical Society Held at Philadelphia for Promoting Useful Knowledge, Vol. 44; Philadelphia: The American Philosophical Society, 1957), pp. 127–34.
16. Isaac Newton, *Opticks: or, A Treatise of the Reflexions, Refractions, Inflections and Colours of Light. Also Two Treatises of the Species and Magnitude of Curvilinear Figures* (London: printed for Sam. Smith, and Benj. Walford, printers to the Royal Society, at the Prince's Arms in St. Paul's Church-yard, 1704), pp. 132–7; Newton, *Optice* (7, n. 14), pp. 293–9.
17. Newton, *Opticks* (1704) (7, n. 16), p. 132; idem, *Optice* (1706) (7, n. 14), p. 293.
18. Idem, *Opticks* (1704) (7, n. 16), p. 133; idem, *Optice* (1706) (7, n. 14), pp. 293–4.

of these earliest queries survived from the first edition through all other editions without any modification of their essential meaning even though Newton added more examples and illustrations later to Queries 8, 10, 11, and 16. How is one to explain that fact, since Newton later added yet other queries that seem quite at variance with the first sixteen?

It is not impossible that Newton's language when he composed those first queries was deliberately designed to conceal his private speculations on the presence of the vegetable spirit in matter. Query 4 is particularly suggestive in that respect, where Newton asked whether the rays of light are not "reflected, refracted and inflected by one and the same Principle, acting variously in various circumstances?"[19] Clarke's Latin is even more suggestive, as the word *vis* (force) translates "Principle."[20] One is reminded of Newton's early description of the alchemical agent: it is a single principle but "the particularities of its method are many, according to the nature of the subject in which it operates." It was the same in all things, the "vital agent diffused through all things... most subtle and wholly volatile" that was responsible for shaping common matter into all its glorious variety. In 1704–06 just such a "Principle" or "force" was "acting variously in various circumstances." "Principle" and "force" were at that time quite probably public euphemisms for the vegetable spirit of the private papers.

One must take into account also Newton's well-known letter to Bentley in 169⅔.

> Tis unconceivable that inanimate brute matter should (without ye mediation of something else wch is not material) operate upon & affect other matter without mutual contact....[21]

The context of the discussion was the force of gravity in Newton's remarks to Bentley, but the statement he made there denying action at a distance was perfectly general: he found it "so great an absurdity" that he believed "no man who has in philosophical matters any competent faculty of thinking can ever fall into it."[22] When Newton raised questions then respecting the apparent effects bodies had on light rays before actual contact was made, it would seem that he was utilizing the queries as a device for suggesting an agent present in all bodies that mediated the interactions between light and matter. To

19. Idem, *Opticks* (1704) (7, n. 16), p. 133.
20. Idem, *Optice* (1706) (7, n. 14), p. 294: "Et Reflectuntur, Refringuntur, atq; Inflectuntur, una eademq; Vi, varie se in variis circumstantiis exerente?"
21. Idem, *Correspondence* (2, n. 14), vol. III, 253–6 (Newton to Bentley, 25 Feb. 169⅔), quotation from pp. 253–4.
22. Ibid., p. 254.

speak of bodies acting where they were not, should, for those with a "competent faculty of thinking," immediately imply the presence of a mediating agent.

The evidence he needed to state his idea explicitly came from Hauksbee's experiments before the Royal Society that were the outgrowth of the work on cold light from "phosphors" that began in 1703.[23] Working his way through the naturally glowing element phosphorus and the "mercurial phosphorus" (i.e., the element mercury that will glow when agitated in a vacuum), Hauksbee discovered that many different materials will glow when rubbed together in a vacuum. By late 1706 Hauksbee had found that he could elicit a light within an evacuated glass globe simply by spinning the globe against his hand.

Newton had obtained the electrical effects he reported in 1675 by rubbing a glass lens, and he rapidly identified the source of Hauksbee's new luminosity as electrical in nature. Perhaps the identification at first was simply based on the experimental fountain of electricity (rubbed glass) that was common both to his earlier "odd motions" of light objects and to Hauksbee's luminosity, but Hauksbee's continued experimentation soon provided additional evidence. Though Hauksbee preferred to think two types of effluvia (electric and luminous) were involved, Newton believed a connection between light and electricity had been firmly established.

Newton's new conviction on that point had not been formulated in time to influence the *Optice* of 1706,[24] but it did soon appear in other papers, both private and public. Scholarly explorations of this important episode appear with some regularity, the most recent recognizing that the many drafts and published formulations that Newton produced between 1707 and 1717 indicate a mind greatly stimulated and enthusiastic, by no means in decline even though moving into its eighth decade of activity.[25]

23. Henry Guerlac, "Francis Hauksbee: Expérimentateur au Profit de Newton," *Archives Internationales d'Histoire des Sciences* 16 (1963), 1113–28; idem, "Newton's optical aether: his draft of a proposed addition to his *Opticks*," *Notes and Records of the Royal Society of London* 22 (1967), 45–57 [the two Guerlac essays reprinted in Guerlac, *Essays and Papers* (2, n. 41), pp. 107–19 and 120–30, respectively]; Heilbron, *Electricity* (7, n. 11), pp. 229–41; idem, *Physics at the Royal Society* (7, n. 11), pp. 54–66.
24. Cf. Guerlac, "Francis Hauksbee" (7, n. 23) for the details of Hauksbee's experimentation that did enter the *Optice* of 1706.
25. R. W. Home, "Force, electricity, and the powers of living matter in Newton's mature philosophy of nature," in *Religion, Science, and Worldview* (1, n. 20), pp. 95–117. See also, idem, "Newton on electricity and the aether,"

The products of that mind came in what one may describe as two waves. In the first, there was the recognition that the electrical force might well be one of the fundamental secrets of the alchemists. In the second, there was the realization that what applied to micromatter might be extrapolated to apply to the cosmos as well.

In the early stages of his reaction to the new information Newton saw the electrical material as the mediator between light and gross matter. As in 1675, he used terminology of aethers, spirits, winds, and effluvia: the electrical substance is an "aetherial spirit or spiritual effluvium"; light objects are "carried about by the Etherial effluvium as with a wind."[26] "Certainly there is some spirit hid in all bodies, by means of which light and bodies act upon each other mutually."[27] "This spirit is . . . the cause of electrical attraction and not only reflects, refracts and inflects light but also emits it. . . ."[28] It is elastic and is capable of being rarified and expanded by friction so as to diffuse into the space around the rubbed object, and even without rubbing may reach to "small distances." It is by the means of this electric spirit that bodies *seem* to act upon light at a distance, but really "electric bodies could not act at a distance without a spirit reaching to that distance."[29] Whereas in 1706 Newton had added

in *Contemporary Newtonian Research* (1, n. 10), pp. 191–213; Joan L. Hawes, "Newton and the 'Electrical Attraction Unexcited,' " *Annals of Science* 24 (1968), 121–30; idem, "Newton's revival of the aether hypothesis and the explanation of gravitational attraction," *Notes and Records of the Royal Society of London* 23 (1968), 200–12; idem, "Newton's two electricities," *Annals of Science* 27 (1971), 95–103; McGuire, "Force, active principles, and Newton's invisible realm" (4, n. 9); Alexandre Koyré and I. Bernard Cohen, "Newton's 'Electric and Elastic Spirit,' " *Isis* 51 (1960), 337; Marie Boas Hall and A. Rupert Hall, "Newton's electric spirit: four oddities," *Isis* 50 (1959), 473–6; Henry Guerlac, "Sir Isaac and the ingenious Mr. Hauksbee," in *Mélanges Alexandre Koyré, publiés à l'occasion de son soixante-dixième anniversaire*, ed. by I. B. Cohen and R. Taton (2 vols.; Paris: Hermann, 1964), I, 228–54; idem, "Francis Hauksbee" (7, n. 23); idem, "Newton's optical aether" (7, n. 23); Heilbron, *Electricity* (7, n. 11); idem, *Physics at the Royal Society* (7, n. 11); Westfall, *Never at Rest* (1, n. 4).

26. Portsmouth Collection MS Add. 3970 (6, n. 93), no. 9, f. 626r, v, quoted in Guerlac, "Newton's optical aether" (7, n. 23), pp. 122–3 of Guerlac, *Essays and Papers* (2, n. 41).
27. Isaac Newton, *De vi electrica*, Portsmouth Collection MS Add. 3970 (6, n. 93), ff. 427–8, published and translated in Newton, *Correspondence* (2, n. 14), vol. V, 362–9, quotation from pp. 362 and 365.
28. Ibid., pp. 363 and 366.
29. Isaac Newton, *Opticks: or, a Treatise of the Reflections, Refractions, In-*

a new query to the *Optice* discussing the movement of rays of light in proximity to glass and other pellucid substances, using the terminology of vacua and action at a distance,[30] to the 1717/18 edition of *Opticks* he added a paragraph to the end of that discussion saying that what he meant there by vacuum and the attractions of light toward glass or crystal was to be understood in terms of three of the "aether" queries also added in 1717/18, where the interactions of heat and light with the "AEthereal Medium" were discussed.[31]

The new electrical medium was active, universal, and responsible for cohesion, attraction, and repulsion as well as the interactions of light and matter: "the electric spirit is a medium most active and emits light;"[32] it is "sufficiently active to emit light at a distance from the gross body if it be there put into a trembling agitation. . . ."[33] It is "a certain most subtle spirit which pervades and lies hid in all gross bodies;"

> by the force and action of which spirit the particles of bodies attract one another at near distances, and cohere, if contiguous; and electric bodies operate to greater distances, as well repelling as attracting the neighboring corpuscles; and light is emitted, reflected, inflected, and heats bodies. . . .[34]

Was it also responsible for vegetation? Yes, at least in part. "This spirit may be also of great use in vegetation, wherein three things are to be

flections and Colours of Light. The Second Edition, with Additions. By Sir Isaac Newton, Knt. (London: printed by W. Bowyer for W. Innys at the Prince's Arms in St. Paul's Church-Yard, 1717; London: printed for W. and J. Innys, printers to the Royal Society, at the Prince's Arms in St. Paul's Church-Yard, 1718), pp. 350–1; idem, *Opticks* (2, n. 15), pp. 375–6; Portsmouth Collection MS Add. 3970 (6, n. 93), f. 241r,v, quoted in Home, "Force, electricity, and living matter" (7, n. 25), pp. 102–3. See also Hawes, "Newton and the 'Electrical Attraction Unexcited' " (7, n. 25).

30. Newton, *Optice* (1706) (7, n. 14), Query 21, pp. 315–19.
31. Idem, *Opticks* (1717–18) (7, n. 29), Query 29, pp. 345–9, esp. 349; idem, *Opticks* (2, n. 15), Query 29, pp. 370–4, esp. 374. Newton's reference was to the new Queries 18, 19, and 20 (pp. 323–4 in the 1717–18 ed., pp. 348–50 in the modern ed.).
32. Isaac Newton, "Draft of the General Scholium for the Principia, MS C," University Library, Cambridge, Portsmouth Collection MS Add. 3965, ff. 361–2, published in Newton, *Unpublished Papers* (2, n. 51), pp. 355–64, quotation from pp. 357 and 362.
33. Portsmouth Collection MS Add. 3970 (6, n. 93), f. 626r, quoted in Guerlac, "Newton's optical aether" (7, n. 25), in *Essays and Papers* (2, n. 41), p. 122.
34. Newton, *Principia* (Motte–Cajori: 2, n. 13), vol. II, 547; (Koyré–Cohen: 1, n. 9), vol. II, 764–5.

considered, generation, nutrition, & praeparation of nourishment."[35] Or again, in a passage that echoes the classic rationale behind the Stoic concept of *pneuma*:

> And for faciliating this assimilation of ye nourishment & preserving the nourished bodies from corruption it may be presumed that as electric attraction is excited by friction so it may be invigorated also by some other causes & particularly by some agitation caused in the electric spirit by the vegetable life of the particles of living substances: & the ceasing of this vigour upon death may be the reason why ye death of animals is accompanied by putrefaction.[36]

Active in itself, the electric spirit further "invigorated" by "other causes" might be an essential element in the life of living substances, of use in generation, in the preparation and assimilation of nourishment, and in the preservation of the living form of the organism. The loss of its vigorous participation would lead to disintegration and "putrefaction."

Was electricity then equivalent to the vegetable spirit for which Newton had searched so long? No, not quite equivalent, but certainly related – a natural active substance that facilitated the activity of light with respect to gross matter and that facilitated the internal activities necessary for a living creature to maintain life. Newton's new appreciation of the electric spirit as a source of activity in matter was undoubtedly the root of his equivocation in his final Query 30 when he suggested that bodies received much but not all of their activity from light.[37] Light kept its function as outside activator, but the electric spirit functioned inside of bodies to organize and "regulate" internal processes.

> Do not all bodies abound with a very subtil active vibrating spirit by w^ch^ . . . the small particles of bodies cohaere when continguous, agitate one another at small distances & regulate almost all their motions amongst themselves as the great bodies of the Universe regulate theirs by the power of gravity?[38]

35. Portsmouth Collection MS Add. 3970 (6, n. 93), f. 235r, quoted in Hawes, "Newton's two electricities" (7, n. 25), p. 96.
36. Portsmouth Collection MS Add. 3970 (6, n. 93), f. 235v, quoted in Home, "Force, electricity, and living matter" (7, n. 25), p. 115. See also a similar passage from f. 241r of the same manuscript, quoted by Home in ibid., p. 111, and a Latin version from Newton's *De vita & morte vegetabili*, f. 237r,v of the same manuscript, quoted in Hawes, "Newton's two electricities" (7, n. 25), p. 98.
37. Newton, *Opticks* (1717–18) (7, n. 29), p. 349; idem, *Opticks* (2, n. 15), p. 374.
38. Portsmouth Collection MS Add. 3970 (6, n. 93), f. 241r,v, quoted in Home, "Force, electricity, and living matter" (7, n. 25), pp. 102–3.

The agent responsible for the regulation of the motions of small bodies "amongst themselves" was indeed what Newton had been looking for, the natural agent that organized the small parts of matter according to the ideas and will of their omnipotent Creator. The reality had turned out to be more complex than the alchemists had led him to believe, however, for both light and electricity were involved, and perhaps something else as well, "some other causes" that invigorated the electric spirit in living bodies.

On the other hand, the electric spirit came as near to being a "spiritual body" as anything one might hope to find in this world. It was tangible, one of the requirements of "body," for Newton had touched it. When "Mr Hawsksby before y^e^e R. Society" rubbed a cylindrical rod of glass or of hard wax, the body emitted "an electric spirit or vapour wch pushes against the hand or face so as to be felt...."[39] But it was also active, as one has just seen, and that was one of the requirements of "spirit." It was, in contrast to grosser "body," "so rare and subtile" that its emission caused "no sensible Diminution of the weight of the electrick Body" even when the electric spirit had expanded through a sphere of over two feet in diameter around its source.[40] All in all, the electric and elastic spirit was a very satisfactory active natural entity, but in the 1713 General Scholium, Newton ended by saying that we do not have enough experiments to demonstrate the laws by which it operates.[41]

In the second wave of Newton's reaction to the new information, the elasticity of the electric spirit seems to have been the critical factor. Omitting almost all reference to electrical phenomena, Newton spoke again in the terminology of aethereal media. In the new queries of the 1717/18 *Opticks* the topic was introduced to the public first as the "refracting or reflecting Medium" of Query 17, operating in optical phenomena in ways analogous to water waves.[42] In Query 18 its utility was

39. Ibid.
40. Newton, *Opticks* (1717–18) (7, n. 29), p. 327; idem, *Opticks* (2, n. 15), p. 353.
41. Newton, *Principia* (Motte–Cajori: 2, n. 13), vol. II, 547; (Koyré–Cohen: 1, n. 9), vol. II, 765. Although Motte's addition of "electric and elastic" to the description of the "spirit" in his translation is not justified by the Latin text of either the 1713 or the 1726 editions of the *Principia*, there is ample justification for it elsewhere. See the critical apparatus for the passage in the Koyré–Cohen edition cited here, also Cohen, *Introduction* (1, n. 9), pp. 26–7.
42. Newton, *Opticks* (1717–18) (7, n. 29), pp. 322–3, quotation from p. 322; idem, *Opticks* (2, n. 15), pp. 347–8, quotation from p. 348. See also Guerlac, "Newton's optical aether" (7, n. 23).

expanded to account for the transmission of heat in a vacuum, for the communication of heat from light particles to bodies, for the intensity and duration of heat within hot bodies, and for the communication of heat from hot to contiguous cold bodies.[43]

So far, clearly, Newton had confined the usefulness of the new medium to short-range phenomena, but at the end of Query 18 he suddenly extrapolated it to the heavens on the basis of its strong elasticity.

> And is not this Medium exceedingly more rare and subtile than the Air, and exceedingly more elastick and active? And doth it not readily pervade all Bodies? And is it not (by its elastick force) expanded through all the Heavens?[44]

Newton's new cosmic "Medium" of Query 18 is related to but not identical with the electric and elastic spirit of the General Scholium. Even though the new aethereal medium thus introduced in the last *Opticks* had properties based on those of the electric and elastic spirit, and was utilized by Newton to explain phenomena that he had explained by the electric spirit only a few years earlier, in some ways this new universal medium is quite different from the electric spirit, as Home has observed.[45]

The remaining "aether queries" of 1717/18 show mixed purposes. In Queries 23 and 24 Newton filled the "Capillamenta" of the nerves with this medium and argued that its vibrations carry both sensory and motor excitations,[46] but Queries 19–22 are devoted to establishing its cosmic qualities and actions.

Queries 19 and 20 are based on the reactions of light in the vicinity of bodies and are an effort on Newton's part to establish the notion that the medium stands rarer inside gross bodies and then grows more and more dense in the "free and open Spaces" outside of them.[47] Then the longer Query 21 introduced the possibility that those density gradients in the aether in the spaces between the sun, stars, planets, and comets "cause the gravity of those great Bodies towards one another" in a manner reminiscent of the aether gradients in the letter to Boyle of 167⁸⁄₉, although in 1717/18 the exceedingly great elastic force of the medium seemed to Newton to be necessarily involved as well, and he offered a

43. Newton, *Opticks* (1717–18) (7, n. 29), pp. 323–4; idem, *Opticks* (2, n. 15), pp. 348–9.
44. Newton, *Opticks* (1717–18) (7, n. 29), p. 324; idem, *Opticks* (2, n. 15), p. 349.
45. Home, "Newton on electricity and the aether" (7, n. 25).
46. Newton, *Opticks* (1717–18) (7, n. 29), p. 328; idem, *Opticks* (2, n. 15), pp. 353–4.
47. Idem, *Opticks* (1717–18) (7, n. 29), p. 324; idem, *Opticks* (2, n. 15), pp. 349–50.

calculation to demonstrate that the aether must be $4.9\ 10^{11}$ more elastic than air in proportion to their respective densities.[48] He concluded Query 21 with the argument that the particles of the aether are smaller even than the particles of light, which smallness contributes to the greatness of the force by which they recede from one another and makes them less able to resist the motion of projectiles; that led him finally in Query 22 into the crucial question of the resistance – or lack of it – experienced by the planets and comets in their motion through this aethereal medium. In Query 22 Newton very carefully distinguished the new aether from the old dense forms, argued that the resistance was so small as to be "inconsiderable," so small that planets would suffer no "sensible alteration" in their motions in 10,000 years. As comparable examples of other improbably rare media he offered the air of the upper parts of our atmosphere, electric exhalations, and magnetic effluvia, and in the last two examples insisted also upon their extreme potency, even though those media were in a "rare and subtile" state.[49]

Newton's final aethereal suggestions for a cause of gravity in those queries have long puzzled scholars on several levels. Newton inserted them into the *Opticks* without removing any of the other ideas that were predicated upon the incorporeal Divine Mind and His "boundless uniform Sensorium," so inconsistencies appear from query to query and, inevitably, questions are raised as to what Newton "really" meant and why he did such a peculiar thing as to create this last aether. Then, too, the last aether itself is a strange hybrid. Since it is not a dense aether acting mechanically by impact, Newton could hardly have hoped that it would satisfy the mechanical philosophers who had objected to his reintroduction of occult principles in the *Principia*. In fact, Newton insisted so strongly upon the great activity of his new medium that this aether itself became in effect one of the "active principles" to which mechanical philosophers objected. Finally, there has been confusion about the relationship between Newton's last active aether of 1717/18 and the electric and elastic spirit of the General Scholium that first appeared in 1713 but that persisted in the 1726 edition of the *Principia*. Newton kept both of them in place in his major publications – both the active aether and the electric spirit – and since sometimes their functions overlapped or were even identical, naturally the two have often been conflated.

From the perspective of this book some of these difficulties disappear.

48. Idem, *Opticks* (1717–18) (7, n. 29), pp. 325–7, quotation from p. 325; idem, *Opticks* (2, n. 15), pp. 350–2 quotation from p. 350.
49. Idem, *Opticks* (1717–18) (7, n. 29), p. 327; idem, *Opticks* (2, n. 15), pp. 352–3.

The electric and elastic spirit came first in time, in response to experimental evidence. Newton understood it to be a new source of activity in matter that worked in conjunction with light. In private papers he assigned it some of the functions of the vegetable spirit for which he had searched so long and in the General Scholium he assigned it broad but more exoteric functions. It was, however, microcosmic, pervading the pores of gross bodies and, when excited by friction, operating short distances away from its source. It never became cosmic.

It did, on the other hand, provide Newton with a model of the "spiritual body" for which he had also been searching. It was body but it was also spirit, and above all it had the property of elasticity. By extrapolating from the properties of the electric and elastic spirit Newton created a cosmic aether that was vigorously elastic, so elastic that it filled the heavens by the efforts its particles made to recede from one another. Its particles were exceedingly small, even smaller than those of light, and therefore exceedingly active. With its rarity, activity, and greater elasticity, the new aethereal medium was even more of a spiritual body than was the electric spirit. Its primary new function was cosmic. Even though it existed within gross bodies and there absorbed some of the functions Newton had earlier assigned to the electric spirit, it did not absorb all of them; on the other hand, it did become responsible for gravity which the electric spirit never did. The active aether thus was conceptually related to the electric spirit but distinct from it.

Since the new aether was active and not mechanical in its operation, it seems most unlikely that Newton even intended it as a concession to the orthodox mechanists. Quite the contrary, its creation was much more likely a response to his own theological anxieties regarding the immense direct involvement with matter he had postulated for the Supreme Deity. Since Newton thought that the Supreme God does nothing by Himself that He can do by others, an intermediary agent for gravity would have seemed to him to be a more appropriate solution. By definition, it was an agent closely assimilated to divinity, by virtue of its inherent activity, but to an Arian theologian it was more acceptable simply because it was an *agent*, a created being that acted for the Supreme God to put His will into effect in the moment-to-moment supervision of matter. But Newton's postulation of this new cosmic aether to manage gravitational affairs in no way precluded the literal omnipresence of God. Since Newton thought God to be present where body is, as well as where body is not, His cosmic presence was assured no matter how many aethers filled the heavens. Therefore, to Newton, in contrast to modern readers, the continued omnipresence of the Divine Mind as the ground of all being, and His continued intimate awareness of all matter in His boundless uniform sensorium, constituted no inconsistency with the new semimaterial spiritual body as a gravitational agent.

Comets, miracles, and compasses

In a stray draft probably associated with the Leibniz–Clarke correspondence of 1715–16, Newton explicitly formulated his ideas about miracles, occult qualities, and laws of nature.

> For Miracles are so called not because they are the works of God but because they happen seldom & for that reason create wonder. If they should happen constantly according to certaine laws imprest upon the nature of things, they would be no longer wonders ↗ or miracles ↙ but might be considered in Philosophy as a part of the Phenomena of Nature notwithstanding that the cause of their causes might be unknown to us. And Occult qualities have been exploded not because their causes are unknown to us but because by giving this name to the specific qualities of things, a stop has been put to all enquiry into the causes ↗ of their qualities ↙ as if they could not be known because ↗ the great Philosopher ↙ Aristotle ↗ was not able to find them. ↙ Mr Leibnitz ↗ alters & ↙ enlarges the signification of the words ↗ Miracles & occult Qualities ↙ that he may make use of them in exploding (I do not say confuting) the Philosophy of Mr Newton so far as it relates to the argument for proving a Deity from the Phenomena of Nature. And at the same time he is propounding Hypotheses (not Quaeres to be examined by experiments but praecarious opinions to [be] believed without any proof) which turn Philosophy into a Romance.[50]

Newton was exceedingly annoyed with Leibniz's caustic remarks on the force of gravity as an occult quality and a perpetual miracle and was

50. Lehigh University Libraries, Bethlehem, PA, Isaac Newton, "MS on Miracles," verso (of its single page); all indication of Newton's deletions omitted, the "be" in square brackets inserted for sense, and a terminal period substituted for Newton's (unopened) terminal square bracket. Arrows up and down indicate Newton's interlineations. The recto and part of the verso of this sheet are concerned with Catherine Barton's inheritance from Lord Halifax. The first part of the quotation given here was also cited in Henry Guerlac and M. C. Jacob, "Bentley, Newton, and providence (The Boyle lectures once more)," *Journal of the History of Ideas* 30 (1969), 307–18, on p. 309; closely related material occurs in Newton, *Correspondence* (2, n. 14), vol. VI, 285–90 (Newton to Conti, 26 Feb. 1716), esp. pp. 285–6.

drafting a response for his own use or perhaps for Clarke's use in one of his (Clarke's) replies to Leibniz.[51]

Newton's sense of outrage against Leibniz, and especially against what he took to be a willful misrepresentation by Leibniz of his (Newton's) position, has often been explored,[52] but one may perhaps glean some new insights by a further examination of Newton's ideas on miracles: "Miracles are so called not because they are the works of God but because they happen seldom," essentially an Augustinian declaration and one that attempted to avoid contemporary quarrels about ordinary and extraordinary providence.

Augustine's views, partly indebted to Philo and the Stoics but also thoroughly biblical, were that ultimately there was only one miracle, creation itself, and that within that original creation God had implanted "seminal reasons" that held all the possibilities for the future. Like the spermatic logoi of the Stoics, Augustine's seminal reasons were hidden within the natural world; when unusual events occurred that seemed miraculous to human beings, those events were really only the working out of the hidden natural causes and were never *contra naturam*, against nature. All events were both "natural" and "miraculous" in this sacramental view of the whole order of creation. The events that happen regularly are "daily miracles," but people become accustomed to them, no longer giving them reverence as manifestations of God's power, so God also allows for unusual events to be drawn out from the *seminales rationes* that provoke wonder, amazement, and awe at the mysterious creative power of God.[53] As did Newton, Augustine thought that both natural law and "Miracles... so called" are the works of God. When events occur that are apparently miraculous, Augustine argued, they are

51. Clarke responded to the issue on miracles in his second, third, fourth, and fifth replies: cf. *Leibniz–Clarke Correspondence* (6, n. 119), pp. 23–4, 35, 53, 114–15.
52. See especially, Alexander, "Introduction," in ibid., pp. ix–lv; Koyré and Cohen, "Newton and the Leibniz–Clarke correspondence" (6, n. 119), and A. Rupert Hall, *Philosophers at War. The Quarrel between Newton and Leibniz* (Cambridge University Press, 1980), pp. 146–67, 183–6, 218–23.
53. Benedicta Ward, *Miracles and the Medieval Mind: Theory, Record, and Event, 1000–1215* (Philadelphia: University of Pennsylvania Press, 1982), pp. 3–6; Robert M. Grant, *Miracle and Natural Law in Graeco-Roman and Early Christian Thought* (Amsterdam: North-Holland Publishing Company, 1952), pp. 27–8, 217–19, 244–5, 263; Reijer Hooykaas, *Natural Law and Divine Miracle. The Principle of Uniformity in Geology, Biology and Theology* (Second Impression; Leiden: E. J. Brill, 1963), pp. 170–1, 206–26.

so only for us, "who have a limited knowledge of nature," but they are not so for God nor are they "against nature." The seminal reasons that God placed in the world when He first created all things have a "double potentiality" and "were created to exercise their causality in either one way or the other":

> by providing for the ordinary development . . . in appropriate periods of time, or by providing for the rare occurrence of a miraculous production . . . , in accordance with what God wills as proper for the occasion.[54]

Augustine's views were enormously influential, but, as Grant has pointed out, Augustine wrote at a time when ancient science and indeed ancient civilization were nearing their nadir, and Augustine's sense of natural law was not strong.[55] In addition, Augustine was really not interested in natural law as such; he was instead fundamentally concerned with the encounter between God and humanity, not with the earthly stage of that encounter. That gave to Augustine's ideas an emphasis upon the psychological impact of "miracle" in creating a sense of wonder and an awareness in human beings of God's activity, and it allowed a wide scope for the definition of events as miraculous.[56]

Although Augustine's views must be held partly responsible for the subsequent medieval readiness to see *mirabilia* and "signs" nearly everywhere, later theologians began to restrict and differentiate. Augustine's approach led to an emphasis on the unusual, but that emphasis was gradually eroded even as Augustine's conviction that God was ultimately responsible for everything was maintained. As early as the eleventh century Anselm of Canterbury distinguished three categories of events: the miraculous, the natural, and the voluntary. Voluntary events happen by the will of a creature, natural ones by powers God has given to nature, miraculous ones by the will of God alone.[57]

The distinction between natural and miraculous events was precisely what Newton wanted to avoid by returning to a more Augustinian position. The centuries between Anselm and Newton had seen more and more events described as natural and the concept of the miraculous become more and more restricted.[58] If secondary causes could be found

54. Augustine, *The Literal Meaning of Genesis* (3, n. 19), VI, 13–14, quotations from VI, 13.24 and VI, 14.25.
55. Grant, *Miracle and Natural Law* (7, n. 53), pp. 218–20.
56. Ibid., p. 118; Ward, *Miracles* (7, n. 53), p. 4.
57. Ibid., pp. 4–9.
58. Paul H. Kocher, *Science and Religion in Elizabethan England* (San Marino, CA: The Huntington Library, 1953), pp. 93–118; Herschel Baker, *The*

for an event, then it was deemed a part of God's ordained power, sustained by his ordinary providence and concourse, as one has already seen. Only when God acted "without means," disrupting the natural order and imposing His will directly on this world, was the event considered miraculous and evidence of God's absolute power and special or extraordinary providence.

Unlike Augustine, the proponents of the new science in the seventeenth century were keenly aware of natural law, and the balance between natural and miraculous had shifted so far toward the natural that the miraculous was virtually excluded. Some in fact denied even the possibility of miracle, and along that path loomed the twin specters of deism and atheism. Newton's position would have resurrected the sacramental view of the whole order of creation: everything that happens is both natural and miraculous. Things that happen seldom arouse "wonder" while things that happen "constantly" do not, yet all "are the works of God."

Nevertheless Newton's stance is subtly different from Augustine's, primarily because Newton was so deeply involved with natural law. After all he had recently discovered (or, in his own view, restored) the law of universal gravitation that operated "constantly" and with majestic regularity to the furthest reaches of the world. Perhaps he had wavered in his Arianism and had believed for a time that God acted directly in the matter of gravity, but even if so the draft of his ideas on miracles was composed in the context of the rapid development of his new active aether that was to serve as a secondary cause for gravity, the aether that appeared in print within a year or two in 1717/18. Newton was strongly inclined to focus on regularity, whether God acted "without means" or through secondary causes, and that emphasis contrasts sharply with the Augustinian emphasis on the unusual event capable of arousing wonder, though Newton knew he must account for the unusual as well, especially

Wars of Truth. Studies in the Decay of Christian Humanism in the Earlier Seventeenth Century (Cambridge, MA: Harvard University Press, 1952), pp. 12–25; Force, *Whiston* (3, n. 74), esp. pp. 32–89, 121–55; R. D. Stock, *The Holy and the Daemonic from Sir Thomas Browne to William Blake* (Princeton: Princeton University Press, 1982), pp. 3–116; R. M. Burns, *The Great Debate on Miracles From Joseph Glanvill to David Hume* (Lewisburg: Bucknell University Press; London: Associated University Presses, 1981), pp. 9–130, 247–51. For more recent arguments in this perennial and continuing debate, see also David Kirk Himrod, "Cosmic Order and Divine Activity: A Study in the Relation between Science and Religion, 1850–1950" (University of California, Los Angeles: Ph. D. dissertation, 1977) and Hooykaas, *Natural Law and Miracle* (7, n. 53).

the unusual events prophesied in Scripture. What Newton needed to lend coherence to his sacramental view of the creation was an agent that operated with majestic regularity and yet was capable of generating the unusual events of natural and sacred history, and he found what he needed in comets.

Newton was convinced from biblical sources that the world would end by fire but would probably be renewed, for in his undergraduate notebook he entered the topic "Of Earth." He had just discussed fire, air, water, and salt in a thoroughly naturalistic chemical fashion, but it was the *end* of Earth that interested him, the destruction of this earthly globe.

> Of Earth
>
> Its conflagration testified Peter 2, Chapter 3, verses 6, 7, 10, and 12. The wicked probably to be punished thereby, Peter 2, Chapter 3, verse 7. The succession of worlds is probable from Peter 2, Chapter 3, verse 13, in which text an emphasis upon the word "we" is not countenanced by the original. Revelations, Chapter 21, verse 1; Isaiah, Chapter 65, verse 17, Chapter 66, verse 22. Days and nights after the Judgment, Revelations, Chapter 20, verse 10.[59]

II Peter 3 is just as Newton said: "the heavens that now are, and the earth, by the same word [of God] have been stored up for fire, being reserved against the day of judgment and destruction of ungodly men." The "succession of worlds" is indeed probable from verse 13 if the word "we" is not countenanced. Revelations 21:1 contains John's vision of "a new heaven and a new earth" after the first earth has passed away. In Isaiah Jehovah Himself speaks: "For, behold, I create new heavens and a new earth; and the former things shall not be remembered, nor come into mind."

When Newton later encountered, among the Stoics, ideas on a final conflagration and a succession of worlds, such ideas would have served to validate Stoicism for him as a system of thought that contained at least part of the Truth. Newton seems to have treated the study of ancient philosophy as a sort of treasure hunt in which the golden nuggets of pure, sacred original wisdom might be recovered. Scripture remained his touchstone, but when the Stoics agreed with Scripture it was evidence that they had succeeded in preserving or recovering a fragment of ancient knowledge.[60] Philosophical *ekpyrosis* in its turn would surely have rein-

59. McGuire and Tamny, *Certain Philosophical Questions* (1, n. 22), pp. 374–7.
60. McGuire and Rattansi, "Newton and the 'Pipes of Pan' " (2, n. 2); John Maynard Keynes, Lord Keynes, "Newton the man," in *The Royal Society Tercentenary Celebrations 15–19 July 1946* (Cambridge University Press, 1947), pp. 27–34. Newton's attitude toward ancient philosophy was quite

forced Newton's interest in the end of the world and pushed him toward incorporating a physical provision for it in his own system, though of course it was a common assumption in the seventeenth century that the physical state of our world paralleled the moral history of mankind, as one saw especially vividly in Burnet's complex frontispiece.

The physical agent for effecting a pyrotechnic end to this world was to be the comet of 1680. In 1724/5, just about two years before his death, Newton explained his conclusion in detail of John Conduitt. "I was on Sunday night, the 7th of March, 1724–5, at Kensington with Sir Isaac Newton, in his lodgings," Conduitt said. Newton had just emerged from "a fit of the gout" and apparently felt talkative. So he told Conduitt "in one continued narration" what he had often hinted before, "viz. that it was his conjecture... that there was a sort of revolution in the heavenly bodies...." His idea was that the vapors and light emitted by the sun, and other celestial matter as well, "gathered themselves by degrees, into a body... and [attracting yet more matter] at last made a secondary planet [such as our moon]," then a primary planet, and finally a comet,

> which after certain revolutions, by coming nearer and nearer to the sun, had all its volatile parts condensed, and became a matter fit to recruit, and replenish the sun (which must waste by the constant heat and light it emitted), as a faggot would this fire, if put into it (we were sitting by a wood fire), and that that would probably be the effect of the comet of 1680 sooner or later....[61]

The comet of 1680 had first been sighted before sunrise early in November of that year, becoming invisible by the end of the month because of its proximity to the rising sun. In mid-December another comet appeared, moving away from the sun in the early evening. Or at least almost everyone thought it was another comet; an exception was the Astrono-

similar to that of Cudworth as Cudworth's work is explicated in Gunnar Aspelin, *Ralph Cudworth's Interpretation of Greek Philosophy. A Study in the History of English Philosophical Ideas* (Göteborgs Årsskrift XLIX 1943:1; Göteborg: Wettergren & Kerbers Förlag, Elanders Boktryckeri Aktiebolag, 1943). See also Danton B. Sailor, "Newton's debt to Cudworth," *Journal of the History of Ideas* 49 (1988), 511–18.

61. John Conduitt, "Memoirs of Sir Isaac Newton, sent by Mr. Conduitt to Monsieur Fontenelle, in 1727," in Edmund Turnor, *Collections for the History of the Town and Soke of Grantham. Containing Authentic Memoirs of Sir Isaac Newton, Now First Published From the Original MSS. in the Possession of the Earl of Portsmouth* (London: printed for William Miller, Albemarle-Street, by W. Bulmer and Co. Cleveland-Row, St. James's, 1806), pp. 158–86, quotations from p. 172.

mer-Royal John Flamsteed who thought the two comets were one and the same, a comet that had simply reversed its direction in the vicinity of the sun. Newton argued with Flamsteed for some little while that there really were two comets, but in 1684, in the early stages of writing the *Principia*, he recognized that comets are subject to the inverse-square law of attraction toward a center of gravitation, which of course made Flamsteed's strange idea seem more probable. By the time the *Principia* was published in 1687 Newton knew that comets describe conic sections focused on the sun.[62]

The taming of the comets, making them more or less domesticated members of the solar system, was not the least of Newton's achievements in the *Principia*. Such a notion was almost unheard of at the time. Comets had always seemed radically alien; they were erratic and ephemeral bodies that portended no good, and had traditionally been taken as "signs" and portents of disaster – not disaster itself, for none had ever been known to crash into anything, but God's fiery sword that slashed across the normal circular motions of the heavens to show mankind that the normal course of earthly events was about to be disrupted in a most unpleasant way. With Newton's work, however, the newly domesticated comets were promoted in status from signs to agents of destruction.[63]

If the conic section described by a comet is an ellipse, the orbit will be closed and the comet will make a periodic return. But Newton fully realized that his new concept of universal gravitation in which all bodies *mutually* attract each other could be expected to disrupt the regularity of a comet's orbit. Over a long period of time, a comet's motion might very well be sufficiently retarded by its proximity to other heavenly bodies so that it would "descend to the sun." In the *Principia* he sounded this note in rather a positive way: "So fixed stars, that have been gradually wasted by the light and vapours emitted . . . , may be recruited by comets that fall upon them; and from this fresh supply of new fuel those old stars, acquiring new splendor, may pass for new stars."[64] To Conduitt he was more explicit: the comet of 1680 "might perhaps have five or six revolutions more first. . . . " He could not say when it might drop into the sun, "but whenever it did, it would so much increase the heat of the sun, that this earth would be

62. Westfall, *Never at Rest* (1, n. 4), pp. 391–7, 402–35.
63. Webster, *From Paracelsus to Newton* (3, n. 80), pp. 25–31, 40–2; Genuth, "Comets, teleology" (5, n. 79).
64. Newton, *Principia* (Motte–Cajori: 2, n. 13), vol. II, 540–1, quotation from p. 541; (Koyré–Cohen: 1, n. 9), vol. II, 756–7.

burnt, and no animals in it could live." The earth already bears "marks of ruin" that cannot be wholly accounted for by Noah's flood, Newton said, suggesting that a similar conflagration had already taken place in the past. When Conduitt asked how the earth might have been re-peopled in that case, Newton answered, "that required the power of a creator." Indeed all of the "revolutions of the heavenly bodies" were "by the direction of the Supreme Being."[65]

So with comets Newton had found a precise mechanism, described by his mathematical principles, by which the dissolution of worlds could be effected from time to time according to the will of their omnipotent Creator. The dissolution would take place through the regular operation of natural law, but built into that regularity were the hidden causes similar to those of Augustine, the working out of which would effect the surprising irregularity, the wonder-causing irruption of God's miraculous activity.

Leibniz sarcastically observed that Newton's God did not seem to have had sufficient foresight to give His machine "a perpetual motion," that He was an unskillful workman "obliged to mend his work and to set it right."[66] Clarke, who is generally understood to have been Newton's spokesman in this exchange, first replied that the "notion of the world's being a great machine, going on without the interposition of God ... is the notion of materialism and fate"; it tends to push "providence and God's government" out of the world.[67] In any case, Clarke added in his next reply, in a response especially appropriate for Newton's use of comets to effect the end of the world, Leibniz's idea of a "correction" or "amendment" to the present system when it becomes disordered is one that stems only from human perceptions.

> In reality, and with regard to God; the present frame, and the consequent disorder, and the following renovation, are all equally parts of the design framed in God's original perfect idea. 'Tis in the frame of the world, as in the frame of man's body: the wisdom of God does not consist, in making the present frame of either of them eternal, but to last so long as he thought fit.[68]

There is an echo there of Newton's statement in his "Vegetation" manuscript that the earth, like all things living, must have its time of flourishing, old age, and death; there is also an echo of the Scriptural destruction of the world in accordance with the inscrutable will of the

65. Conduitt, "Memoirs" (7, n. 61), pp. 172–3.
66. *Leibniz–Clarke Correspondence* (6, n. 119), pp. 11–12.
67. Ibid., p. 14.
68. Ibid., p. 23.

Almighty God. Against the true wisdom of Stoic and Judeo–Christian traditions Leibniz had set the impossible (and false) notion of an eternal world that ran perfectly and independently on and on without decay – or so it seemed to Newton. The ideas Leibniz expressed were anathema to Newton; he had fought against them all his adult life. Leibniz, on the other hand, never seems to have understood that Newton really wanted a system of the world that *required* God's active participation and governance, no matter how inscrutable that God's plans might be.[69]

The destruction of worlds was not the only function Newton found for comets; he assigned them restorative functions as well. Indeed, even though this earth will be burned and all the animals in it will perish when the comet of 1680 falls into the sun, the sun itself will be restored and replenished and shine with fresh splendor. In the meantime the tails of comets will provide humidity and active principles to replenish planetary fluids and conserve a cosmic balance of matter, as both Genuth and Schaffer have amply demonstrated in their recent comprehensive studies on this point.[70]

Genuth in particular found that the restorative and conservative functions of the comets' tails filled the place in Newton's cosmology left empty by the demise in the 1680s of his earlier circulatory aethers. Beginning in 1687 the tails provided for "the production and nourishment of vegetables," replaced the fluids used up in vegetation and putrefaction, and perhaps furnished the vital aerial spirit as well. Newton later realized that the tail of a comet is extremely tenuous despite its brilliance, for one can see stars shining through it, and qualified the *Principia* statement somewhat in private though not in print. The *Principia* had seemed to suggest that the comets (and vapors from the sun and fixed stars) actually supplied water to the earth along with the vegetable and aerial spirits, and that probably is what Newton believed at first, for he had in 1680/81 thought the cometary atmospheres to be quite thick. But in 1705 Gregory recorded that the *Principia* was not to be understood that way:

69. For a provocative and enlightening argument that the experience of the numinous always incorporates fear and awe of the tremendous, mysterious Other and His incomprehensible providential arrangements, see Stock, *The Holy and the Daemonic* (7, n. 58).
70. Genuth, "Comets, teleology" (5, n. 79); Simon Schaffer, "Newton's comets and the transformation of astrology," in *Astrology, Science and Society. Historical Essays*, ed. by Patrick Curry (Woodbridge, Suffolk: The Boydell Press, 1987), pp. 219–43.

> This is not to be understood of the real fluid water so restored, for its certain there are not such rains at or after a comet as this would inferr; & the tail of a comet is too thin for this, a fixd starr of the smaller size being visible thorough such an immense thickness thereof; but of that subtile Spirit that does turn Solids into Fluids. A very small Aura or particle of this may be able to doe the business.[71]

So after reflection and more observation, Newton seems to have limited the restorative fluid supplied by comets' tails to the vegetable spirit alone, a "very small Aura or particle" of which could turn solids into fluids right here on earth.

But the great cosmic interchange of matter continued in its familiar circulatory pattern as cometary, solar, and stellar vapors and spirits condensed for the uses of the planets and were wasted and replenished, and sun and stars themselves were nourished by the comets that plunged into them. As Genuth has argued, the circulatory changes in which Newton's comets participated are very similar to the lesser but still cosmic circulation of the aether of 1675. Comets had indeed assumed many of the functions of the lost aether. As seen in Chapter 5, in the alchemical literature the vegetable spirit always had a cosmic dimension, even though primarily operative in micromatter, and Newton needed to maintain a cosmic source of supply. The cosmic part of the vegetable spirit thus continued to be supplied by the tails of comets, and that was almost certainly the source of those "other causes" that Newton supposed worked with light and electricity to stir up and maintain the life of earthly animals, vegetables, and minerals.

It was no easy matter to establish the periodicity of the comets upon which their various functions in God's cosmic plan depended. Schaffer's study in particular explores the immense effort put into their work on comets by both Newton and Halley, especially after the spring of 1694. Newton had a graphical method for computing parabolic approximations to closed elliptical orbits, but it was important to both men to demonstrate definitively that the highly eccentric orbits of their comets were truly elliptical and truly closed. Only then could comets become permanent members of the solar system and participate in their several important cosmic roles. "Halley's Comet," still famous, was that of 1682; its period was established at about 75–6 years. Its return has now been

71. Hiscock, Ed., *Gregory's Memoranda* (2, n. 16), p. 26. Genuth, "Comets, teleology" (5, n. 79), pp. 45–6, n. 57, examines the inconsistencies this change would have introduced into Newton's cometology and argues that Gregory was recording his own opinion, not Newton's, contrary to my own argument that Gregory was recording Newton's changed opinion.

duly recorded four times and since 1758 has often been cited for the definitive establishment of Newtonian principles. Newton's favorite, however, was the comet of 1680, which he and Halley concluded had a period larger than 500 years. It has not returned, but it was of course the one destined, according to Newton, to usher in the apocalypse after five or six more revolutions. Schaffer argues that only heavy commitment to the idea of comets as permanent bodies could have sustained the "tortuous analysis" by which Newton and Halley moved from nonperiodic parabolic approximations to closed elliptical orbits with definite periodicities.[72]

In addition to recognizing the vast importance of comets to Newton in their apocalyptic and vitalizing roles, Schaffer has located the significance of comets in Newton's program for the restoration of true religion through the restoration of true natural philosophy. Comets had been falsely forced down below the orb of the moon as sublunary meteors in the Greek system of solid heavenly spheres, Newton had observed in 1685, even though the more ancient Chaldeans had earlier known them to be true celestial bodies. Restoring the comets to their rightful place in the free spaces of the heavens in his own system meant to Newton a step toward restoring the Truth in religion that was ultimately based on the Truth in natural philosophy. Rescuing comets from corrupt astrological practices and false priestcraft, where a form of idolatry prevailed that wrongly attributed God's powers to intermediate material bodies, meant returning His powers to the one true God and according Him a proper worship for His cosmic activity.[73] As Newton's disciple William Whiston later remarked, quoting the famous Pauline apocalyptic "restitution" text of Acts 3:21,[74] the publication of the *Principia* marked "an eminent prelude and preparation to those happy *times of the restitution of all things, which God has spoken of by the mouth of all his holy prophets since the world began.*"[75]

In his work on comets Newton brought together many of the variegated strands of his studies. His mathematics was instrumental in establishing their periodicity, a necessary prelude to establishing their true cosmic functions. With their periodic returns demonstrated, he then used comets as the hidden causes that account in a "natural" way for the "miraculous" congruence of natural and sacred histories, especially for the confluence of those histories at the end of time in the final conflagration of a world

72. Schaffer, "Newton's comets" (7, n. 70), esp. pp. 227–31.
73. Ibid., pp. 237–43.
74. I am indebted to Robert S. Paul for help in locating this citation.
75. Quoted in Schaffer, "Newton's comets" (7, n. 70), p. 234.

"stored up for fire, against the day of judgment and destruction of ungodly men." In the meantime, before the day of judgment arrives, the comets serve God's will in providing, along with solar and stellar emanations, a cosmic source for the divine activation of micromatter, to enliven it and shape it into the so many beautiful and well crafted forms God has ordained. "A very small Aura or particle of this [cometary effluvium] may be able to doe the business," serving as the heavenly quintessence that links divine will with matter and stirs up the active principles of earthly cohesion and fermentation, of electricity and light. Finally, the restoration of a true natural philosophical understanding of comets led humanity one step closer to the restoration of true religion and the true worship of the one true God, a Deity who desired to be worshipped not for His essence but for His activity, the issues of His will.

However, one comes full circle back to the Augustinian attempt to balance miracle and natural law in a sacramental view of nature. Augustine's emphasis upon the unusual natural event, and his relative neglect of regularly acting natural law, had perhaps contributed to the eagerness with which subsequent generations searched for unexpected irruptions of divine activity in the natural world. Newton's emphasis upon regularly acting natural laws, laws that in their regular activity could be counted on to supply the occasionally unusual event as well as "daily miracles," perhaps contributed to the eagerness with which generations subsequent to his own took up the search for natural law.

The fate of the divine compasses of medieval iconography illustrates the result of that shift in emphasis. God's creative compasses were taken in the Middle Ages to be emblematic of His wonderful ordering of nature by number, weight, and measure; they carried an aura of the original divine miracle of creation itself.

Newton had understood the divine compasses in a similar fashion: the symbol of natural mathematical law in its miraculous Augustinian sense. So he had understood the dimensions of ancient temples in their cosmic significance; so he had understood his own work in which he had found the mathematical law of gravity written by a God very skilled in geometry and mechanics.[76] So he had understood the creative vegetable spirit itself which was after all, he hoped, to be brought under the rubric of natural law also. So he had found comets, and especially the comet of 1680, to be the hidden causes built into nature to work the creative will of the omnipotent Deity in unusual

76. Newton, *Correspondence* (2, n. 14), vol. III, 233–8 (Newton to Bentley, 10 Dec. 1692), esp. 235.

Plate 11. William Blake, *Portrait of Sir Isaac Newton* (1795). The Tate Gallery, London. Reproduced by permission of The Tate Gallery, London/Art Resource, New York.

and even nonmechanical ways, according to their natural operations under the mathematical law of gravity.

But by the end of the eighteenth century, as observed in Chapter 5, the divine compasses had become in the work of William Blake the symbol not of miraculous creativity but of restriction. Newton himself had become for Blake an exemplar of Urizen (Your Reason), in Blake's mythology a personification of the reasoned search for natural law that was the foe of every divine creative impulse. In Plate 11,[77] Blake's *Newton* gazes not toward a wonder-working creator God in His eternal heaven; rather he bends downward toward the earth with cold concentration. In his hand the compasses, no longer divine, trace only noncreative, mechanistic, earthly, and temporal law.[78]

77. William Blake, *Portrait of Sir Isaac Newton* (1795), The Tate Gallery, London.
78. Stock, *The Holy and the Daemonic* (7, n. 58), pp. 359–70; Blunt, "Blake's 'Ancient of Days' " (5, n. 103).

Newton had in fact engaged both heaven and earth, had surveyed both past and future, had studied eternity as well as time, and divine creativity as well as mechanism, in order to do what he had done. But *regularity* was that aspect of the Newtonian system that most captivated eighteenth-century thinkers, especially the regular operation of gravity that brought the return of Halley's Comet in 1758 and so dramatically confirmed Newton's mathematical principles. The unusual manifestations of divine will in the system that Newton had expected and provided for were minimized or eliminated as the eighteenth century focused more and more on the regular mechanical operation of natural laws that did not require divine supervision. Thus by the end of the century Newton's sacramental view of natural law and miracle, in which all was the will of God, had been so thoroughly subverted that Blake could blame Newton for the destructive, restrictive triumph of cold human reason. That result was hardly what Newton had intended. "When I wrote my treatise about our Systeme I had an eye upon such Principles as might work with considering men for the *beleife* of a Deity," Newton said,[79] and the almost total misperception of Newton's system in the eighteenth century is surely one of history's greater ironies.

Final credo

Sometime late in his life – perhaps as late as the 1720s but certainly not earlier than 1710 – Isaac Newton attempted to formulate a creed that he could recommend for all Christians. Although drawn from his biblical and theological studies and reflecting the terminology appropriate to those fields, parts of this creed also reflect Newton's final convictions on the relationship of the Supreme Deity to Christ and to the created world.

The initial paragraph of the creed offered a partial restatement of the God of the General Scholium and in addition explicitly differentiated the Supreme Deity from other spirits.

> 1 That ↗ God ↙ the ffather is ↗ an ↙ infinite, eternal, omniscient, immortal, ↗ & ↙ invisible spirit whom no eye hath seen nor can see, ↗ & ↙ all other spirits ↗ are ↙ sometimes ⟨appearing, *deleted*⟩ visible.[80]

79. Newton, *Correspondence* (2, n. 14), vol. III, 233–8 (Newton to Bentley, 10 Dec. 1692), quotation from p. 233, emphasis added.
80. Yahuda MS Var. 1, Newton MS 15.4 (2, n. 50), f. 67v. Arrows up and down indicate Newton's interlineations in this and the following quotations.

Only God the Father bears the traditional theological attributes of infinity and immortality, is all-knowing and eternal. He is furthermore completely inaccessible to human senses even though other spirits are not so, and are indeed sometimes visible. Similarly, Newton said in the General Scholium that God is eternal, infinite, omnipotent, and omniscient, and that "He is utterly void of all body and bodily figure, and can therefore neither be seen, nor heard, nor touched. . . ."[81]

In the next succeeding paragraphs of the creed Newton struggled to define the relationships that exist, or should exist, between and among Father, Son, and humanity. The God who decreed in the Fourth Commandment that we should have no other gods before Him "is the first author or father of all things." God Almighty, God the First Author, and God the Father are "equipollent" expressions, Newton added, "& therefore this God has no equalls." We are to worship Him alone, and idolatry consists in the worshipping of a false god, "A God who is not what your worship supposes him to be," a fictitious god, a "vanity." There is, however, an obedience and worship due to kings, "suitable to" the power and dominion God has lodged in them; likewise there is a worship due to Jesus as our Lord and King, the Messiah, the Prince. For, Newton continued with many deletions and interlineations, Jesus "was obedient to death even the death of the cross." Therefore God "highly exalted him & set him at his own right hand far above all principalities & power & might & dominion" and gave him "a name above every name not only in this world but also in that wch is to come." At the name of Jesus every knee should bow, but all worship of Jesus has as its ultimate purpose the glory of God the Father. The Supreme Deity

> is to be worshipped as the ffather almighty, the supreme potentate, the first author of all things, the Lord God omnipotent: & other ⟨*illegible word, deleted*⟩ ↗ Potentates ↙ are to be worshipped according to the power & dominion wch he has given them over us & the benefits wch we receive from them & wth such a worship ↗ only ↙ as he has granted them in order to his own glory ffor all things should be done ↗ in due order ↙ to the glory of God the father.[82]

Newton said we may give the name of king to viceroys, continuing his effort to refine the relationship between Jesus and the Supreme Deity, but we may *not* say that a king and his viceroy are *both* kings. Neither may we do that with gods. Denying the status of mediators to angels or

81. Newton, *Principia* (Motte–Cajori: 2, n. 13), vol. II, 545; (Koyré–Cohen: 1, n. 9), vol. II, 762–3.
82. Yahuda MS Var. 1, Newton MS 15.4 (2, n. 50), f. 68r.

the souls of dead men, however, Newton accorded that status uniquely to Jesus: as there is but one God, so "there is but one Mediator between God & man, the man Christ Jesus." Each – both God and Jesus – has his proper worship: God the Father "of whom are all things & we in him" and Jesus "for redeeming us wth his blood." And on the basis of all his deliberations Newton concluded that it was proper to direct our prayers to God in the name of the Lamb (Jesus Christ) but not to the Lamb in the name of God.[83]

One could hardly wish for a clearer statement of Newton's anti-Trinitarianism than that carried by those several statements on the subordination of Jesus to the Supreme Deity in all things. The worship of Jesus for himself would be idolatry; we are to worship him because his death redeemed us, and that obedience unto death, "even the death of the cross," led God to exalt him mightily, "above all principalities & power & might & dominion." But all worship accorded to Jesus is to be done "to the glory of God the father." Newton would have no Christian adhere to the doctrine that Jesus shares equally in the eternity and dominion of God the Father. So far was Newton from the orthodox Trinitarian position (that Father, Son, and Holy Ghost constitute One God in Three Persons) that he did not even find it necessary to consider the Holy Ghost in these passages.

But in the sixth and seventh paragraphs of the creed the cosmic grandeur of the Arian Christ appeared. The emphasis on the Atonement, on the sacrificial Lamb, on the humanity of "the man Christ Jesus" as Mediator – all those aspects above that point to the subordination of Jesus to God the Father are submerged before the vision of Jesus as the viceroy with whom God has shared tremendous power.

> 6 That as the father hath life in himself & hath given the Son to have life in himself so the father hath ↗ knowledge & ↙ wisdom & power ↗ & will & counsel & substance ↙ in himself & hath given the son to have knowledge & wisdom & power will ↗ & counsel & substance ↙ in himself.[84]

That attribution of divine, indeed godlike, powers to the Son "in himself" reflects the tensions of Arian Christology. Even though the Son was subordinate in Arian doctrine, he was certainly more than human. Denied coeternity and consubstantiality with the Father, the Arian Son nevertheless had "in himself" the divine attributes because the Father had

83. Ibid., f. 67v.
84. Ibid. In the original, Newton reordered the attributes of the Son by placing numbers above the words to yield the following list: (1) knowledge, (2) wisdom, (3) counsel, (4) will, (5) power, (6) substance.

given them to him and had invested the Son with the office of viceroy. Although many Christian doctrines demonstrate a certain tension between Christ as man and Christ as god, that dichotomy appears in particularly stark relief in Newton's formulation of Arianism and was of the utmost importance to Newton in his efforts to understand the relation between his Creator God and the created world.

It is in the next paragraph that the great significance of Arian Christology in Newton's system of the world becomes fully apparent. Here he explicitly called Christ God's "Agent" and implied that the Christ was responsible for all or almost all divine interaction with this world and the next.

> 7 That Jesus ⟨*illegible word, deleted*; the first & the last, *deleted*⟩ ↗ ⟨&, *deleted*⟩ was beloved of God before the foundation of the world ↙ & had glory wth the father before the world began & was the principle of the creation ⟨of God, *deleted*⟩ ↗ of God, the Agent ↙ , by whom God created ⟨all things in, *deleted*⟩ this ⟨*illegible word, deleted*⟩ world & ↗ who ↙ is now gone to prepare another place or mansion for the blessed; for in Gods house there are many mansions, ⟨*illegible word, deleted*⟩ ↗ & God does nothing by himself wch he can do by another. ↙ [85]

Jesus existed before the creation of the world, was even then beloved of God and shared His glory. For creation itself Jesus was "the principle," acting as God's agent, and it was he "by whom God created this world," just as even now he is preparing "another place" – presumably one to be inhabited by "the blessed" after the apocalypse. As a general rule Newton insisted that God does nothing by himself that "he can do by another." All of those statements are in full accord with the remote and transcendental Supreme Deity of Arianism, but Newton avoided the trap of the absentee Deity of deism by utilizing the cosmic role of the Arian Christ as mediator between God and world, as the "Agent" who put God's will into effect in the creation.

With the assistance of this Arian creed from Newton's last years, the historian may reconstruct that system of the world Newton had hoped to demonstrate in an irrefutable fashion – its beginning, its present state, its end, its renewal. The Supreme God, eternal and infinite, was always and everywhere the ground of all being and His presence constituted duration and dimension for the created world – Newton's absolute time and space. Present with the Father before the beginning of the world was Jesus – the Son, the Word, the Christ – not of the same substance with the Father but the first created being,

85. Ibid., f. 67r.

beloved of the Father before all worlds and His agent and viceroy for the creation of this world and its governance, as well as Mediator between God and humanity through his redeeming Incarnation and death. The Son or Word in his cosmic capacities, all freely given to him by the Father, organized the present world from its initial chaos, and remains active in the continued organization of passive matter into all the lovely forms of minerals, plants, and animals that derive ultimately from the ideas and will of the Father of all things, the Son carrying divine intent and design into the created world, operating through the natural agencies of light, electricity, and cometary effluvia. Operating also through the agency of the active gravitational aether, the Word organized the heavenly bodies into their beautiful system, the mathematical system designed according to number, weight, and measure by a God very skilled in geometry and mechanics, and also managed the comets in such a way that they would both sustain the system in its present state and also in the end produce its pyrotechnic destruction as the Father had willed in His original perfect plan. In the meantime Jesus had spoken for God to the prophets, allowing to them some prevision of things to come according to God's plan, not that human beings might be enabled to foretell the future but that after the prophecies were fulfilled humanity might recognize events as God's willed activity and His continued governance and providence be made known. The true knowledge of God and the proper worship of Him having been lost or corrupted, however, it was required that human beings make every effort to restore the Truth, and the best way to do that was first to restore Truth in natural philosophy, success there leading directly to a restoration of Truth in religion and to the proper worship of God in this physical world, which is His real and true temple, where His activity may be seen. This happy time of the restitution of Truth ushering in the End of Days (for moral and physical histories run together in God's plan), the apocalypse will insure the destruction of the godless and idolators while the blessed find renewed life under a different heaven and in a different mansion that is even now being prepared by the Son somewhere in God's house.

Conclusion

The beginning of the last period of Newton's life saw him rich in the honors and recognition of this world but still dissatisfied with his own life's work. He had not succeeded in defining nature's obvious laws and processes in vegetation, and his first solution to the problem of a cause for gravity had left a residue of theological anxiety. Granted that the *Principia* and the *Opticks* explicated *some* of God's laws better than they had been since original wisdom had been lost, and granted that gravity as an active and divine principle constituted a powerful argument against deism and atheism, still his system as a whole was not entirely coherent.

The most dramatic changes in the system that Newton put in place, more or less, in his final years stemmed from electrical experiments and demonstrations at the meetings of the Royal Society, where he himself occupied the presidential chair. Concluding that a connection had been made between light and electricity, Newton came to see the electrical effluvium as a new source of activity in micromatter, one that worked with light to stir up and organize the particles of passive matter in living forms – in short, as another component of the alchemical vegetable spirit. Then, extrapolating from the pronounced elasticity of the electric spirit, he created a new version of the aether as an explanatory cause for gravity – an aether that was not only exceedingly elastic but also exceedingly active, a substance intermediary between the incorporeality of God and the full corporeality of body, a substance that helped to allay Newton's theological anxieties by serving as a cosmic mediator between God and matter in the operation of gravity.

Though one can hardly argue that Newton's speculations upon electricity and upon the new active aether achieved full scientific fruition, they certainly did serve to bring more coherence to his physical system and to align it more fully with his theological system. If Newton had wavered in his Arianism during the period when he thought the Supreme Deity subsumed the operations of gravity directly, the new aether allowed him fully to reinstate his Arian convictions. In Newton's final *credo* God Almighty once more has His agent, His viceroy, by whom He creates and governs and through whom humanity has been redeemed, a Mediator between Himself and the world and between Himself and human beings. The Mediator, one must assume, is in charge of all the natural active entities, the "other spirits" of the creed, of light, electricity, cometary effluvia, and the active gravitational aether of the last *Opticks*. As God's viceroy, having life, power, wisdom, will, knowledge, counsel, and substance "in himself," the Mediator brings divine ideas into the world and insures that they are embodied in all their glorious variety. The Mediator also insures the stately gravitational motions of the heavenly bodies, bodies moving regularly as long as God so wills but destined in His good time to demonstrate the irregular irruption of His willed activity in their apocalyptic end and renovation. It was a fully sacramental view of God and nature that Newton achieved, and, although not a fully demonstrated one as he had hoped, certainly a coherent one in which law and miracle, natural principles and divine ones, all worked together to the greater glory of God.

8

Epilogue

It was my intention when I first began work on this book, on 1 October 1974, to complete my study of Newton's alchemy, and to do only that. My first book having been focused on the "foundations" of Newton's alchemy and restricted primarily to the earlier alchemical manuscripts in the Keynes Collection, King's College, Cambridge, I was only too well aware of the great quantity of Newtonian alchemical material of later periods both at King's College and elsewhere. It seemed necessary to me to become much more familiar with those other manuscripts in order to obtain some sense of closure on the subject. Would that I had been able to stay with that intention and only that, because then this book would have been completed many years ago.

It would, however, have been a very different book. It would have been a book focused much more narrowly on Newton's alchemical interests, though one hopes even so that it would have contained some of the general cultural context, presented here primarily in Chapters 2 and 3, that made the alchemical enterprise seem so important to quite a number of significant thinkers in early modern Europe. Especially in the linkage between Christ and the philosopher's stone, in Reformation expectations of a general *renovatio* in which matter as well as humanity was to be redeemed, in Newton's studious conviction that alchemy (and the divine activity in micromatter to which it spoke) should prove to be the necessary corrective for the overly mechanized system of Descartes – especially, in short, in the religious ambience of alchemy, may one locate its great contemporary appeal and so make Newton's devoted study of it more comprehensible. The seventeenth century was, after all, still a very religious age. But, had that been the book I wrote, it would still have been a book focused almost exclusively on Newton and alchemy.

My slow recognition that alchemical studies held religious significance for Newton himself was one of the turning points in my thinking that led me on to quite a different book. Sixteen years ago I was imperfectly detached from modernist convictions and from our general cultural perception of Newton as the founder of modern science. Even though I was willing to entertain the heretical notion that Newton's alchemy was worthy of scholarly examination, I was not willing to entertain a religious interpretation of it. Religious sentiments are both more acceptable and more perceptible in this postmodern era in which religious revolutions profoundly affect many parts of the globe, which may perhaps help to explain why I now perceive Isaac Newton so differently. I have apologized above for my previous attitude to Mary S. Churchill, of whose argument for the religious significance of Newton's alchemy I was at one time quite dismissive. My specific retraction may be found in Chapter 1, but this entire book may also be considered in that light.

The other turning point in my thinking, and one that really forced me into a more comprehensive approach, came with my close study of Dibner MSS 1031 B, Newton's treatise on nature's obvious laws and processes in vegetation (Appendix A). As had most scholars who had studied Newton, I had fallen into the pattern of focusing almost entirely on one category of his voluminous manuscripts. Having divided up the scholarly work of studying Newton into areas congruent with modern academic interests, we had inadvertently divided up Newton. Dibner MSS 1031 B insisted that I rethink those divisions, for there, in what was primarily an alchemical treatise, Newton himself discussed both gravity and God. It is entirely understandable that the study of Newton's papers had been parceled out among the various fields of modern expertise, for the papers lend themselves to that approach: Newton pursued each area of study as if it were the only one of importance to him and almost always adhered to the terminology and the concepts embedded in that particular field. Dibner MSS 1031 B alerted me to the fact that that was not always the case and encouraged me to find other approaches to certain problems.

The first instance of that sort concerned the problem of gravity. The speculative aethereal gravitational system in Dibner MSS 1031 B is similar to the one in Newton's student notebook of the mid–1660s, yet differs from it. It is in addition similar to the one in his "Hypothesis" of 1675, yet it differs from that one also. The similarities, and the differences, are such as to suggest an evolutionary sequence: (1) student notebook, (2) Dibner MSS 1031 B, (3) "Hypothesis," a sequence set out in Chapter 4. The recognition of that sequence helped date Dibner MSS 1031 B to the early 1670s; eventually it led me into an attempt to discover an evolutionary pattern in all of Newton's thinking about gravity.

The pattern I found in the matter of gravity is delineated in Chapters 4, 5, 6, and 7. Especially in Chapter 5, my predilection for the evolutionary chronological sequence of development in Newton's thinking about gravity, that had been suggested to me by the sequence of the

1660s and 1670s, led me to emphasize Newton's apparent commitment to aethereal gravitational systems well into the 1680s and forced me to argue for a radical redating of *De gravitatione* to 1684 or early 1684/5 – to a period when Newton was actually engaged in writing the *Principia*. For it was only the discoveries that constitute part of the *Principia* itself that finally forced Newton to make a revolutionary break with Descartes, and with every contemporary mechanical system, on the matter of gravity. Having broken with mechanical causation for gravity, however, Newton renewed his search for its cause. So far was he from accepting the nineteenth and twentieth centuries' argument that the mathematical laws of gravity suffice that Newton spent the rest of his life searching for a new causal explanation: his first one is described in Chapter 6, his second one in Chapter 7. Both of his solutions constitute strong arguments for the unity of Newton's thought, for theological constructs seem to have been fundamental to each of the two he later offered and indeed to have been also determinative of his shift from the first to the second.

My excursion into the realms of gravity may seem anomalous in a study purporting to examine the role of alchemy in Newton's thought, but that too was conditioned in a very general way by Dibner MSS 1031 B. In addition to presenting in that manuscript a mechanical gravitational system predicated upon the circulatory motion of the aether, Newton there had his active alchemical agent, the "vegetable spirit," thoroughly entangled in the gravitational aether and following its motion. Clearly, in Newton's mind, the mechanical aether and the active principle of alchemy were intimately related in some way, and it became incumbent upon me to follow his evolving thought on their *connection*, as well as to follow their separate paths. That I have attempted to do in Chapters 4 through 7, often juxtaposing gravity and alchemy in ways that may seem surprising to the modern reader, in order to emphasize the point that there was some connection in Newton's mind between the causal principle for gravity and the causal agent in alchemy. Their relationship varied dramatically from one period to another of Newton's life, however. His first gravitational system was set out before he began to study alchemy. In the second, in Dibner MSS 1031 B, the alchemical and gravitational agents were closely united, as also in the "Hypothesis" of 1675. But after that they were apparently separated again, as in Newton's letter to Boyle of 1678/9. In 1684 the mechanical gravitational aether disappeared completely, and the only residue of aethereal particles left in Newton's cosmos – those very few widely scattered ones with great empty spaces between them – are best understood as the necessary cosmic source of supply for an alchemical agent that worked primarily in micromatter but that had always had such a cosmic dimension. That, then, was their relationship – or lack of relationship – for two to three decades after 1684, as Newton struggled to formulate an incorporeal cause for gravity but still maintained his belief in a corporeal active principle at work in micromatter, one that still had its cosmic source of supply. Finally, in the last period of his life, the two kinds of agents came somewhat closer together again. Basing his new speculations on experimental evidence of electrical activity, and on the connections he thought he had grasped among light, electricity, and the corpuscles of matter, Newton suggested in private papers that a semicorporeal electricity, a sort of "spiritual body," had some of the functions of the active alchemical agent. Then, extrapolating from the extraordinary elasticity of that microcosmic actor, Newton created a final version of the macrocosmic gravitational aether – one that was itself active in a way his earlier aethers had never been. At somewhat the same time he called upon the tails of comets to serve as a cosmic source of supply for the microcosmic agent, even as his new cosmic aether absorbed some of the functions of the "electric and elastic spirit." It is to be hoped that discussion of these connections and relationships among Newton's various speculative aethereal systems will serve to alleviate some of the scholarly problems in the field, for heretofore it has been tacitly assumed that Newton's aethers were generally unifunctional, and that does not now seem always to have been the case.

The connections in Newton's mind between gravity and alchemy, however dramatically they may have varied from time to time, constitute yet another argument for the unity of Newton's thought. He simply did not ultimately divide his studies into the logic-tight compartments that most recent scholars have deemed appropriate. Not only was there a connection between gravity and alchemy, there was also one between gravity and God and one between alchemy and God. The God passages in Dibner MSS 1031 B set out Newton's adherence to theological voluntarism, in which God's will is His primary attribute; the shift in the meanings of the word "protoplast" from Dibner MSS 1031 B to the later "Hypothesis" spelled out the effects of Newton's acceptance of theological Arianism during the period around 1673, a position in which God's will is always put into effect in the world through the agency of the Arian Christ. Again, Dibner MSS 1031 B proved to be the culprit that forced me into an examination of yet another area of Newton studies, the theological. Later I noticed other documents from Newton's pen that emphasized Newton's conviction that God desires to be worshipped for His activity, the issues of His will, and also his conviction that humanity may learn of God's activity by a study of "the frame of nature" and by a study of prophecy and its historical fulfillment.

The ultimate unity of Newton's thought slowly centered itself for me on my growing realization that Newton supposed himself to be studying God's activity in every area – in micromatter, in cosmic order, in history. Although the connections between and among Newton's various pursuits are often far from obvious to the modern reader, I have attempted in this book to suggest the connections as it seems Newton saw them. First, last, and always, I have tried to demonstrate those connections on the basis of Newton's own writings – both public and private – and in Chapter 7 I have argued that, with the assistance of Newton's late Arian *credo*, the historian may indeed reconstruct the outlines of the system of the world Newton had hoped to construct. It was a system *not* just of "the mathematical principles of natural philosophy." On the contrary, it was to have been a grand unification of natural and divine principles, and it included a vision of God's activity not only *in* this world as we know it but also at the world's beginning and at its apocalyptic end and renovation. It was a vision in which the Arian Christ, as God's "Agent" throughout time, always putting the will of the Supreme God into effect, kept God intimately connected both with the physical world and with humanity: that was Newton's ultimate answer to the twin specters of deism and atheism that had always haunted him. It is also a vision that forces one to the conviction that one must give a religious interpretation not only to Newton's alchemy but to *all* of Newton's work, including the *Principia* and the *Opticks*, since Newton himself was apparently motivated to study "the frame of nature" in order to learn of God's activity.

As I have argued in Chapter 1, one result of this more unified understanding of Newton is to suggest a reevaluation of his methodology. If Newton really did accept the doctrine of the unity of Truth, and if he really did believe that all of his many studies were capable of yielding access to at least a partial Truth, then his methodology in all his work was surely much broader than we had previously supposed. All the evidence from his papers seems to me to argue for the broader method. Especially in Chapter 6, as one watches him look backward in time for a cause for gravity, one reaches the conclusion that Newton respected the possible validity of ancient natural and divine knowledge. How did the ancients explain gravity, he asked again and again, and finally he found ideas there that he could and did use. He was of course highly selective in his use of ancient material; he never accepted anything uncritically in any area of study. Rather he sought always to find a synthesis into which his partial truths might coalesce to give a larger truth. Any knowledge from antiquity had to be tested against the things he knew from other sources. Just so did mathematical knowledge require to be tested and balanced by observational and experimental knowledge; just so did rational, logical constructs require to be tested and balanced by the biblical and historical records; just so did mechanism require to be tested and balanced by alchemical concepts. The doctrine of the unity of Truth may well have been one factor that made Newton so flexible and creative in the published work that subsequent generations have so admired and still admire.

If the religious interpretation offered here of all of Newton's work be accepted by historians of early modern Europe, it may ultimately have some bearing on a major historical problem, that is, the rise of modern science. As is well known, the scientific enterprise that was to become such an integral part of Western civilization arose in Europe in the sixteenth and seventeenth centuries, at no other time and in no other place. Why was that the case? What were the necessary historical ingredients for that revolution? Why did those ingredients, whatever they were, coalesce in early modern Europe and never elsewhere or at another time? Many factors have been and are considered relevant to those questions – factors associated with intellectual, social, technological, political, economic, and institutional developments. I do not wish to discount or minimize the importance and relevance of such factors. But in a period when societal support for the study of nature was quite slender, and when the study of nature had so far produced virtually none of its later ability to control and manipulate nature, the question of *motivation* for a study of nature has seemed to me to remain elusive. I would like to suggest, on the basis of this study of Newton, that one motivational factor may be located in Judeo–Christian doctrines of long standing: the doctrine of the divine creation of the world and its associated doctrine that the attributes of the Creator are reflected in the nature of the world He created. "The heavens declare the glory of God," the Psalmist had said, or, as Newton put it, there is no way to come to a knowledge of the Deity except by revelation or by "the frame of nature." In the post-Reformation turmoil of early modern Europe, when interpretations of revelation were all in doubt, a religious hunger for knowledge of the Deity could be satisfied by a study of "the frame of nature." That religious rationale for the study of nature may in turn have sustained and validated the nascent scientific enterprise in a still Christian Europe until the time arrived when science no longer had need of such support.

APPENDIX A

"Of natures obvious laws & processes in vegetation"

Dibner MSS 1031 B (part)
Dibner Library of the History of Science and Technology
Special Collections Branch
Smithsonian Institution Libraries
Smithsonian Institution, Washington, D.C.

Editorial note

Dibner MSS 1031 B was lot no. 113 in the Sotheby sale of Newton manuscripts in 1936. It was later purchased by Dr. Bern Dibner and designated as Burndy MS 16 in The Burndy Library, Norwalk, CT. A generous gift to the nation from The Burndy Library in 1976 placed it in the Dibner Library.[1]

1. The first public notice of the existence of this manuscript appeared in *Catalogue of the Portsmouth Collection*, (1, n. 27), Sect. II, (1), No. 31, p. 12, where it was described as an "Abstract of a treatise on Nature's obvious laws and processes in vegetation." Returned to Lord Portsmouth at that time, it was sold at auction in 1936: *Catalogue of the Newton Papers*, (1, no. 27), lot. no. 113, p. 17, where it was described as "[Vegetation of Metals: draft of a Short Treatise, *incomplete* (in English)] *about* 4500 *words*, 12 *pp*. autograph, with many corrections and alterations, *sm. 4to*." Dibner purchased it in London during or soon after World War II; when I pressed him for details about that transaction, he assured me that it seemed more important at the time to save it from the blitz than to keep records of that sort of thing, a sentiment with which I heartily concur since this manuscript more than any other has provided the clues for decoding the role of alchemy in Newton's thought. Through the Burndy Library's gift to the nation in 1976 the manuscript is now available for study in Washington, D.C., in the Dibner Library of the History of Science and Technology, Special Collections Branch, Smithsonian Institution Libraries, Smithsonian Institution. There it is described as: "*Vegetation of metals* [between 1660 and 1727]. [12] p.; 22 cm., Holograph notes on vegetation and the generation of minerals, as well as air, heat, fire, cold and God. Last page only in Latin," *Manuscripts of the Dibner Collection*, (1, n. 30), No. 80, p. 7. See also Dobbs, "Newton manuscripts at the Smithsonian Institution," (1, n. 30).

Table A-1. *Symbols used by Newton*

Symbol	Meaning	Symbol	Meaning
∇	water, H_2O	✳	ammoniac, NH_4Cl
☿	mercury, Hg	🜍	sulfur, S
⊖	common salt, NaCl	⊕	world
⦶	niter, KNO_3	B.M.	balneum mariae, a water bath

The manuscript consists of three small quarto booklets of four pages each, twelve pages in all, all closely written with many corrections and alterations. Almost all of it is in English, but at the end there is an inverted Latin section separated from the English by a slash. That is, the Latin begins on f. 6v and ends on f. 6r, where the last of it is contiguous to but upside down with respect to the end of the English section.[2] Although the handwriting is virtually identical in the English and Latin sections, probably the English was written first. Then later, but not much later, Newton elaborated in Latin on a few of the ideas in the main body of the composition, fitting the new sentences in the little space that was left at the end of the last booklet. The manuscript is untitled, but its incipit makes an appropriate title. All of the lengthy English section is transcribed here.

The handwriting indicates a date in the early 1670s. In addition, its content is closely related to the content of "An Hypothesis explaining the Properties of Light, discoursed of in my severall Papers," a document Newton sent to the Royal Society in 1675.[3] A date of about 1672 thus seems appropriate; probably Newton had these papers on his desk when he was drafting the "Hypothesis."

The overriding concern in editing the manuscript has been to adhere as closely as is feasible to the original. Neither Newton's erratic spelling nor his equally variable punctuation has been corrected. Parentheses and square brackets are Newton's own, and the reader will soon realize that they are not always closed in the most appropriate place, or indeed closed at all. However, all bars on terminal consonants and all ligatures have been silently expanded. Probable readings, for places in which the manuscript is virtually illegible, are placed in angle brackets, as are Newton's deletions. Editorial comments are also placed in angle brackets, but they are italicized as well to distinguish them from Newton's words. Newton's interlineations are indicated by arrows up and down. The symbolic notations of the original are retained in the transcription but are translated in Table A-1 in order of their appearance.

2. Folio 6v is reproduced in *Manuscripts of the Dibner Collection* (1, n. 30), Plate VII, facing p. 46. The entire manuscript is reproduced in facsimile in B. J. T. Dobbs, *Alchemical Death & Resurrection*, (1, n. 30), Appendix.
3. Newton, *Correspondence*, (2, n. 14), Vol. I, 362–82 (Newton to Oldenburg, 7 Dec. 1675).

[f. 1r] 1 Of Natures obvious laws & processes in vegetation:

2 That ⟨*illegible word and* vegetation of metalls is described to be don by the same laws by ye universall consent of the magi, *all deleted*⟩

2 That metalls vegetate after the same laws. ↗ Proved transitorily fr ↙ ffrom ye circumstances observed by miners, ↗ more fully from ↙ The consent of ye Sophy wth one another & wth natures processe, & ye strange distractions of all other chymists from both nature & one another. And ye corruptibility of all things

⟨3 That vegetation may be *and illegible word* ↗ though ↙ promoted by art is naturall., *all deleted*⟩

⟨4 That *and illegible word* natures process in vegetation are ↗ is ↙ best understood in ye simplest p, *all deleted*⟩

3 A description of their vegetation in the earth

4 A description of their vegetation in a glasse. & that this is as much naturall as tother

5 The circumstances in wch they agree wth plants & animalls. And of met trees by nature & Art

6 That vegetation ⟨proceeds from y^e^, *deleted*⟩ is ye sole effect of a latent spt & that this spt is ye same in all things only discriminated by its degrees of maturity & the rude matter This instancd in metalls &c. In fermentation of wines in Autumn, in Antypathys in ye contagiousnes of putrefaction ↗ In crocus metallorum. ↙

7 Of ye actions & passions of grosser matter & how far that is common.

8 Of the ⟨effects produced by, *deleted*⟩ the degrees of maturity in all kingdoms ⟨*illegible word, deleted*⟩ ↗ mixture ↙ putrefaction ⟨conjunction, vegetation &c How a *illegible word* might conserve its specie, *all deleted*⟩ ↗ conjunction vegetation &c. & that this is only observable universalls ↙

9 How things conserve their species & how a tree might bee conserved & nourished In the glass & that its probable those metalline trees in the earth grew after this manner ↗ & why the continuall mutations in mans body are insensible. Note in ye sam minerall are found divers metalls or ores ↙

10 Of seed & propagation in number bulk & quality. Why ye ejection of it debilitates. Why the product is somtimes male & somtimes female

11 Of protoplasts y^t^ nature can onely nourish, not form them, Thats Gods ⟨work, *deleted*⟩ ↗ mechanism ↙ y^t^ these natures

12 Why the two Elixirs are the most ⟨nourishing, *deleted*⟩ amicable & universall medicine to all beings what ever

⟨*Space of about ½ inch left in MS before next paragraph.*⟩

Notes ↗ of agreement ↙ . Y^e^ less diffence in maturity the quicker union & work, in mineralls; & hence animalls easy transmute &c. 2 Y^e^ lesse nourishment ye quicker & safer concoction. 3 Tis a safer work to imbibe gradually then give ye nourishment at onece. 4 heterogeneous impuritys are hurtful 5 a gentle heat. 6 nature ever begins wth putrefaction or fermentation wherby there is an intimate union & exertion of sp^ts^ ↗ & purgation of impuritys ↙ . 7 After putrefaction ye work is pretty secure, in y^t^ is ye main danger of miscarige. Hence diseases cheifly in the stomack & blood ↗ miscarring after first coition. Y^e^ egg

seldom miscarys ↙ 8 putrefaction exerts a spirit ↗ purgeth feces ↙ , 9 & makes an intimate mixture. ↗ & stink like that of a carcasse ↙ 10 After the term of conjunction vegetation admits of noe interruption. by cold ye compositum dyes, & by excessive heat. 11 Vegetation must bee performed in humido. 12 the motion stays at the maturity of ye first movers 12 suddein heats & colds prejudice maturity. 13 putrefaction seperates feces. ⟨*illegible word, deleted*⟩ hence urin gall &c 14 Totall putrefaction makes a black stinking rottennes 15 after conjunction the matter is apt to grow into all figures & colours though transitory because ye motion is not yet terminated. ⟨*illegible letter or figure, deleted*⟩ 16 In ye same Oare severall metalls are found all wch vegetate distinctly. 17 That salt cheifly excites to vegetation 18 That in ye first days of ye stone green is ye only permanent colour & so in ye least mature vegetables. 19 That sometimes metalls grow like trees in ye earth. 20 That nothing has so great power on animalls as mineralls witnes not only the Alkahest to destroy ↗ & ↙ ye Elixir to conserve but but their operations in common chymicall physick & in springs &c & therefore since wee live in ye air where their most subtile vapours are ever disperst wee must of necessity have a great dependance on them, witness healthfull & sickly yeares, ye barronnes of grownd over mines &c
[f. 1v] They therefore unite wth o^{r} bodys & become p^{t} of y^{m} wch they could not doe if they had not a principle of vegetation in them & that of the same disposion wth ye matter of ye rest of o^{r} bodys. therefore o^{r} bodys vegetate as they doe in a glas. 21 They are conjoyned like male & female. 22 The female seed is ever in greatest proportion 23 They are debilitated by emission of seed. 24 Y^{e} ⟨*illegible letter, deleted*⟩ Elixir is multiplyd in vertue & ye child is more vigorous & long livd y^{n} ye parents 25 There is a term of conjunction after fermentation & long decoction 26 at that terme ye matter dys not though ye matter cool for then birds lay their eggs. 27 After y^{t} term it dys if it cole

Dissimilitudes. 1 They retaine noe constant form, o^{r} bodys a constant one. 2 They are augmented in vertue as well as bulk, wch o^{r} bodys are not. 3 They grow not wthout a totall death & putrefaction, wch is ⟨*illegible word, deleted*⟩ beyond ye pour of nature to make o^{r} bodys doe. ↗ Resp: so is an infant generated from ye mixture of male & female seed ↙ 4 They grow wthout air, wch nothing els can doe 5 by groth they attain a faculty of communicating their vertue to others of wch their can bee noe instance in other vegetables. 5 they attain a supereminent fixity in their whole substance wch noe other vegetable can. 6 They can convert ⟨*illegible word or figures, deleted*⟩ ↗ 2 or 3 nay ↙ 10 ↗ or more ↙ times their owne weight ⟨*illegible word, deleted*⟩ of nourishment at once. Resp: they are mutually nourishment or ferment each to other. Also there ought to bee most of ye female seed. 7 ⟨They, *deleted*⟩ Every p^{t} of the whole is sperme, in other animalls a very small portion

⟨*Space of about ½ inch left in MS before next paragraph.*⟩

Of ye production of ye upper region from mineralls. 1 How mettalls dissolv in divers liquors to a saline or vitriolate substane. 2 Much more ought theire fumes while in as subtile & volatile a condition as water it selfe, ⟨to unite wth, *deleted*⟩ when they happen to pervade water to concret wth its p^{ts} ⟨into ye form of, *deleted*⟩ & impregnat it after the manner of salt or vitrioll. 3 They are imediately deprived of vegetation by the coldness of ye water. 4 Hence they are much more alienate from metalls then solutions of metalls & cannot be reduced to them ↗ by destillation as the other may ↙ , being concrete in a highly volatile & anomalus condition. 5 ↗ * ⟨*Newton's asterisk: see below for section to be inserted here*⟩ ↙ ⟨by their lying in ye water ⟨*illegible word, deleted*⟩ in vicissituds of heat & cold they are ↗ congealed ⟨indurated, *deleted*⟩ into clusters compacted & ↙ changed in som measure from a ↗ spirituall to a corporall from a ⟨*illegible word, deleted*⟩ ↙ volatile to a fixt condition. This instaned in Gur & minerall concrets in Corall & stony juices in Niter & Tartar in salts laid in the sun, yea & in vitriolls. ↗ in the christallizing of salts wch show a propensity to concret. & hence salt concrets in sand ↙ , *all deleted*⟩ 6 By this accidentall hardeing they are so much more alienated from a metalline state & made so difficult to be reduced back that they may well be thought as differing as salts and & vitriolls. Since therefore they are not vitrioll they must be salt. Thus is ye sea impregnated wth salt, & that most towars the equator where most fumes are raised. And this conclusion seems necessary for &c. 7 This salt is a concret of all sorts ⟨*illegible word, deleted*⟩ of metalline fumes w^{t}ever becaus all sorts pervade ye sea & are mixed promiscuously ↗ Hence the difficulty of precipitating it ↙ . 8 Because the sea is perpetually repleinshed wth fresh vapours it cannot bee freed from ↗ a ↙ salin tast by destillation that salt arising wth ye water wch is not yet ⟨indurated, *deleted*⟩ concreted to a a grosser body. * [5 moving up & down in the water they associate into little masses as they chance to meet & compact, & ⟨*illegible word or letter, deleted*⟩ so chang in some measure from a spirituall to a ⟨corporall, *deleted*⟩ ↗ grosse ↙ , from ↗ a ↙ volatile to a fixt condition such as is sea salt ↗ hence in destillation there is not only a gross salt left behind but also a subtil sp^t not yet coagulated ariseth wth y^e ▽ ↙ . And this will appear more then a conjectur by considering that these ⟨confus, *deleted*⟩ ↗ saline ↙ clusters have a propensity still to associate into greater clusters for by evaporation they will convene into pretty larg christalline cubes wch indurate more & more in the sun till they becom of a strong concrescense insoluble in water. Their propensity to ⟨concur may be also noted, *deleted*⟩ associate wth one another ⟨or any sollid substance they, *deleted*⟩ [f. 2r] may appear also by the edulcifiing ⟨*illegible word, deleted*⟩ ↗ of ↙ water in its passag through sand ffor as salt will ⟨sooner, *deleted*⟩ christalliz upon sticke put into water especially if ye water bee gently agitated, when it would not otherwise christallize, soe salt which in the maine sea would not chistallize yet by the waters motion through sand it is brought by little to settle on the sand ⟨& at last to eq, *deleted*⟩ & that the spirits as well as the grosser salt. Analogous to this is the generation of Tartar out of wine by the graduall association of subtler parts into grosser clusters till they at last coagulat into that stony concrete wch conteins copiously the most fixed of salts. ⟨], *deleted*⟩ Another instance is in vapors coagulating into clouds.]

8 In like manner the said ⟨watry, *deleted*⟩ vapors meeting wth water or watry vapours in mountains & cavitys of ye earth may concret into the same or other sorts of salt according to ye nature of the fumes ascending there. ⟨But if those that, *deleted*⟩ And after ye same manner they may associate themselves wth vapors in the aire & impregnate rain water ⟨but, *deleted*⟩ And where they meet wth

convenient substances to rest upon as some sort of rock & walls & ⟨*illegible word or letter, deleted*⟩ most earths in ye upper crust of the earth there they may associat in grosser ↗ christallin ↙ forme ⟨ & constitut, *deleted*⟩ (noe other way then they would coagulate ⟨and of, *deleted*⟩ ↗ in ↙ water ⟨*illegible letter, deleted*⟩ upon ye sides of ye vessel or sticks thrown into it) & so constitute ye salts wee find in such places

9 But ⟨as ☿, *deleted*⟩ it ⟨*illegible word, deleted*⟩ is to be observed y^{t} ⟨*illegible letter, deleted*⟩ fumes ↗ may ↙ unite wth fumes after another more intimate manner than ↗ they ↙ will doe wth liquors as may bee seene in subliming many things together thus ☿ ⟨unites wth y^{e}, *deleted*⟩ sublimed wth salts unites in another manner wth their spirit then if you should distill of y^{e} ⟨gross &, *deleted*⟩ spirits alone & dissolv it in the liquor So these minerall fumes meting wth ⟨y^{e}, *deleted*⟩ subtile invisible vapors ↗ leisurely & by degrees ↙ must concret wth them after another manner then they ⟨would, *deleted*⟩ doe ⟨w^{n}, *deleted*⟩ wth wth water or the same ⟨fumes, *deleted*⟩ ↗ vapors ↙ when they begin to concrete into water & becom visible ⟨*illegible word, deleted*⟩ clouds ↗ by wch they are at once overwhelmed & drowned ↙ . And these concretions ought in reason when they associate into a saline form to bee of a more open & subtile ⟨constitution, *deleted*⟩ ↗ ⟨constitu, *deleted*⟩ ↙ texture & constitution such as is Niter ↗ ⟨& as y^{t} sp^{t} being the ferment of fire & blood &c, *deleted*⟩ ↙ Nor is it strange y^{t} so slight causes should produce so ⟨*illegible word, deleted*⟩ different salts as ⊖ & ⦶ if wee consider y^{t} ye fixt salt left in ignition returns to ⦶ by dissolution, y^{t} ⦶^{s} vertue is impaired by refining, & ⟨y^{t} y^{t} y, *deleted*⟩ it is further confirmed by ye affinity of y^{t} sp^{t} wth niter. ↗ y^{t} is ↙ y^{e} ⟨*illegible word, deleted*⟩ ferment of fire & all vegetables ye other most apt to take fire & most ⟨*illegible word, deleted*⟩ promoting vegetation of all salts. Hence it appeare why ⦶ never found wthout Common salt ⟨becaus ye aire is ever replenished wth crasse, *deleted*⟩ ↗ ⟨the sam humidity while in its most subt, *deleted*⟩ ↙ ↗ becaus ye aire is ever replenished wth ⟨vap, *deleted*⟩ ↙ vapors (y^{t} is small drops of water) as well as subtil ⟨vapors, *deleted*⟩ ↗ ones ↙ & ye same humidity while subtile causes ⦶ & when it thickens ⊖. Hence also little or noe ⦶ is in ye sea⟨for its *illegible word* same *illegible word* insensible quantitys there may be in it *illegible word, all deleted*⟩ becaus ye grosse water stifles all or ye far greatest p^{t} of the exhalation the aire indeed is replenishd wth this exhalation from neibourig regions & so may impregnate rain water wth wth niter & so it may ⟨replenish, *deleted*⟩ receive niter from rivers but ⟨*illegible words, deleted*⟩ ye proportion is inconsiderable compared to all those vapors y^{t} arise into it. And all this will appear more then conjectur by considering 1 fums do arise ↗ plentifully ↙ , 2 ⟨and fumes, *deleted*⟩ they will abide wth water in a pellucid form & ↗ 3 ↙ therefore appear in evaporation of a saline forme. ↗ 4 ↙ they must therefore produce somthing like salt copiously 5 there are noe such products but ⊖ & ⦶ generally found. 6 These are generally washed down by ye descent of water hence ⦶ is most copious in houses & dry places, hence also the sea is salter y^{n} the earth 7 these salts would therefore soon ⟨*illegible word, deleted*⟩ vanish if they were not constantly new generated [f. 2v] & this is further confirmed by their bee plentifully produced in places where there was none before & where they could not bee had but out of y^{e} ⟨*illegible word, deleted*⟩ vaporous air nay

that it descends wth rain yet in that saline form ⟨*illegible word* ascend wth it, *deleted*⟩ ↗ it descends ↙ is two gros to ascend wth it [that tis noe stranger for it to praecipitate out of vapors upon rock y^{n} out of waters upon the sides of a vessell.] They ↗ are ↙ therefore constantly generated & that out of a most subtil vapor y^{t} ascends wth as little heat as water. ⟨*illegible word, deleted*⟩ Nor is it to bee thought that this vapor or these salts have noe other fountain but ye upper crust because y^{t} would soon bee exhausted. Since therefore ⟨*illegible word, deleted*⟩ minerall fumes will wth water coag into a salin form & salts are constantly made fumes & noe fumes so copiously ascend as minerall ones nor noe salts are so copiously generated as ⟨*illegible word, deleted*⟩ ⊖ & ⦶ w^{t} can one bee but ye product of ye other. [Note sea salt praecipitates on sand & soe is constantly washed down as new is generated. & y^{t} w^{t} water descend wth ⊖ ascends wth out it therefore ye sea grows noe salter.

Of sal ✳ . Sal gemmae &c. alume &c

How salt turns to ⟨*illegible words* stones chrystall & *illegible word, all deleted*⟩ a hard stony pellucid stony substance insoluble in water whence sand christall & all pellucid stony concrescences are produced & glorious gemms also if the matter be pure & pervaded by pure & variously coloured metalline fumes before ⟨*illegible word or letter, deleted*⟩ it be hard. This instanced in ye converting christall to rubys or other pretious stones by ye elixir. W^{t} water converts to stone of Gur & coral ↗ Petrification of wood by by water wch congealed in ye pores to stone ↙

That clay wch is a great ingredient of this upper crust is nothing but stone poudered. That salts may ↗ putrefy & ↙ by putrefaction ↗ will ↙ generate another sort of black ↗ ish ↙ rotten ↗ fat substance ↙ substance the ⟨*illegible word, deleted*⟩ most fertile part of this upper crust & ye nearest matter out of w^{ch} ⟨ye gr, *deleted*⟩ vegetables are extracted & into wch after death they returne. And this confirmed in that nothing promotes fermentation & putrefaction more y^{n} salts ⟨(though they *and illegible words, deleted*⟩ where they are incited to it, they are alone as I may say dead & have noe active principle of vegetation in them till they bee incited to it by other substances that are in a live & vegetating state of what they are mixed wthall & thence it is that ⟨they, *deleted*⟩ where nature is not powerfull enough to incite them to action they on the contrary ↗ retard & ↙ hinder her working And thence salted meats are so slow in putrefaction & so hard to digest. But if their latent principle can bee once exerted it shows it selfe more vigorously: Hence ⦶ seemes ye most fertile ⟨*illegible words, deleted*⟩ ↗ & inriching of land & ↙ lesse powerfull to conserve meats because not so close lockd up & difficult to putrefy. Here note whither ye praeparation of salts by putrefaction if it can bee truly attained would not prove a noble way of Physick 2 If ⊖^{ts} wch are a meane twixt ye the minerall & other kingdoms will vegetate why may not metalls & that as much more powerfully than ⦶^{ts} as ⦶^{ts} doe y^{n} other ⟨mine, *deleted*⟩ earth. 3 nay since metalls may putrefy into a black fat rotten stinking substanc why not ⟨*illegible word, deleted*⟩ ↗ salts ↙ also. And soe much for salts, stones, clay, & fat mold, the cheif ingredients of this upper ⟨*illegible word, deleted*⟩ earth

As for water it is to bee observed how the Alcahest (a minerall sp^{t} of ye same root wth that wch constantly ascends & pervades all things, only prepared by ye philosopher ⟨*illegible word, deleted*⟩ &c resolvs all these ↗ upper ↙ substances into water ↗ yea metalline ☿ it selfe ↙ . whence it appears not only that they have one common matter but ye ye minerall sp pervading all things may doe the like in som measure to them. 2dly what was dry ⟨*illegible word or letter, deleted*⟩ & grosse may a great p^{t} of it by putrefaction relent to water A carcas if distilld before putrefaction will leave a great deale of fixed earth but if it bee laid in the warm son & open air to putrefy it wil resolve & exhale ⟨most or all, *deleted*⟩ ↗ almost ↙ of it in tim into fumes [f. 3r] And putrefaction resolvs substances not only into ⟨sp^{t}, *deleted*⟩ water but oyle also as may be seen by their fatnesse & sp^{t} as in ye fermentation of beer [yea & aire two as is evident from ye swelling & bubling]

⟨Now *illegible word*⟩ the changes of ye minerall spirit being done by into salt stones water &c being done *illegible word or letter* all or most of them by gross mechanicall ways wthout vegetation & only by severall transpositions of p^{ts}, they seem to bee somany violations done to ye metalline nature (whose propper action in its species is only *illegible word or letter* vegetation) & so render the reduction of them back to that nature a very difficult work. Tis a work not to be done by vegetation, *entire paragraph, deleted*⟩

Now these things thus produced ⟨by y^{e}, *deleted*⟩ salt stones earth water &c seeme so aleinate ⟨& sto, *deleted*⟩ from ye metalline nature y^{t} one would scarce ⟨*illegible words, deleted*⟩ think they took their rise thence, ⟨*illegible letter, deleted*⟩ Nay they are at *p*fect enmity wth & if mixed doe hinder or destroy ye work. but ye reason's manifest. ffor being changed into these substances not by vegetation but for ye most part ↗ onely ↙ by a gros mechanicall transposition of *p*ts, they are ⟨not, *deleted*⟩ to bee reduced back by the same way, not by vegetation but by the same mechanicall ⟨restitution, *deleted*⟩ ↗ transposition ↙ till they bee reduced back to their first order & frame, ⟨*illegible word or letter, deleted*⟩ Since therefore vegetation is ye only naturall work of metalls & the reduction of these is besides y^{t} work & yet these cannot vegetate as they doe till they bee reduced, they must of necessity hinder their working & so be counted heterogeneous. for w^{t} will not comply will disturb ye acting.

Yet ye reduction of these ⟨*illegible letter, deleted*⟩ is possible to bee performed by ⟨*illegible words, deleted*⟩ mechanicall ways unravelling their production. Water by ye suns heat & by assention & descention will yeild earth as hath been tryed by distilling it often, Also ⟨*illegible word or symbol, deleted*⟩ ↗ standing water w ↙ will putrefy by ye suns heat, corrupt & let fall a foeculent earth & that ⟨suff, *deleted*⟩ successively wthout period. Out of these earths may bee extracted a salt. This salt may be brought to putrefy & ye minerall sp^{t} thereby set loose from ye water wth wch it was concreted & so returnes to ye same state it had ⟨when, *deleted*⟩ at ⟨*illegible letter, deleted*⟩ its first ascent out of ye earth y^{t} is to ye nearest metalline matter & (though debilitated by these changes) yet ⟨*illegible words, deleted*⟩ if pervading ye earth where other metalls vegetate might enter them ↗ receive metallick life ↙ & by degrees recover ⟨*illegible letter, deleted*⟩ ↗ their primitive ↙ metalline forme.

⟨*Remainder of f. 3r (about two inches) left blank.*⟩

[f. 3v] Air (by wch I mean ⟨y^{t} w^{ch}, *deleted*⟩ not vapours) but y^{t} wch cold will not condens to water may be generated 1 Out of water by freezing it. 2dly out of saline or vitriolate spirits by their ebullition when poured together. 3 out of salts & vitriolls in the drawing of their spirits, 4 out of metalls ↗ & som other substances ↙ by corroding them wth acid liquors as aqua fortis. (hence poyson swells a man) 5 by fermentation. Hence ebullition flying of bottle beer &c. swelling after a stroake. In generall by any meanes wher y^{e} ⟨internall, *deleted*⟩ parts of a body are set a working among themselvs. (W^{ch} seems to argue an agent in freezing.) that thereby ye constringed aire may bee let loos wch intimate ⟨ye origen of, *deleted*⟩ ↗ the ↙ earthly substances to be ⟨*illegible word, deleted*⟩ but AEthereall concretions that they so easily approach ⟨*illegible letters, deleted*⟩ ↗ towards ↙ it again. Of ye reduction of aire to a gross body I know but one instance & y^{t} in ye stone wher during ye firs solution much air is generated, enough to burst a weak glas & wch yet returns to ye stone againe

By minerall dissolutions & fermentations there is constantly a very great quantity of air generated wch perpetually ascends wth a gentle motion (as is very sensible in mines) ⟨*illegible letter, deleted*⟩ being a vehicle to minerall fumes & watry ⟨exhalations, *deleted*⟩ vapors, boying up the clouds & still (protruded by ye air ascending under it) riseth higher & hiher till it straggle into ye ethereall regions, Carriing also wth it many other vapors & exhalations & whole clouds too when they happen to bee so high as to loos their gravity. The quantity of air constantly generated may be aestimated by ye quantity of rains y^{t} fall of whose ascent we are as insensible as of ye airs. But better by the quantity of air compared to its ascent. As if it bee supposed to rise a mile in 3 or 4 days wch it may doe wth a very gentle ⟨motion, *deleted*⟩ & insensible motion: that would amount to 5 foot water in depth round ye earth. This constantly crowding for room ye AEther will becom prest thereby & so forced continually to descend into ye earth from whence ye air cam & there tis gradually condensed & interwoven wth bodys it meets there ↗ & promotes their actions beeing a tender fermet ↙ . but in its descent it endeavours to beare along w^{t} bodys it passeth through, that is makes them heavy & this action is promoted by the tenacious elastick constituon whereby it takes ye greater hold on things in its way; & by its vast swiftness. Soe much AEther ought to descend as air & exhalations ascend, & therefore ye AEther being by many degres more thin & rare then air (as air is y^{n} wather) it must descend soe much the swifter & consequently have soe much more efficacy to drive bodys downward then air hath to drive them up. And this is very agreeable to natures proceedings to make a circulation of all things. Thus this Earth resembles a great animall ↗ or rather inanimate vegetable ↙ , draws in aethereall breath for its dayly refreshment ↗ & vital ferment ↙ & transpires again wth gross ⟨*illegible letter, deleted*⟩ exhalations. And according to ⟨*illegible word, deleted*⟩ the condition of all other things living ought to have its times of beginning youth old age & perishing [This is the subtil spirit wch searches ye most hiden recesses of all grosser matter which enters their smallest pores & divides them more subtly then any other ↗ materiall ↙ power w^{t} ever. (not after ye way of common menstruums by rending them ↗ violently ↙ assunder &c) this is Natures universall agent, her secret fire, ⟨ye materiall soule of all matter, *deleted*⟩, y^{e} ⟨sole, *deleted*⟩ ↗ onely ↙ ferment & principle of ↗ all ↙ vegetation. The material soule of all matter wch being constantly inspired from above pervades & concretes wth it into one form & then if incited by a gentle heat actuates ⟨it & makes it vegetate, *deleted*⟩ & enlivens it but so tender & subtile is it wthall as to ⟨*illegible letter, deleted*⟩ vanish at ye least excess and (having once begun to act) to cease ↗ acting ↙ for ever & congeale in ye matter ⟨*illegible word, deleted*⟩ at ye defect of heat ↗ unless it receive new life from a fresh ferment ↙ . And thus perhaps ⟨*illegible words, deleted*⟩ a great pt if not all the moles of sensible matter is nothing but AEther congealed & interwoven into various textures whose life depends on that ⟨*illegible letters, deleted*⟩ pt of it wch is in a middl state, not wholy distinct & lose from it like ye AEther in wch it swims as in [f. 4r] a fluid nor wholly joyned & compacted together wth it under one forme in som degree ↗ condensed ↙ united to it yet remaining of a much more rare ↗ tender ↙ & subtile disposition & so this seems to bee the principle of its acting to resolve ye body & bee mutually condensed by it & so mix under one form ↗ being of one root ↙ & grow together ⟨*illegible word or letter, deleted*⟩ till ⟨they attain, *deleted*⟩ the compositum attain ye same state wch ye body had before solution. Hence 1 ye earth needs a constant fresh supply of aether. 2 Bodys are subtiliated by solution ⟨*A short dash drawn by Newton below this last paragraph separates it from what follows.*⟩

Note that tis more probable ye aether is but a vehicle to some ⟨*illegible word, deleted*⟩ more active spt. & ye bodys may bee concreted of both together, they may imbibe aether as well as air in generation & in y^{t} aether ye spt is intangled. This spt perhaps is ye body of light 1 becaus both have a prodigious active principle ↗ both are perpetuall workers ↙ 2 because all things may bee made to emit light by heat, 3 ye same ⟨heat, *deleted*⟩ cause (heat) banishes also the vitall principle. 4 Tis suitable wth infinite wisdom ⟨to derive effects from y^{e}, *deleted*⟩ not to multiply causes wthout necessity 5 Noe heat is so pleasant ↗ & brigh ↙ as ye suns,⟨*illegible letter, deleted*⟩ 6 light & heat have a mutuall dependance on each other & noe generation ↗ wthout heat ↙ . heat is a necessary condition to light & vegetation. * ⟨*Newton's asterisk: see below for section to be inserted here*⟩ 6 Noe substance soe indifferently, subtily & swiftly pervades all things as light & noe spirit searches bodys so subtily percingly & quickly as ye vegetable spirit. [* heate exites light & light & light exites heat, heat excites ye vegetable principle & that ⟨excites, *deleted*⟩ increaseth heat.]

⟨*Space of about one inch left in MS before next paragraph.*⟩

Of heat. agitates aether & that agitated stirrs up light & heat, aether is agitated by rushing in twixt p^{ts} or by sudden extrusion

Of light.

⟨*Space of about 1/4 inch left in MS before next paragraph.*⟩

Of fire is either ⟨*illegible word and* or flame, *deleted*⟩ a gross body ⟨or vapor. the first *and illegible word, all deleted*⟩ whose parts are in a vehement motion among themselves & excite it to emit light or flame. the first either wants a principle of praeserving that state as hot iron or borrows it from an externall agent as coale or hath it wthin it selfe as the sun. &c. The last is nothing but ye parts of fumes turned to coales.

⟨*Space of about ½ inch left in MS before next paragraph.*⟩

Cold & freezing have divers principles because a thing will freez wthout growing colder & grow colder wthout freezing. ↗ also becaus things freez not proportionably to their fluidity as in water oyle ☿ spt ↙ Cold is only rest, freezing is by an agent as fumes of lead coagulate ☿. Congelation is made when any agent ⟨enters, *deleted*⟩ ↗ settles on ↙ the *p*ts & makes y^{m} ⟨porus &, *deleted*⟩ rough or ↗ rather ↙ adhaeres to their out side & acquiesces by cold.

ffluidity is preserved by ye aethers agitation & smallness of *p*ts. ☿ composed of hard metalline globules.

Hardness & union of *p*ts by aethers extrusion & their roughness.

Volatility & fixity.

Action of Salts dissolving, promoting fusion, fighting, praecipitating.

Divers liquors coagulated & liquefyed by divers causes as whites of eggs by heat & lime, corall &c

propension & aversion to mix.

Pellucity opacenese. Elasticity. Expansiveness

[f. 4v] Of ye contrivance of vegetables & animalls. Of sensible qualitys. Of ye soules union

⟨*Space of about ½ inch left in MS before next paragraph.*⟩

Of God. What ever I can conceive wthout a contradition, either is or may ⟨effected, *deleted*⟩ ↗ bee made ↙ by something that is: I can conceive all my owne powers (knowledge, ⟨*illegible word, deleted*⟩ activating matter &c) wthout assigning them any limits Therefore such powers either are or may bee made to bee.

Example. ↗ All the dimensions imaginable are possible. ↙ A body by accelerated motion may ⟨becom infinitly long or, *deleted*⟩ trancend all ⟨space, *deleted*⟩ distance in any finite tim assigned. ↗ also it may becom infinitly long ↙ This if thou denyest tis because thou apprehendest a contradiction in the notion & if thou apprehendest none thou wilt grant it ⟨*illegible word, deleted*⟩ ↗ in the ↙ pour of things.

Arg 2. The world might have been otherwise then it is (because there may be worlds otherwise framed then this) Twas therefore noe necessary but a voluntary & free determination y^{t} it should bee thus. And such a voluntary [cause must bee a God]. determination implys a God. If it be said ye w^{ld} could bee noe otherwise y^{n} tis because tis determined by an eternall series of causes, y^{ts} to pervert not to answer ye 1st prop: ffor I meane not y^{t} y^{e} ⊕ might have been otherwise notwth-standing the precedent series of causes, but y^{t} ye whole series of causes might from eterity have beene otherwise here, ⟨because they as well as, *deleted*⟩ ↗ because they may be otherwise ↙ in other places

⟨*The remainder of f. 4v (about four inches) is blank, as well as about ½ inch at the top of f. 5r.*⟩

[f. 5r] Nothing can be changed from w^{t} it is wthout putrefaction. ↗ Of ☿. violent seperations & coalitions ↙

no putrefaction can bee wth out ⟨changing, *deleted*⟩ ↗ alienating ↙ the thing putrefyed from what it was

Nothing can bee generated ↗ or nourished ↙ (but of putrefyed matter) ⟨wthout *p*^{cedent} putrefaction, *deleted*⟩

All putrefyed matter is capable of having somthing generated out of it ↗ & in motion towards it ↙

All natures opperations are twixt things of differing dispositions. The most powerfall agent acts not upon it selfe

Her first action is to blend & confound mixtures into a putrifyed Chaos

⟨*illegible words, deleted*⟩ Then are they fitted for new generation or nourishment

⟨Shee never works but wth easy heat., *deleted*⟩

All things are ⟨ge, *deleted*⟩ corruptible

All things are generable

Nature only works in moyst substances

And wth a gentle heat

Art may set nature on work & ⟨*illegible word, deleted*⟩ promote her working in ye production of any thing what ever. * ⟨*Newton's asterisk: see below for section to be inserted here.*⟩ Nor is ye product less natural then if nature had produced it alone. Is ye child artificiall because ye mother took physick, or a tree less naturall ⟨beca, *deleted*⟩ which is planted in a garden & watered then that which grows alone in ye feild. or if a ⟨beast, *deleted*⟩ ↗ carcass ↙ bee put in a glasse & kept warm in B. M. y^{t} it may putefy ↗ & breed ⟨worms, *deleted*⟩ insects ↙ are not those ⟨worms, *deleted*⟩ ↗ insects ↙ as naturall as others bred in a ditch wthout any such artifice. * Thus an oak may stand 100 years wthout rotting But if it bee scraped thin & kept twixt moyst & dry it may soon by art be brought to dirt & praepared for a new generation. Thus metalls though in a massy body & above grownd where minerall humidity is but weak & thin are in mans memory observed to rot & though they may long persist in the earth wthout corruption yet duly ordered ⟨*illegible letter, deleted*⟩ and mixt wth ⟨an active, *deleted*⟩ ↗ due ↙ minerall humidity; may by art soon rot & putrefy & consc ⟨*last word broken off*⟩

Natures actions are either ⟨seminall, *deleted*⟩ ↗ vegetable ↙ or ↗ purely ↙ mechanicall (grav. flux. meteors. vulg. Chymistry)

The principles of her vegetable actions are noe other then the ⟨seeds, *deleted*⟩ ↗ seeds or seminall vessels ↙ of things those are her onely agents, her fire, her soule, her life,

The seede of things is all that substance in them that is attained to the ⟨*illegible word, deleted*⟩ ↗ fullest ↙ degree of maturity that ⟨that, *deleted*⟩ is in that thing ⟨*illegible letter, deleted*⟩ so that then being nothing more mature to act upon then they acquiesce.

Vegetation is nothing else but ye acting of w^{t} is most maturated or specificate upon that wch is ⟨*illegible letter, deleted*⟩ less specificate or mature to make it as mature as it selfe And in that degree of maturity nature ever rests.

The portion fully mature in all things is but very small, & never to bee seene alone, but only as tis inclothed wth watry humidity. The whole substance is never maturated but only that *p*t of it wch is most disposed ⟨during wch action, *deleted*⟩ The maine bulk being but a watry insipid substance in wch rather then upon wch the action is performed

Putrefaction is ye reduction of a thing from y^{t} maturity & specificateness it [f. 5v] had attained by generation

⟨*Here Newton drew a short line to separate this last paragraph from what follows.*⟩

1 All vegetables have a disposition to act upon other adventitious substances & alter them to their one temper & nature. And this is to grow in bulk as the alteration of ye nourishment may bee called groth in vertue & maturity or specificateness.

2 When y^{e} ⟨*illegible letter, deleted*⟩ nourishment has attained the same state wth the species transmuting the action ceaseth

3 And then is that body thus maturated able ↗ in like manner ↙ to act upon any new matter & transform it to its owne state & temper. Hence ye more to ye lesse mature is as agent to patient

⟨But it is not the whole substance of this nourishment that is transformed or of ye species y^{t} transforms but a very small substance in them both wch if it were seperated ye mass would bee but a ↗ mixture of ↙ dead earth & insipid water., *entire paragraph deleted*⟩

⟨*A short dash drawn by Newton separates the above from what follows.*⟩

7 Tis ye office ↗ therefore ↙ of those grosser substances to bee medium or vehicle in wch rather then upon wch those vegetable ⟨potentates, *deleted*⟩ ↗ substances ↙ perform their actions

8 Yet those grosser substances are very apt to ⟨bee, *deleted*⟩ put on various external appeanes according to the present state of the ⟨*illegible word, deleted*⟩ invisible inhabitant as to appeare like bones flesh ⟨*illegible words, deleted*⟩ wood fruit &c Namely they consisting of ⟨heterogeneous, *deleted*⟩ ↗ differing ↙ particles watry earthy saline aery oyly spirituous &c those parts may bee variously moved one among another according to the acting of the ⟨*illegible word, deleted*⟩ ↗ latent ↙ vegetable substances & be ⟨put, *deleted*⟩ variously associated & concatenated together by their influence

4 All these changes thus wrought ⟨*illegible words, deleted*⟩ in ye generation of things so far as to sense may appeare to bee nothing but ⟨severall *illegible word, deleted*⟩ mechanisme ⟨*illegible words, deleted*⟩ or severall dissevering & associating the *p*ts of ye matter acted upon & that becaus severall changes ⟨*illegible word, deleted*⟩ to sense may be wrought by such ways wth out any interceding act of vegetation. Thus acid two pouders mixed each to a 3^{d} colour, ↗ ye unctuous pts in ↙ milk by a little agitation concret into one mass of butter Nay all ye operations in vulgar chemistry (many of wch to sense are as strange transmutations as those of nature) are but mechanicall coalitions ↗ or seperations ↙ of particles as may appear in that they returne into their former natures if reconjoned or (when unequally volatile) dissevered, & y^{t} wthout any vegetation.

5 So far therefore as ye same changes may bee wrought by the slight mutation of the tinctures of bodys in common chymistry & such like experiments many may judg that ⟨there is noe other cause that will, *deleted*⟩ such changes made by nature are done ye same way that is by ye sleighty transpositions of ye grosser corpuscles, for upon their disposition onely sensible qualitys depend. But so fast as by ⟨generation, *deleted*⟩ ↗ vegetation ↙ such changes are wrought as cannot bee done wthout it wee must have recourse to som further cause And this difference ⟨is seen clearest in fossile substances, *deleted*⟩ is vast & fundamentall because

nothing could ever yet bee made wthout vegetation wch nature useth to produce by it.

[note ye instance of turning Iron into copper. &c.]

6 There is therefore besides ye sensible changes wrough in ye textures of ye grosser matter a more subtile secret & noble ⟨change wrought, *deleted*⟩ ↗ way of working ↙ in all vegetation which makes its products distinct from all others & ye immeadiate seate of thes [f. 6r] operations is not ye whole bulk of matter, but rather an exceeding subtile & ↗ inimaginably ↙ small portion of matter diffused through the masse wch if it were seperated there would remain but a dead & inactive earth. And this appeares in that ⟨*illegible words, deleted*⟩ vegetables are ⟨by any, *deleted*⟩ deprived of their vegetable vertue ⟨either by the fligh, *deleted*⟩ by any small ⟨*illegible word or letter, deleted*⟩ excesse of heat, the tender spirit being either put to flight or at least corrupted thereby (as may appear in an egg) whereas those operations wch depend upon ye texture of ye grosser matter (as all those in common chemistry do) receive noe dammage ⟨there, *deleted*⟩ by heats far greater. Besides if we consider an ⟨chicken, *deleted*⟩ ↗ egg ↙ , noe doubt but when it is first sat upon the whole matter in wch ye vegetive vertue resides ⟨*illegible word, deleted*⟩ is put into action ⟨*illegible word, deleted*⟩ wch if it were ye whole substance ye rudiments would be spread all over wch yet is begun but in a very little space.

This vegetable spirit is radically the same in all things & differs ⟨*illegible word, deleted*⟩ but in degre of digestion or maturity from the state of corruption ⟨(as metals, *deleted*⟩ differ ⟨*differ not deleted but Newton probably meant it to be*⟩. or as it is applyed to gros matter (viz as metalls differ in both respects)

And it has but but one law of acting that as when two vegetables spirits are mixed of unequall maturity they fall to work, putrefy mix radically & so proceed in perpetuall working till they arrive at the state of the les digested & if nothing hinder they still proced to ye state of the more digested where they infallibly stop

Hence it appears how nutrition is made yea by salves outwardly applyed

As metalls in divers states of digestion put on severall forms & unite after divers manners wth grosser matter so other vegetable powers

In animals &c the putrefaction is not sensible 1 becaus it is continuall ⟨*illegible words, deleted*⟩ 2^{d} quickly finished, 3dly but of a very small portion at once, ⟨a greater, *deleted*⟩ ↗ a greater ↙ portion at the same time being in motion to maturity & a far greater portion then that allready mature, by whose contrary dispositions tis allayed as to ye production of any eminent sensible qualitys. ↗ & that all together seems but one continued action of growing ↙ 4 Tis not like ye putrefaction of a carcass ⟨*illegible word, deleted*⟩ a confusion of the whole where all opposite qualitys are alike powerfull to destroy each other but ⟨rather, *deleted*⟩ much more mild ↗ the matter having already been reduced by ye putrefactions in vivo ↙ & its methodized by ye great potency of the body both in quantity & power to convert it to its owne temper. Tis more like ye putrefaction or fermentation of wine or bread or mault. 5 its putrefaction or fermentation in turning to blood is insensible yea & that more grosse in the stomak 6 Wthout such putrefaction how can vermin breed in ye body & yet y^{t} is in insensible

There are exhalations dispersed from every *pt* of the body ⟨to every *pt*, *deleted*⟩ as well as into ye aire & in this sense it may be said to bee totum in toto &c like ye foetus in sperm. And these emanations are the fountain of their sperm & perform ye same office to ye body wch sperm doth in the production of an infant, praeparing & distributing nourishment for they being of the same nature wth ye body (excepting order) they must act after the same maner upon adventitious matter that is prepare it.

APPENDIX B

"Hermes"

Keynes MS 28
King's College, Cambridge

Editorial note

Keynes MS 28 was lot no. 31 in the Sotheby sale of Newton manuscripts in 1936, described then as "Hermes Trismegistus. Tabula Smaragdina [et Commentarium] and a Translation into English of the Tabula, 3½ *pp.* Autograph *sm. 4to.*" Acquired by John Maynard Keynes, 1st Baron Keynes, this manuscript and also between one-third and one-half of all the Newton manuscripts dispersed by the 1936 sale were left by Lord Keynes at his death to King's College, Cambridge, where they now comprise the Keynes Collection.[1]

In its present form Keynes MS 28 consists of an English version of the *Emerald Tablet* of Hermes Trismegistus, a Latin version of the same material (with some variation), and a Latin *Commentarium* on the *Emerald Tablet* that is Newton's own composition. By permission of the Provost and Scholars of King's College, Cambridge, all three items are here transcribed (omitting from Keynes MS 28 only two title pages that each bear the single word "Hermes") and are followed by an English translation of Newton's *Commentarium*. I have expanded the Latin abbreviations for "que" and for final consonants and omitted all indication of ligatures; otherwise the transcriptions adhere as closely as possible to the original, following Newton's spelling and punctuation and indicating Newton's interlineations by arrows up and down. His deletions are placed in angle brackets; editorial comments are also placed in angle brackets but are italicized to distinguish them from Newton's words. The annotations that follow the English version of the *Emerald Tablet* and also follow the *Commentarium* are Newton's own.

The foliation of Keynes MS 28 seems peculiar at first, but the difficulties disappear with the recognition that part of another item in the Keynes Collection, Keynes MS 27, was once combined with part of Keynes MS 28. Keynes MS 27 was lot no. 30 in the Sotheby sale of 1936, described as "[Hermes Trismegistus]

1. Two items in the earliest description of the Newton alchemical papers probably refer to the papers now known collectively as Keynes MS 28: *Catalogue of the Portsmouth Collection* (1 n. 27), Sect. II, (1), No. 9 ("Hermes. Tabula smaragdina et commentarium") and No. 10 ("The same in English"), p. 11. No English commentary survives, however, and, unless lost between 1888 and 1936, may never have existed. See also *Catalogue of the Newton Papers* (1, n. 27), lot no. 31, p. 5; Munby, "The Keynes Collection" (6, n. 13); Dobbs, "Newton's *Commentary*" (1, n. 5).

'The Seven Chapters' [A Treatise on Transmutation] *about* 4750 *words*, 19 *pp.* Autograph, with many alterations and re-writings *sm. 4to.*"[2] Although Keynes MSS 27 and 28 do not constitute all of his work on Hermes, they both were part of Newton's long-term study of the Hermetic texts, as a step-by-step examination of the two manuscripts will make clear.

In Keynes MS 28, ff. 1r–4v form one physical unit, made by folding a folio sheet in four and cutting the fold between f. 1 and f. 2 to make a little booklet. Folio 1r bears the word "Hermes" as a title. The English *Emerald Tablet* in the bold confident handwriting of the late 1680s and early 1690s follows on f. 2r,v, with annotations at the end of it (f. 2v) in a later hand. Ink stains on ff. 2v–3r of Keynes MS 28 match stains on a part of Keynes MS 27.

The ink stains in Keynes MS 27 (the English *Seven Chapters*) are on ff. 2r and 13v, which is the second half-folio of Keynes MS 27. The stains indicate quite clearly that the English *Seven Chapters* at one time fitted in just after the English *Emerald Tablet* of Keynes MS 28, and the English *Seven Chapters* is in the same bold confident handwriting as that of the English *Emerald Tablet*. The English versions of the two Hermetic texts were thus written about the same time (late 1680s or early 1690s) and are Newton's own translations from French versions found in the two-volume *Bibliothèque de philosophes [chymique]* (Paris: 1672–8).[3]

The part of Keynes MS 27 that was once included in the first physical unit of papers in Keynes MS 28 was the first part of the present Keynes MS 27 to be written. It includes Newton's entire translation of the *Seven Chapters*. However, at some later time, presumably toward the end of the 1690s or even later, Newton separated his two translations. He then gave the *Seven Chapters* a new cover (now ff. 1r, v and 14r, v of Keynes MS 27), wrote in a late hand on the front of the cover (f. 1r of Keynes MS 27) "The Contents of ye 7 Chapters," and provided his translation with an analytical table of contents. Probably Newton added the late annotations to his English *Emerald Tablet* at about the same time.

To summarize, parts of both Keynes MSS 27 and 28 (the translations) were written in the late 1680s or early 1690s. Parts of both manuscripts (the annotations to the English *Emerald Tablet* and the analytical table of contents for the English *Seven Chapters*) were written late in the 1690s or even later, probably at the time the two translations were physically separated from one another.

The remainder of Keynes MS 28, however, was written at an earlier period. The first physical unit of papers, ff. 1r–4v, is blank after the ending of the English *Emerald Tablet* and the later annotations on f. 2v, except of course for the telltale inkstains; that is, there is no writing on ff. 3r–4v of Keynes MS 28. But there is

2. *Catalogue of the Newton Papers* (1, n. 27), lot no. 30, p. 5; *Catalogue of the Portsmouth Collection* (1, n. 27), Sect. II, (2), No. 34, p. 14 ("The seven chapters of Hermes, with part of an unfinished letter on the back"). The letter was later described more fully by Sotheby's: "At the end is the Autograph Draft of a Letter (unsigned) regarding the use of the circle in constructing equations with 3 real roots. (9 *lines*)."
3. Cf. supra, Chapter 6 at notes 21–39.

a second unit of papers, consisting of five half-folios folded together, Keynes MS 28, ff. 5r–10v. The outermost (ff. 5r,v and 10r,v) serves as a cover for the entire second unit. It is blank except for the single word "Hermes" on the front (f. 5r) in a large, elaborate, undatable "title-page" script. The Latin *Emerald Tablet* and the *Commentarium* come next (ff. 6r–7r), and the remainder of the unit is blank, except that the interlineated word "generant" spills over in one place (from f. 6v to f. 9r, which two folios are a part of the same sheet of paper, the second one in the unit). The handwriting is smaller and earlier than that of the translations and the later added material in Keynes MSS 27 and 28 previously discussed. In fact, the handwriting of the *Commentarium* is indistinguishable from that of the various versions of *De motu* and other papers of Newton associated with the early stages of the writing of the *Principia* in 1684 and early 1684/5.[4] A date of 1680–4 for the Latin material in Keynes MS 28 seems appropriate on the basis of the handwriting.[5]

A plausible chronological reconstruction of the history of the papers in Keynes

4. I am grateful to Dr. Michael Halls, Modern Archivist of King's College Library, Cambridge, for providing a xerox copy of Newton's *Commentarium* from Keynes MS 28 that enabled me to make this direct comparison with the tracts on motion and on gravity held by University Library, Cambridge, in particular with *De motu Corporum*, ULC MS Add. 3965.5, f. 21r, called the "initial revise" by Whiteside and dated by him to the winter or early spring of 1684–85. Cf. Newton, *Mathematical Papers* (5, n.15), vol. VI, 92–3. The handwriting of the *Commentarium* is also identical with that of ULC MS Add. 4003, *De Gravitatione et aequipondio fluidorum et solidorum in fluidis scientiam*, dated by the present writer to 1684 or early 1684/5 on the basis of Newton's conceptual development and the conceptual relationships of that manuscript with a large number of other documents: cf. supra, Chapter 5, esp. at notes 41–63.
5. Cf. Alan E. Shapiro, "Beyond the Dating Game: Watermark Clusters and the Composition of Newton's *Opticks*," (in press). Shapiro has systematically studied the watermarks in the surviving papers used by Newton in his many fields of study, utilizing papers datable by other means to establish the probable dates for different watermarks, then utilizing the watermarks to help date otherwise undatable papers. The three half-sheets in question, Keynes MS 28, ff. 5r.–10v, have only the countermark and not the full watermark that would help establish the date of writing more fully. However, Shapiro has concluded with some certainty that these sheets from Keynes MS 28 are from the same batch of paper Newton utilized for a series of chemical experiments dated "Aug. 1682" (Cambridge University Library, Add. MS 3973.4, ff. 13–16). Most of the paper with that watermark, however, was used by Newton in the early 1690s, and Shapiro comments as follows: "Since only four sheets of this paper seem to have been used before the early 1690s, it is tempting to believe that after using a few sheets in 1682, the stack of paper was buried in his [Newton's] laboratory and finally uncovered a decade later." Though Shapiro's argument is clearly not conclusive for the dating of Keynes MS 28, ff. 5r.–10v, he has assured me that his conclusion is consistent with my dating "and can fairly be taken as supporting it" (personal communication, 16 November 1989).

MS 28 can now be made. In the early 1680s Newton copied a Latin version of the *Emerald Tablet*, now on f. 6r.[6] He then wrote his commentary on it, now ff. 6v–7r, and, putting a cover around these pages, laid them aside. A decade or so later, "ye ffrench Bibliotheque" (f. 2r) having fallen into his hands, he made translations of the Hermetic material he found there and put another cover around those sheets. Later still, perhaps another decade later, he reorganized the two sets of material, putting both versions of the *Emerald Tablet* with the commentary and adding a few annotations to the English version. The two sets of papers in Keynes MS 28 remain physically distinct and were treated as separate items in the nineteenth-century catalogue.[7] Combined in lot 31 of the Sotheby sale, however, they now constitute Keynes MS 28.

[f. 2r] Tabula Smaragdina.
Hermetis Trismegistri
Philosophorum patris.

Tis true without lying, certain & most true.

That wch is below is like that wch is above & that wch is above is like y^{t} wch is below to do ye miracles of one only thing

And as all things have been & arose from one by ye mediation of one: so all things have their birth from this one thing by adaptation.

The Sun is its father, the moon its mother, the wind hath carried it in its belly, the earth is its nourse. The father of all perfection in ye whole world is here. Its force or power is entire if it be converted into earth.

Separate thou ye earth from ye fire, ye subtile from the gross sweetly wth great indoustry. It ascends from ye earth to ye heaven & again it descends to ye earth & receives ye force of things superior & inferior.

By this means you shall have ye glory of ye whole world & thereby all obscurity shall ⟨*illegible word, deleted*⟩ fly from you.

Its force is above all force. ffor it vanquishes every subtile thing & penetrates every solid thing.

So was ye world created.

From this are & do come admirable adaptations whereof ye means ↗ (or process) ↙ is here in this.

[f. 2v] Hence I am called Hermes Trismegist, having the three parts of ye philosophy of ye whole world

That wch I have said of ye operation of ye Sun is accomplished & ended.

See ye ffrench Bibliotheque. Theatrum Chemicum vol. 6. p. 715. & Vol 1 p 362 et p 8 et p 166 ↗ & p 685 ↙ et vol 4 p 497

6. I have been unable to trace the exact source of this version of the *Emerald Tablet*, but by 1680 Newton had had access for over a decade to two very similar versions in the six-volume *Theatrum chemicum* that he purchased in 1669. Cf. *Theatrum chemicum* (3, n. 44), vol. I, 8, and vol. VI, 715; Harrison, *Library* (1, n. 6), item 1608, p. 249, now missing.
7. Cf. App. B, n. 1 supra.

[f. 6r] Hermetis Trismegisti
opera Chemica.
Tabula Smaragdina.

Verum est sine mendacio, certum et verissimum. Quod est inferius est sicut id quod est superius et quod est superius est sicut id quod est inferius ad perpetranda miracula rei unius. Et sicut res omnes fuerunt ab uno meditatione et consilio unius: ita omnes res nascuntur ab hac una re adaptione. Pater ejus est sol, Mater ejus est Luna Portavit illum Ventus in ventre suo, Nutrix ejus est Terra. Pater omnis perfectionis totius mundi est hic. Vis ejus est integra si versa fuerit in terram. Separabis terram ab igne, subtile a spisso suaviter magno cum ⟨diligentia et, *deleted*⟩ ingenio. Ascendit a terra in coelum iterumque descendit in terram & recipit vim superiorum & inferiorum. Sic habebis gloriam totius mundi et fugiet a te ⟨tenebrae et, *deleted*⟩ omnis obscuritas. Haec est enim totius fortitudinis fortitudo fortis. Nam vincet omnem rem subtilem omnemque solidam penetrabit. Sic Mundus creatus est. Hinc erunt adaptiones mirabiles quarum modus est hîc. Itaque vocatus sum Hermes Trismegistus habens tres partes philosophiae totius mundi. Completum est quod dixi de opere solari.

Commentarium.

Quae sequuntur verissima sunt. Inferius et superius, [a] fixum et volatile, sulphur et argentum vivum similem habent naturam et sunt una res ut vir et uxor. Nam solo digestionis et maturitatis gradu differunt ab invicem. Sulphur est argentum vivum maturum, et argentum [f. 6v] vivum est sulphur immaturum; ⟨ad eoque, *deleted*⟩ ↗ et propter hanc affinitatem ↙ coeunt ut mas & foemina et agunt in se invicem ↗ et per actionem illam transmutantur in se mutuo & prolem nobiliorem generant ↙ ad perpetranda miracula hujus rei unius. Et sicut res omnes ex uno Chao per consilium Dei unius creatae sunt, sic in arte nostra res omnes id est elementa quatuor ex una hac re quae nostrum Chaos est per consilium Artificis & prudentem rerum ⟨dispositionem, *deleted*⟩ adaptionem nascuntur. Est et ejus generatio humanae similis, nimirum ex patre et matre qui sunt Sol et Luna. Et quando per horum coitum Infans concipitur, ⟨*illegible word, deleted*⟩ gestatur is in ventre ⟨venti nostri, *deleted*⟩ ↗ ⟨Draconis, *deleted*⟩ venti ↙ ad usque nativitatis horam, & post nativitatem nutritur ad ubera ⟨Terrae nostrae cum lacte virginalis, *deleted*⟩ ↗ ⟨Latonae dealbatae, *deleted*⟩, ⟨albae foliatae Terrae fo, *deleted*⟩, Terrae foliatae ↙ donec adolescat. ↗ Hic ⟨Draco, ventus, *deleted*⟩ ventus est balneum Solis et Lunae, & Mercurius, & Draco et Ignis qui tertio loco succedit ut operis gubernator: et ⟨Latona, *deleted*⟩ ↗ terra ↙ nutrix est Latona abluta ⟨et foliata alba vel rubea, *deleted*⟩ & mundificata, quam utique AEgyptij pro Dianae et Apollinis id est tincturae albae et rubrae nutrice habuere. ↙ Fons omnis perfectionis totius mundi est hîc. Vis ⟨*illegible word, deleted*⟩ et efficacia ejus est integra ⟨si versa fuerit in terra, *deleted*⟩ & perfecta si per decoctionem ad rubedinem et multiplicationem & fermentationem vertatur in terram fixam. Debet autem prius mundificari separando elementa suaviter et paulatim absque violentia & faciendo ut materia tota per sublimationem ascendat in coelum, ac deinde per sublimationis reiterationem descendat in terram: qua ratione acquiret ⟨naturam tam, *deleted*⟩ ↗ et vim penetrantem ↙ spiritus ⟨ob subti, *deleted*⟩ & vim fixam corporis. Sic habebis gloriam totius mundi et ⟨claritate summa fulgebis, *deleted*⟩ ↗ fugiet a te omnis obscuritas et omnis inopia et aegritudo. ↙ . Nam haec res ubi per solutionem & congelationem ascenderit in coelum et descenderit in terram evadet omnium rerum fortissima. Vincet enim & coagulabit omnem rem subtilem et omnem solidam penetrabit ac tinget. Et quemadmodum mundus ex Chao tenebroso per lucis productionem ↗ ⟨creatus fuit et per, *deleted*⟩ ↙ & separationem firmamenti aerei et aquarum a terra, ⟨creatus fuit, *deleted*⟩ ↗ creatus fuit ↙ sic opus nostrum ex Chao ⟨tene, *deleted*⟩ nigro & materia sua prima per separationem elementorum et illuminationem materiae originem ducit. Unde oriuntur adaptiones et dispositiones mirabiles in opere nostro quarum modus est hic in mundi creatione [f. 7r] adumbratus. Ob hanc artem vocor Mercurius ter maximus habens tres partes philosophiae totius mundi, ut significetur Mercurius philosophorum qui ex substantijs tribus fortissimis componitur habetque corpus animam et spiritum & est mineralis vegetabilis et animalis & in regno minerali vegetabili et animali dominatur.

Annotationes.

[a] Avicenna in ↗ cap. 4 ↙ Tractatuli sui ⟨*illegible word, deleted*⟩ ↗ qui extat ↙ in Artis Auriferae vol. 1. p. 268. Senior in Th. Ch. v. 5. p. 222.

Translation of the *Commentarium*

Commentary

The things that follow are most true. Inferior and superior,[a] fixed and volatile, sulfur and quicksilver have a similar nature and are one thing, like man and wife. For they differ one from another only by the degree of digestion and maturity. Sulfur is mature quicksilver, and quicksilver is immature sulfur; and on account of this affinity they unite like male and female, and they act on each other, and through that action they are mutually transmuted into each other and procreate a more noble offspring to accomplish the miracles of this one thing. And just as all things were created from one Chaos by the design of one God, so in our art all things, that is the four elements, are born from this one thing, which is our Chaos, by the design of the Artificer and the skillful adaptation of things. And this generation is similar to the human, truly from a father and mother, which are the Sun and the Moon. And when the Infant is conceived through the coition of these, he is borne continuously in the belly of the wind until the hour of birth, and after birth he is nourished at the breasts of foliated Earth until he grows up. This wind is the bath of the Sun and the Moon, and Mercurius, and the Dragon, and the Fire that succeeds in the third place as the governor of the work: and the earth is the nurse, Latona washed and cleansed, whom the Egyptians assuredly had for the nurse of Diana and Apollo, that is, the white and red tinctures. This is the source of all the perfection of the whole world. The force and efficacy of it is entire and perfect if, through decoction to redness and multiplication and fermentation, it be turned into fixed earth. Thus it ought first to be cleansed by separating the elements sweetly and gradually, without violence, and by making the whole material ascend into heaven through sublimation and then through a reiteration of the sublimation making it descend into earth: by that method it acquires the penetrating force of spirit and the fixed force of body. Thus will you have the glory of the whole world and all obscurities and all need and grief will flee from you. For this thing, when it has through solution and congelation ascended into heaven and descended into earth, becomes the strongest of all things. For it will constrain and coagulate every subtle thing and penetrate and tinge every solid thing. And just as the world was created from dark Chaos through the bringing forth of the light and through the separation of the aery firmament and of the waters from the earth, so our work brings forth the beginning out of black Chaos and its first matter through the separation of the elements and the illumination of matter. Whence arise the marvellous adaptations and arrangements in our work, the mode of which here was adumbrated in the creation of the world. On account of this art Mercurius is called thrice greatest, having three parts of the philosophy of the whole world, since he signifies the Mercury of the philosophers, which is composed from the three strongest substances, and has body, soul, and spirit, and is mineral, vegetable, and animal, and has dominion in the mineral kingdom, the vegetable kingdom, and the animal kingdom.

[a] Avicenna, Chapter 4, his little tracts published in *Artis Auriferae* vol. 1, p. 268. Senior in *Theatrum chemicum* vol. 5, p. 222.

APPENDIX C

"Out of La Lumiere sortant des Tenebres"

Yahuda MS Var. 1, Newton MS 30
Jewish National and University Library
Jerusalem
and
Babson MS 414 B
The Sir Isaac Newton Collection
Babson College Archives
Babson Park, Massachusetts

Editorial note

"Out of La Lumiere sortant des Tenebres" contains Newton's translated abstracts from *La lumière sortant par soy même des tenebres*,[1] a book written pseudonymously in Italian verse, translated anonymously into and commented upon in Latin, and finally translated anonymously into French (both text and commentary).[2] The original in Italian verse was said to be by "Marc-Antonio Crassellame Chinese," evidently a pseudonym. However, two copies of the Latin version of the work, which are now in the Bibliothèque Nationale, contain manuscript notes treating the pseudonym as an anagram and deriving from it "Otto Tacchenio chilense starmoso."[3] Consequently, many libraries now assign the authorship of the verses to Otto Tachenius, whose *Hippocrates chimicus* (Venice, 1666, with many subsequent editions and translations) was well known in the seventeenth century.

The translation here is into English, and there is no reason to think that the translation is not Newton's own. He owned the book, and it is still extant, much dog-eared, among the Newton books in the Wren Library of Trinity College,

1. *La lumière* (1, n. 37).
2. Ferguson, *Bibliotheca Chemica* (3, n. 43), Vol. I, 180–1.
3. *Catalogue Général des Livres Imprimés de la Bibliothèque National. Auteurs* (Ministère de l'Éducation Nationale; 231 vols. in 232; Paris: Imprimerie Nationale, 1897–1981), CLXXXI, columns 599–601, esp. col. 600, items R.42399 and R.42398.

Cambridge.[4] Newton did a number of translations from French to English or Latin during the later 1680s and early 1690s, the manuscripts of which show peculiarities similar to those of "Out of La Lumiere sortant des Tenebres": little wavy lines used as space fillers, alternative translations interlineated, and changes in the text that are best interpreted as attempts to improve the translation.[5]

Two sets of Newtonian notes on the book are extant: one in Newton's bold, confident handwriting of the late 1680s and early 1690s and the other in a later hand. The sheets comprising the two sets seem to have been inadvertently mixed at some time, even before they were catalogued by the nineteenth-century syndicate appointed to examine the papers for the University of Cambridge, because in that catalogue they are listed as "Extracts 'ex lumine de tenebris' " and "Out of 'La Lumière sortant des Ténèbres'...and commentary thereon (1 f.), but incomplete."[6] At the time of the Sotheby sale in 1936, however, two folio sheets (rather than the single one of the syndicate's description) were sold as lot no. 40: "Lumina de Tenebris. 'Out of La Lumiere sortant des Tenebres' [Abstracts, in English], *about* 1700 *words*, 4 *pp*. Autograph."[7] The two folio sheets of Sotheby lot no. 40 constituted the first part of the earlier set of notes, and they now comprise Yahuda MS Var. 1, Newton MS 30, Jewish National and University Library, Jerusalem. The first folio bears the title as given here, and the second folio continues Newton's abstracts in orderly fashion; but the second folio failed to complete this earlier set of notes. Two more folio sheets in an identical hand continue the same set of notes, but they were not sold in Sotheby lot no. 40. Probably they had been placed with the other manuscript that depends on *La lumière*, in the nineteenth century or perhaps even before. In any case they were sold in 1936 as a part of Sotheby lot no. 41: "Lumina da Tenebris [Notes and Abstracts, in English] *about* 2500 *words*, 5 *pp*. Autograph."[8] Sotheby lot no. 41 consisted of two parts. The first, a single folio sheet closely written front and back in a later hand, is entitled "Lumina de tenebris" and contains abstracts of Newton's original abstracts; it is now Babson MS 414 A, The Sir Isaac Newton Collection, Babson College Archives, Babson Park, MA. The second, now Babson MS 414 B at the same location, contains the last two folios of the original translated abstracts of the late 1680s or early 1690s.[9] Reproduced here is the complete set of the original translated abstracts that thus derive from two sources: Yahuda MS Var. 1, Newton MS 30, and Babson MS 414 B. The most likely date

4. Harrison, *Library* (1, n. 6), item 1003, p. 184, now Trinity College NQ.16.117.
5. This series of translations is discussed in detail in Chapter 6.
6. *Catalogue of the Portsmouth Collection* (1, n. 27) Sect. II, (5), 6 and 20, respectively, pp. 17–18.
7. *Catalogue of the Newton Papers* (1, n. 27), p. 8.
8. Ibid.
9. *A Descriptive Catalogue of the Grace K. Babson Collection of the Works of Sir Isaac Newton and the Material Relating to Him in the Babson Institute Library, Babson Park, Mass. With an Introduction by Roger Babson Webber* (New York: Herbert Reichner, 1950), p. 191. The Babson description is somewhat misleading: "Most of p. 2 has been rewritten on p. 3 and half of p. 4, on better paper and in more legible writing." In actuality, pp. 3–4 (Babson MS 414 B) were written first, as noted above.

Table C-1. *Symbols used by Newton*

Symbol	Meaning
☿	mercury
☉	gold
⊕	vitriol (a sulfate, perhaps of iron or copper)
🜍	sulfur

for this set of papers lies between 1687, when the book was published, and about 1692, when Newton's handwriting changed its form.

Although most of the material in these abstracts is not of Newton's own composition, since the material is abstracted rather than simply transcribed, it contains the parts of the book that Newton found important. Then, too, there are a few places that do contain Newtonian material: The interpretations in square brackets are Newton's own insertions. And probably at least in part because the original publication has been filtered through Newton's mind, but also in part because of the relative rationality of the original, this set of notes presents, at least in a few places, a remarkably clear exposition of late seventeenth-century alchemical theory. Most alchemical literature is deliberately arcane, hiding in cryptic language or in idiosyncratic symbolism the very information it claims to convey. *La lumière*, the commentary on it, and Newton's abstracts, on the other hand, at least occasionally express their ideas in a rationalistic form that we can comprehend even as we take note of its errors.

Editorial devices are designed to reproduce the original manuscript pages as accurately as possible. Newton's deletions are indicated by angle brackets as are editorial comments, but the editorial remarks are also italicized. Newton's corrective interlineations are indicated by arrows up and down. In one place (f. 1r of Yahuda MS Var. 1, Newton MS 30) Newton placed a possible alternative translation above the line, and that also is indicated by arrows up and down. Grammar, spelling, abbreviations, symbols, the lines in the text, and the occasional italics are Newton's, as is the material in square brackets, with the exception that I have expanded Newton's abbreviations for "que." Symbols are left untranslated in the transcription but are translated into words in Table C–1 in order of their appearance in the text.

[Yahuda MS Var. 1, Newton MS 30, f. 1r]

Out of La Lumiere sortant des Tenebres.

Cant. 1.

Stanz. 1, 2, 3, 4 The Philosophic work resembles ye creation of ye world being made out of the same matter wch matter conteins all things necessary to ye Art.

Stan. 5, ⟨6, 7 *deleted*⟩ This matter is a living universal innate spirit, wch in form of an aereal vapour perpetually descends from heaven to earth to fill its porous belly & is afterwards born among impure sulphurs & in growing passes from a volatile nature to a fixt one, giving it self the form of an humidum radicale.

Stan. 6 If o^{r} oval vessel is not sealed by winter it will not retein the pretious vapour, & o^{r} fair infant will dye at his birth if he is not readily succoured by an industrious hand & ye eyes of a Lynx: for otherwise he cannot be further nourished by his first humour, as a man wch after his being nourished of impure blood in his mothers womb, lives of milk when he is born.

Cant 2.

Sta. 1, 2 They that relye on ye sound of words work on vulgar gold & mercury thinking to fix y^{e} ☿ by a slow fire. But ☉ & ☿ vulgar are destitute of the universal fire wch is the true agent & wch agent or spirit forsakes ye metalls as oft as they are exposed to ye violent flames of a furnace, & thence ye metall out of ye mine being deprived thereof is no more then a dead & immoveable body.

Sta 3. Tis another Mercury & Gold of wch Hermes speaks, a mercury moist & hot & always constant ith' fire, a Gold wch is all fire & all life. The vulgar are dead bodies deprived of spirits, o^{rs} are corporeal spirits always living.

Sta 4. O grand mercury of Philosophers, 'tis in thee that Gold & Silver unite when they are brought out of power into act. A mercury all Sol, a Sol all Luna A triple substance in one & one in three, mercury Sulphur & Salt, three substances in one Substance.

Sta 5. But where is this aurific mercury wch resolves into salt & sulphur becoming ye radical humid of metalls & ↗ their ↙ animated seed? Tis imprisoned in so strong a prison that nature herself cannot draw it from thence unless indoustrious art faciliate the means.

Sta 6. What then does Art? It purifies by a vaporous flame the ways to the prison, having no better guide nor surer means then that of a sweet & continual heat to assist nature & give her place ↗ leave ↙ to breake ye bonds where o^{r} mercury is tied

S^{t} 7 This is ye mercury wch you must seek because in it alone you may find all that is necessary to ye wise. In it are found in nearest power Luna & Sol w^{ch}, wthout Gold & Silver vulgar, being united become the true seed of Silver & Gold.

St. 8 But this seed is unprofitable unless it rot & become black for corruption always precedes generation & we must make black before we can whiten.

Cant 3

Stanz. 1, 2, 3, 4. The sages use not violent flames nor burning coales like ye vulgar Chemists but imitate Nature & work wth her fire: a fire vaporous & yet not light, a fire wch nourishes & devours not, a fire natural & yet made by art, dry yet causing rain, moist & yet drying, a water which quenches a water wch washes the body & yet moistens not the hands [Yahuda MS Var. 1, Newton MS 30, f. 1v] ~~ Art wch would imitate nature must work wth such a fire & therewith supply what nature wants. ⟨Art, *deleted*⟩ ↗ Nature ↙ begins, ⟨nature proceeds, *deleted*⟩ Art finishes & alone purifies what nature could not purify. And unless art make plane the way nature will stop.

Sta. 5, 6. And as ye fire is one so is the matter, & poor men can have it as well as rich. Tis unknown to ye whole world, & ye whole world has it before their eyes. Tis despised as dirt by the vulgar ignorant & sold at a vile price, but is pretious to ye Philosopher who knows its value because it conteins all wch he desires. In it are found together the Sun & Moon, not ye vulgar wch are dead. In it is shut up ye fire whence metalls have their life. This is it wch gives ye fiery water & also ye fixt earth & what ever is necessary to a cleare understanding.

St 7, 8, 9, 10. Why then do senseless Chymists instead of one fire sweet & solary & one vessel & one only vapour wch thickens by degrees use a thousand various ingredients. Tis not with soft gumms or hard excrements or blood or humane sperm or unripe grapes, or quintessences of hearbs or Aqua fort'es or corrosive salts or Roman vitriol, nor with dry Talc or impure Antimony or sulphur or Mercury nor with vulgar metals themselves that an able Artist ought to work Our science comprehends y^{e} ↗ whole ↙ Magistery in one sole root wch conteins in it two substances wch have one only essence & these substances ⟨*illegible word, deleted*⟩ wch at first meeting are not gold & silver but in power become at last gold & silver in act provided we can well equalize their weights. These substances are made gold & silver actually & by the equality of their weight the volatile is fixed in sulphur of Gold, the true gold animated & foundation of ye Art.

⟨*Here Newton drew a line all the way across the page.*⟩

Out of the Commentator on La Lumiere sortant de Tenebris.

Preface p. 5. Nature offers us but only subject on w^{ch} ⟨*illegible letters, deleted*⟩ we ought to work & uses but one only vessel & one only fire & one only furnace.

The matter is but one not only in specie but in number, a certain Mineral, ↗ wch is o^{r} Hyle ↙ a Chaos conteining all ye three principles lightly conjoyned in due proportion ↗ so y^{t} ye Artist is only to untie, purify & reunite them by ye rules of nature for producing ye true Chaos wch is a new heaven & new earth. ↙ p. 107, 108, 195, 201, 262, 263, 274.

This mineral is of no use but in the philosophic work & therefore not vendible in Shops, accounted of no value, despised of all ye world, handled by many but rejected through ignorance as happened also to ye author. p. 265, 268. Tis that wch is least mentioned in receipts. Snyders.

It has a metallique splendor & a metallique essence & in things wch want this splendor the sperm of metalls is not to be found. p. 264, 267, 277, 279.

It is in part congealed & fixt but its fixity is only potential it being inveloped wth many volatil vapours wch make it easily fly away & vanish in the air. ffor when in one & ye same subject the volatil part surmounts the fixt they become both volatile p. 266.

It conteins many feces & vulgar eyes see nothing in it but feces & abominations p 100, 250. ↗ ffor by impurities ye 3 principles are kept open & ununited so y^{t} they are found more open in iron then in silver the iron be more coct & most open in this 1st matter p 156, 157, 158 ↙ It's full of impure Sulphurs. ffor the philosophic fire wch putrefies & performs all things is principally found in it (p. 101, 250) & this fire is usually inveloped wth suphureous excrements because it desires a hot nature & invests it self wth a saline attire. Which is the cause y^{t} where the earth is full of sulphurs metalls easily generate by reason of other ↗ material ↙ causes intervening. p. 255.

It is not vulgar gold nor vulgar mercury ↗ (107, 148, 153, 154, 155, 156, 157, 158, 159, 160, 161, 162, 163, 164) ↙ nor any vulgar metall nor any of their oars, nor vulgar sulphur nor Antimony (p. 106, 107, 147, 148, 154, 176,

177, 178, 198, 279, 280.) nor extracted out of earth ↗ or ↙ rain or dew (p 149) nor is it vitriol, but approaches nearer to the metallic nature then vitriol [Yahuda MS Var. 1, Newton MS 30, f. 2r] tho Vitriol be a salt wch conteins in it the spirits mercurial & sulphureous & there is nothing in nature wch conteins ↗ the sulphur ↙ so plentifully & visibly as Vitriol & whatsoever is of the nature of Vitriol. ffor ↗ ⊕ of nature ⟨ ⊕ , *deleted*⟩ is ye first coagulation of the vapor of the elements & ↙ all metals were originally Vitriol being generated of a vapour wch arises from Vitriol. p. 172, 173, 278.

Our aforesaid matter is the true gold of the Philosophers & possesses all that is necessary to the Art. 'Tis a body brought by nature to a certain term of perfection, wch body being once known it's certain that after many errors one might find ye right preparation. p. 28, 100, 148, 281.

As the world was formed of a Chaos or Materia prima wch was voyd & wthout form with darkness upon the face of the deep, in wch the four elements were confusedly blended together & God said Let there be light & a firmament wth waters above & below & dry land & seas & a Sun & Moon &c: so in o^{r} work ye Stone is made of a chaos or materia prima wch by prutrefaction is voyd wth-out form & black or dark & a black or dark cloud hangs over the face of the deep or black putrid matter & this matter conteins in it the four elements confusedly mixed together & then light is separated from ye darkness & the superior waters or vapours by destillation are separated from ye inferior & the inferior are gathered together in one place & ye dry land appears, & then the earth vegetates & at length the Sun & Moon are formed. pag. 59 & inde ab usque p. 102. Item p. 242, 243. But this knowledge is for them that wth the eyes of an Eagle can behold the Sun from its nativity & touch with their hands the Son of ye Sun, know the houre of its nativity from the septenary number, draw it from its darkness, separate the after birth & other superfluities, wash it nourish it & bring it to age of maturity & can know & adore Diana his true Sister, & having Jupiter favourable in their nativity, are as Apes of the creator in ye work of the stone p. 99, 100, 129, 130, 131.

Out of a little confused mass of this o^{r} matter is drawn an humidity dark & ⟨obscure, *deleted*⟩ mercurial conteining all that is necessary to the work the waters superior & inferior (or volatil & fixt) in wch all the Elements are enclosed, wch are to be extracted by a *second separation* & perfectly purified & then led on to generation by means of putrefaction &c. p. 100, 101, 102. This matter is a Chaos or Hyle conteining in it all ye Elements confusedly, wch are to be industriously separated purified & reconjoyned for producing o^{r} true Chaos, that is a new heaven & new earth. p. 108.

Every body according to its species conteins its own seed (wch seed is not found in divers things) & o^{r} body conteins all things necessary to ye art p 148, 149. So vulgar gold has its own seed very perfect & digested but resembles a fruit perfectly ripe & separated from the tree & of ⟨so close, *deleted*⟩ a composition so close & united that if to multiply it you put it in the [Philosophic] earth it will require much time & pains to bring it to vegetation but if instead thereof you take a slip or root of the tree (on wch gold grows) & put it in the earth [terra alba foliata, o^{r} ☿] it will vegetate & bring forth fuit in a short time p 158.

You may expect success when you know how by a crude spirit [materia prima] to extract a ripe one out of the body dissolved & then to unite it with the vital oyle [Yahuda MS Var. 1, Newton MS 30, f. 2v] for working the miracles of one thing. Or to speak more plainly when you can with yor vegetable & mineral menstrues united ↗ [dissolve o^{r} body &] ↙ extract a third essential menstrue, that you may afterward with these three menstrues wash the earth & after washing exalt it &c p. 151, 152.

Vulgar gold & mercury are not at all fit for o^{r} work because they want a thing wch is absolutely necessary in o^{r} art, that is, the proper agent. I speak not of the internal agent but of the external wch excites the internal & leads it from potentia to act. This agent leaves gold in the end of its coction as is described in *Margarita pretiosa.* And so *Zachary* says well that vulgar argent vive wants this agent & *Bernard* excludes metalls alone that is metals deprived of this agent as is explained in *Arca aperta.* ⟨p. 162, 163, *deleted*⟩ Therefore vulgar Gold & Mercury ought not to enter the work neither in whole nor in part. p. 162, 163, 164, ↗ 106, 107. ↙ All the metals have this spirit or agent in the mines except Gold wch it forsook in its final decoction & Mercury to wch it was never joyned; but in fusion they lose it. p. 166, 167.

Metals are thus generated of sulphur & mercury. The vapour of ye Elements wch is very pure & almost insensible & conteins in it the spirit of fire or light, wch is the form of the Univers & being so impregnated wth the spirit of ye universe represents the first Chaos conteining all things necessary to ye creation, that is the matter universal & form universal. And in descending & becoming sensible it first puts on ye body of ye air wch we breath & becomes inclosed in it to nourish & vivify all nature. And that it may act more easily upon the grosser bodies of vegetables & minerals it becomes still denser & insinuates it self into the water. This water is dispersed through all nature & becomes something saline & upon new commotions of ye Elements or fermentations caused by that of the air rarefies by the action of ye included vapor of ye Elements, & this aqueous and mercurial vapor meeting with sulphureous vapors, they mix & circulate together in the matrix of the saline water & being unable to get out of it they joyn with the salt of this water & put on the form of a lucid earth wch is properly ye Vitriol of nature. ffor Vitriol is nothing else then a salt in wch are shut up the spirits mercurial & sulphureous & there is nothing in nature wch conteins the sulphur so plentifully & visibly as vitriol & whatever is of the nature of vitriol. p. 167, 168, 169, 170, 171, 172, 173. ffrom these vitriolique waters by a new commotion of ye Elem^{ts} (or fermentation) caused by y^{t} of ye air, ascends a new vapor neither mercurial nor sulphureous but of ye nature of both & in ascending carries up with it some part of the salt but ye most pure & lucid, & accordingly as it settles in places more or less pure & dry or moist & joynes wth various substances it engenders divers sorts of minerals. And if this double vapor comes to a place where the fat of sulphur adheres they unite & make a glutinous substance from whence by ye action of the 🜍 upon ye humid vapor a metal is formed. If the places & vapors be pure this metal will be gold from wch the proper agent will separate in ye end of the decoction. But if the decoction be not perfected nor the sulphur wholy separated there will be engendred divers imperfect metals according to ye various impurity of the vapor & of the place. Vulgar Quicksilver is also engendred of this vapor when in its ascent from ⊕ the proper agent or spirit is evaporated ↗ & separated ↙ by too sudden a motion as happens to the ↗ spirit of the ↙ other metals in fusion. p. 173, 174, 175. Whence it appears how far off is Vitriol in the generation of the metals & how they are illuded who work upon it as ye true matter of the Stone in w^{ch} ⟨resides, *deleted*⟩ the metallique essence ought to reside. One sees also that metals whilst in their mines have their proper agent but by fusion lose it & then are not to be wrought upon. And if you take the Oars [Babson MS 414 B, f. 1r] of Metals they have many impurities in separating of w^{ch} ⟨they, *deleted*⟩ ↗ you ↙ will lose their spirit or agent. You must therefore take another subject prepared by nature of wch we will treat in an express chapter as clearly as is possible. p. 176, 177, 178.

Our living Gold is that in wch principally consists the true foundation of the Theory & practise. It is really Gold in essence & substance but more perfect then ye vulgar The sulphur or fire of Gold ingendred in ye mines of Phers fixt, pure, balsamic a corporal spirit diffused through all nature, the principle of all vegetation, life, attraction, sympathy & motion, a composition of salt sulphur & mercury, the fire of mercury & most digested part thereof, the form informing all things, the innate heat of ye Elements the lawfull son of ye Sun & ye true Sun of nature p. 180, 181, 182, 183, 184, 185, 192, 193. It is found in its own house wch house is mercury & where mercury principally resides & abounds there is this sulphur to be found. p 193, 194. Philosophers therefor seek their stone in minerals thinking there to find a nature fixt, they being of a nature most compact & (especially the principles of metals) most replenished wth this ethereal spirit. p 195, 232.

'Tis by Philosophers only that o^r Mercury can be drawn from *potentia* into act, Nature alone cannot compass it. p 197. And when Art by sublimation has purified the Mercury or vapor of ye Elements then it must be united with living Gold or Sulphur also purified whereby ye work is shortned & the tincture augmented p 199, 200. This mercury is o^r Chaos conteining all things necessary to ye art, our body, the subject of ye Art, the full Moon, argent vive animated, And because it conteins the three principles ↗ united ↙ in due proportion 'tis sometimes called Vitriol. In effect there is seen the marriage of the Sun & Moon, the King in his bath, Joseph in his prison & the Sun in his Sphere p. 201, 202.

Our Sulphur is inclosed in the radical humor but imprisoned in so hard a shell that it cannot be raised into the air but by extreme industry of art. ffor nature has not in the mines an agreable menstrue capable to dissolve and deliver this Sulphur for want of local motion. And accordingly as ye vapor rises or stays below, all that is of ye first composition rises also or stays below. But if again she could dissolve putrefy & purefy the metallique body without doubt she her self would give us the physical stone, that is a sulphur extracted & multiplied in vertue. The grain wch is not put in the grownd fit to putrefy it will not be multiplied. If one takes ye physical grain & puts it in its earth well aired well purged from impure Sulphurs & brought to perfect purity certainly it will putrefy & the pure will separate from the impure by a true dissolution & go on to a new generation more noble. If you can find this earth you will have little left to do for attaining the ⟨work, *deleted*⟩ perfection of the work. This earth is not the vulgar but a virgin earth, not that found in the earth we tread on but that wch is often raised over our heads upon wch the Sun terrestrial has not yet imprest its actions. This earth is infected with pestilential vapors & deadly poisons from wch it must be purged wth much pains & art & acuated by its crude menstrue to give it more vertue for dissolution. ~ This earth is a female wch desires the embraces of the male or solary seed & is called mercury by the Philosophers. p. 203, 204, 205, 206, 207.

The Mines are cold as to sense, yet minerals continue in their motion of growing. ffor nature uses another sort of heat imperceptible to sense, a warmth of the nature of Spirits wch are always in motion & by motion have an innate [Babson MS 414 B, f. 1v] faculty of warming. Corrosive waters have such a faculty but you must not use them for dissolving metalls & other matters but ye same fire wch nature uses wch must be well acuated to render it more active & more agreeable to ye nature of ye compound. Its construction is very ingenious & therein consists almost the whole physical secret, philosophers having said nothing or but very little thereof. p. 210, 212.

In the feminine sperm of animals the passive elements are predominant & in the masculine the active, whence arises the mutual action between them in order to generation, to wch the acid quality of the menstruum or feminine sperm contributes & so in the other kingdoms. p 222, 223, 224, 228 ffor ye feminine sperm is sharp & pontic ↗ & by its nitrous sharpness & crudity introduces putrefaction ↙ p. 161, 223, 225.

Without putrefaction tis impossible to attain yor end, vizt the deliverance of the sulphur or seed shut up in the prison of the elements. This is the only mean: for if the seed be not cast into the earth to putrefy it remains unprofitable. Now this corruption is not compassed but by a proper menstruum. In animals this menstruum is in the womb. In vegetables tis in the earth. In minerals tis in the proper matrix wch is taken for earth. But as the matrix must be comforted for animals & the earth cultivated for vegetables so in minerals the aurific seed must be cast into an earth well prepared, otherwise the matrix will be infected with stinking vapours & impure sulphur. Be carefull therefore of cultivating this earth, then cast your seed into it & it will bring forth much fruit. p. 233, 234, 235.

Authors say there are three sorts of solution in the Physical work the first is the solution or reduction of the crude & metallique body into its principles to wit sulphur & argent vive, the second is the solution of the physical body & the third is the solution of the mineral earth. These solutions are so inveloped in obscure termes that it is impossible to understand them without the assistance of a faithful master. The first solution is made when we take our metallique body & draw thence a mercury & a sulphur. Here it is that we have need of all our industry & of o^{r} occult artificial fire to extract out of o^{r} Subject this Mercury or this vapour of the elements, to purify it after having extracted it & afterwards by the same natural order to deliver out of its prisons the sulphur or essence of sulphur: wch cannot be done but by ye sole meanes of solution & ↗ of ↙ corruption wch must be perfectly known. The signe of this corruption is the blackness, that is to say one must see in the vessel a certain black fume wch is engendred of ye corrupting humidity of yor ↗ natural ↙ menstruum, for it is ↗ of ↙ this that in the commotion of the elements this vapour is formed. If therefore you see this black vapour, be certain that you are in the right way & that you have found the true method of working. The second solution is made when the Physical body [o^{r} matter] is dissolved joyntly with the two substances above mentioned & that in this solution the whole is purified & takes a celestial nature, that is when all the elements subtilized are prepared fundamentally for a new generation: & this is properly the true philosophique chaos & the true first matter of the Philosophers (as Compt Bernard assignes) for it is only after the [Babson MS 414 B, f. 2r] conjunction of the female & male the mercury & sulphur that it ought to be called the first matter & not before. This solution is the true reincrudation by wch one has a ⟨very pure, *deleted*⟩ seed very pure & multiplied in vertue. ffor if the grain remains in the earth wthout being reincruded & reduced into this first matter the Labourer expects the desired harvest in vain. All sperms are unprofitable for multiplication if they be not first reincruded. Hence it is very necessary to know this reincrudation or reduction into the first matter by wch alone one may make this second solution of the Physical body. As to the third solution it is properly this humectation of the earth or sulphur physical & mineral by wch the infant augments its forces. But as this chiefly relates to multiplication we refer the Reader to what Authors have said on that subject. Ib. p. 241, 242, 243, 244.

APPENDIX D

"Experiments & observations Dec. 1692 & Jan. 169²⁄₃"

Portsmouth Collection MS Add. 3973.8
University Library
Cambridge

Editorial note

When Newton's papers were examined by the syndicate for Cambridge University in the nineteenth century, his chemical–alchemical notebook and a number of loose sheets recording his experimentation in these areas were accepted by the University as "scientific," presumably because of the obviously empirical nature of the material. The manuscript transcribed here was among the loose sheets and was precisely dated in the title by Newton himself.[1] In fact, later entries toward the end of the manuscript, also dated by Newton, carry it forward to June 1693.

Newton's interlineations have been indicated in the following transcription by arrows up and down, but I have omitted notice of his deletions, which seem to have no particular significance in this manuscript. His spelling, abbreviations, and punctuation have been retained, as well as his symbolic representations for the chemicals in his experiments. The probable verbal equivalents of the symbols are given in Table D-1.

[f. 1r] Experiments & Observations Dec. 1692 & Jan 169²⁄₃.

Barm works best when new & loses its vertue in two or three days of hot weather or ↗ in ↙ a week of cold: but if dried upon a stick or the spriggs of a bush it will keep much longer. Two quarts of barm will ferment a hogshead of wort (brewed with 3 bushels of mault for small beer or almost five for strong) & the fermented liquor will yeild two or 3 gallons of new barm. Strong beer yeilds more barm then small & ye barm is stronger The Brewers heate their water till it begin to boile, then put mault to it & after it has stood about 6 hours draw it off from ye grains & boile it two or three hours & when it is blood hot put

1. *Catalogue of the Portsmouth Collection* (1, n. 27), Sect. IV. No. 8, pp. 19–20.

Table D-1. *Symbols used by Newton*

Symbol	Meaning
♀	copper
♆	bismuth
♄	lead
✳	sal ammoniac, ammonium chloride
♁	antimony: either the ore, probably antimony trisulfide, or the metal
⊕	vitriol, a sulfate, usually of iron or copper
♃	tin
♆	bismuth ore
Æ	aqua fortis, nitric acid
♂	iron
☿	mercury
🜍	sulfur
♂	iron ore
♀	copper ore
♃	tin ore
♄	lead ore
♁	antimony ore

barm to it. They have no certain rule for a just proportion of barm but tend ye liquor ↗ wth hot & cold worts ↙ to keep it in a due degree of working till ye barm begin to rise. ffor if it work too fast they check ye working by pouring in cold wort, if too slow they excite it by pouring in hot wort. By working too fast ye beer gets a tang or ill relish in drinking. The faster it works ye better it is so it get not this tang. The froth wch rises the first day falls down again in liquor, but after ye working has continued about 30 hours, it begins to slacken & the liquor being now grown more tenuious the barm ↗ a clammy viscous froth ↙ begins to rise & continues to rise for about 12 hours more. This barm tasts bitter of the hops. And note that ye juices of macerated vegetables & particularly must or new wine ferment of themselves & probably barm had its first rise from thence. ffor in destilling new wine before fermentation the flegm rises first & then ye spirit, but after fermen[f. 1v]tation the spirit rises before the flegm: as it is in wort.

Salt of metalls distilled from decrepitated salt lets go its spiritual part in form of a saline liquor & the metalline part remains below melted under the ↗ melted ↙ common salt in form of a brittle compact metallick substance shining a little wth a metallick colour.

In making ye salt of metals, salts of Mars & spelter must not be mixed wth salt of ♀ . ffor they cause a fermentation & precipitation in ye volatized saline liquor & hinder ye volatizing of ye salt of ♆ very much & produce a salt of metals wch holds not enough of ♄, but in melting in a glass over the fire flies away too easily in ye form of white saline flowers & leaves a substance below about 1/3d or $1/4^{th}$ of y^e whole & not of a dark colour & sweet tast but tastless & of a light ↗ pearly ↙ colour. Whereas y^e salt of metalls prepared wth s^a of ♀ alone being melted in a glass vial luted sends up only a very little white acrimonious fume & then remains below ponderous, sweet & of a dark metalline colour. In ye melting there arises a fume wch at first colours all ye upper part of ye salt wth a reddish colour & gives it an acrimonious ↗ styptick ↙ ill tast, but by continuing the fusion two or three minutes longer this fume sublimes to ye upper part of ye glass & leaves all ye matter below of a ↗ uniform ↙ dark colour & sweet tast, & then is it prepared. ffor this salt must be thus melted (not wth a strong fire but wth a heat almost red wch will only make it soft like birdlime) to drive away all y^e ✳ & heterogeneous acrimony. Before I learnt this way of preparing it, I broke my egg-glass by ye volatile stiptick, vitriolick salts wch arose during the digestion & stopt up the neck of the glass.

[f. 2r] Note that sublimate of unmelted ♁ volatizes ⊕ something better then that of melted ♁ tho ye difference be not much: & that if ♃, Ψ, ♆ be as 9, 30, 1 there must be as much Æ to 60 of ⊕ volatized as will dissolve 45 of ♃, Ψ, & ♆ colliquated & poudered. But if there be less ♃ & ♆, there may be as much Æ as will dissolve 48 or 50 of ye colliquated metal. ↗ Note also that 4 parts of subl. of ♀ & Ψ give 5 of sal metallorum. ↙

I digested ♁ made of Reg ♂^{tis} 120^{gr} wth y^e ☿^{ial} water of ♄ 120^{gr} for a day. They melted down into a compact ↗ fluid ↙ mass wthout bubbling & sent up a little fume into the neck of ye egg-glass wch stood melted like 🜍 & if I misremember not congealed ↗ afterwards ↙ in ye cold. I put to ye melted mass 2^{gr} of ye red precipitate of ♂ but it caused no fermentation. Upon breaking the glass I found the ♁ & ☿ of ♄ mixed in an uniform solid mass wthout bubbles or grains or metallick splendor It broke smooth like glass or resin ↗ but wthout glittering ↙ & was of a darkish pale colour looking almost like a black flint newly broken but not altogether so dark. It inclined also something more to a pale blue then a flint does. I have seen salve almost of ye same colour. That wch sublimed like 🜍 into ye neck being taken out was soft & moist & tasted very acid like ye salt of ♁, & tho ye flegm was evaporated till the matter became dry yet it tasted as acid as before. I washt away ye acidity & dried ye residue upon a piece of glass. After washing it looked white but after drying it inclined to ye colour of 🜍. I held it to ye side of ye flame of a candle & it did not take flame as 🜍 would [f. 2v] do but yet fumed away & ye fumes made the side of ye flame looke blue so that there was a little 🜍 in it & ye greatest part of it was of another kind. Hence I conclude y^t y^e ↗ ♄^{al} ↙ ☿ before it be used ought to be melted in a glass vial to free it from its acid & sulphureous heterogeneity.

Antimony made of Reg ♂^{tis} & sublimed 240^{gr} red precipitate of ♂ 120^{gr} colliquated in a crucible grew fluid like melted ♁ & upon flowing began to boile, & being presently cooled by dipping ye crucible in water the matter when broken looked black & glossy like pitch. It ↗ had ↙ bubbles in it, but ye bubbles were larger & ye solid compact ↗ black ↙ matter between ye bubbles was thicker then happens in fermentation, & whilst in flux it was more fluid by much then fermented stuff. ffor that is viscous like bird lime & does not flow like a liquor but swells like past & is stiff & viscous like bidlime so it will scarce run out of ye crucible but must be scraped out wth a rodd.

I melted (in a vial of glass luted) mineral ♁ 60^{gr} & red precipitate of ♂ 22^{gr} & the matter melted together & had some bubbles in it but did not rise like puff past nor take up ↗ sensibly ↙ more room then before melting. Only the interspersed vacuities were upon fusion turned into bubbles & the bubbles perhaps a little enlarged by some new vapors. The bubbles were as large as in ye former exp^t [f. 3r] & the matter between them as thick or thicker: but here it looked not black (as there) but of the colour of ♁ shining wth a pale white metallick splendor as ♁ doth but without grains, & was something less brittle & friable then ♁ : Whereas crude ♁ 60^{gr} melted wth iron ore 30^{gr} swelled like puff past till it took up ↗ above ↙ 4 or 5 times the first room & then was every where full of little bubbles like barm & very brittle So then ye ferment lies only in ye ore of iron & not in ye red precipitate.

♁ sublimed precipitated edulcorated & melted in a glass vial luted returns into a fine sort of ♁ in colour & grain like other ♁ . Of this ♁ 60^{gr} iron ore 30^{gr} heated red hot in a glass phial, would neither melt nor grow soft like birdlime ↗ nor swell ↙ but continued in a red hot pouder clotted together as crude ♁ & iron ore doe for some time before they melt into a viscous swelling liquor. And tho ye fire was made bigger & continued much longer then was necessary for crude ♁ & iron ore yet they would neither melt nor flow, ↗ but when cold looked like a dark red pouder clotted together ↙ So then artificial ♁ is void of a fermental virtue or at least is not so viscous & active as the natural.

♁ 2^{pts} melted wth iron ore 1^{pt} & sublimed & precipitated. This precipitate melted in a luted glass phial ↗ as easily as ♁ & ↙ became as liquid. At first there arose a few great bubbles, such arise upon melting the precipitate of ♁ alone wch being broke ye liquor continued in a flux wthout any more bubbles rising so y^t there was no fermentation but letting ye glass fall & breaking it, the liquor spurtled like molten pitch or ♁ & some of it (by reason of its fluidity) ran into round globules. When cold it brake ↗ smooth ↙ like an opake glass of a dark red colour & looked like a broken flint but ↗ more red & ↙ not ↗ altogether ↙ so darke. It had no metallick splendor but was rather a dark red opake glass. ↗ Upon repeating ye exp^t with ½ an ounce of matter in a bigger heat, a considerable quantity of ye matter sublimed & ye rest melted into a very brittle glass of a dark red colour wch brake as smooth, glossy & shining as any other glass but wthout a metallick splendor. ↙

ffour parts of ye same precipitate being melted wth one of iron ore, did not grow fluid as in ye former exp^t but yet melted into a viscous matter as iron ore & crude ♁ does & swelled a little by a fermental vertue & when cold was every ↗ where ↙ very full of small bubbles looking all over like a sponge: but it not rise & swell so much by far as iron ore & crude ♁ would have done nor were ye bubbles half so great in diameter. So then this precipitate ↗ seems ↙ more susceptible of ferment y^n y^t of ♁ alone.

Antimony 2^{pts} melted wth ♃ 5^{pts} amalgams in calido wth ☿ not very difficultly; wth 8^{pts} it amalgams very easily.

Rete 3, ♃ 1, ♄ 1, ♈ 1 confusa et imbib cum ↗ totius ↙ nona parte aceti ♁^{lis}, non liquescebant in igne candente. Item Rete 3, ♃ 1. ♈ 2 similiter imbib. non liquescebant. In priore imbibitione ♄ ebullitionem creabat, in posteriore nulla erat ebullitio.

Reg. ♃ ↗ 6^{parte} ↙ imbib. cum septima parte aceti liquescebat & una illa pars aceti convertebat 4 partes Reguli in ♁ & manebant tres partes reguli colliquati in fundo.

Reg ♃ 4^{ptes} imbib cum 1 parte aceti & colliquat cum 1 parte calcis rubri martialis fiebat massa viscosa quae cum ♃ 4 + ♈ 8 liquefactis misceri nolluit.

⚦ 1, ♀ 1, ♃ 1, ♄ 1, ♁ 4 liquefacta fluebant cum fermentatione notabili, sed cum affusis ♄ 4 + ♃ 6 + ♈ 12 ⟨*illegible word*⟩ miscebantur. Igitur metalla alba liquefacta cum mineris in antimonia liquefactis aeque miscentur nisi metalla illa in [f. 4r] ♁ reducantur.

♈ facilius fluit quam ♄, ♃ aeque facile ac ♈ aut paulo facilius, ♈ 2 + ♃ 1 ad huc ↗ multe ↙ facilius, ♈ 3 + ♃ 2 ↗ ad huc ↙ paulo facilius & ♈ 12 + ♃ 7 paulo facilius, ♄ 3 + ♃ 3 + ♈ 5 ↗ paulo ↙ facilius ♄ 1½ + ♃ 3 + ♈ 5 aeque facile aut paulo facilius ♄ 2 + ♃ 3 + ♈ 5 ad huc facilius et quantum sentio omnium facillime aut potius ♄ 5 + ♃ 7 + ♈ 12 fluit omnium facillimè.

Mercury 4^{pts} sublimes wth 9^{pts} of distilled salt of ♁ & requires a dark red heat to raise it. It sublimes into a white heavy saline substance, apt to emit strong fumes after sublimation upon opening ye vessel, & grow moist in the air. Upon ⚦ 1, ♂ 1, sublimatum & praecipitatum metallorum 1, sal metallorum 3 digested 2 days I poured this salt gradually 6^{pts} & it fermented long & much & sent up a thick white fume & volitized a good part of ye matter below & y^{e} ☿ precipitated & flew up into the neck of ye retort ↗ in a running form ↙ , & that matter wch after ye operation was ended remained below was tastles & fusible on a red hot iron & volatile in good measure & would not work upon ye net

April 1693. Reg of any metal imbibed & melted is very apt to vitrify wth too much heat. Rete imbibed melts difficultly & can scarce be imbibed wthout vitrifying.

June 1693. The two serpents ferment well wth salt of ♄ ♃ & ♀ better wth salt of ♄ & ♀ best wth salt of ♄ alone. ♀ added ferments much more in all three cases & volatizes ye mass: but better in ye 2^{d} case y^{n} in ye 1st & best in ye last. To ye 2 serpents 24^{gr} ↗ I added ↙ ☿ of ♄ 24^{gr} added by degrees & when ye fermentation was over I added ☿ 16^{gr} & ye matter swelled ↗ much ↙ wth a vehement fermentation ↗ then before ↙ & in two or three hours sublimed all to ye top except 3 grains wch remained below ↗ spongy ↙ in form of a ↗ dark ↙ cinder, & there was 9¼^{gr} of running ☿ besides a little that stuck in ye neck of ye glass wch might amount to a grain more so y^{t} ye 2 matters dissolved about 1/8 of their weight of ⟨*illegible word*⟩.

APPENDIX E

"Praxis"

Babson MS 420 (part)
The Sir Isaac Newton Collection
Babson College Archives
Babson Park, Massachusetts

Editorial note

Newton's "Praxis" was first described in the nineteenth century as "Chemical nomenclature of the Egyptians, and a praxis of alchemy extracted from various authors, with a duplicate folio partly cancelled."[1] Sold in 1936 as Sotheby lot no. 74,[2] it is now Babson MS 420 and constitutes a part of The Sir Isaac Newton Collection.[3]

The manuscript in its present form has three distinct parts. The most important is a more or less complete treatise entitled "Praxis," of eighteen small quarto pages that is formally divided into five chapters. Although the treatise relies heavily on the work of earlier alchemical authors, the conceptualization and organization of the work are certainly Newton's own, and he entered it in this final form in most of the space provided by three small quarto booklets of eight pages each, booklets such as he often made by folding a folio sheet in quarters and cutting through the single fold at the top of the first page. All of the eighteen treatise pages have marginal space reserved for annotations and additions. They have been numbered pp. 3–20 by the Babson College Archives.

The second distinct portion of Babson MS 420 is much shorter. Although

1. *Catalogue of the Portsmouth Collection* (1, n. 27), Sect. II, (3), no. 21, p. 15.
2. *Catalogue of the Newton Papers* (1, n. 27), p. 12: "Praxis [A Treatise in Five Chapters with the following heading: 'Cap I. De Materie Spermaticis'; 'Chap. 2. De materia prima'; 'Chap. 3. De Sulphure Phõrum'; 'Cap. 4. De agente prima'; 'Chap. 5. Praxis', in English] with very numerous alterations and re-writings, *about 5500 words, 26 pp., in a wrapper of which 2 pp. contain notes on the derivation of the names and symbols of the metals from the Egyptian Gods, the Planets, etc.*, All Autograph sm. 4to."
3. *Descriptive Catalogue of the Grace K. Babson Collection* (App. C, n. 9), p. 193.

not closely related conceptually to the treatise on "Praxis," this second portion is physically a part of one of the treatise booklets. On the first page of the first booklet and also on its verso, pages designated as pp. 1–2 by the Babson College Archives, Newton entered several notes and reflections. Some of these notes spill over from p. 2 onto the verso of the last page of the booklet. In addition to these notes and reflections, Newton saw fit to enter the last part of the treatise on "Praxis" on the fifth and sixth pages of the first booklet (pages now designated as pp. 19–20); otherwise the rest of the first booklet is blank.

Finally, there is another small quarto booklet of eight pages that contains an earlier draft of parts of Chapters 4 and 5 of the treatise. These pages also have reserved marginal space, and they have been numbered pp. 11a–18a to show their relationship with pp. 11–18 of the treatise itself. The draft section constitutes the third distinct part of Babson MS 420, and I have followed the Babson numbering of this complex manuscript in the partial transcription here and in all other references.

Citations in the text to *Le triomphe hermetique* and to a letter from Nicolas Fatio de Duillier date Babson MS 420 unequivocally in the later years of Newton's alchemical career, certainly no earlier than 1689.[4] However, the handwriting seems to be later than that because it is in general somewhat smaller than the very bold writing of the late 1680s and early 1690s. The treatise was probably written later in the 1690s; one may set a date of about 1696 for the final version,[5] which would place it at or near the end of Newton's most serious alchemical work. Although there is evidence that he continued to study alchemy and to rework his alchemical papers,[6] there is no evidence that he continued to experiment with alchemy after he left Cambridge for London and the Royal Mint in 1696.[7]

I have not transcribed the earlier draft of the "Praxis" (pp. 11a–18a) or the notes that precede the treatise (pp. 1–2). The earlier draft is heavily canceled and rewritten; the formal treatise contains in a more legible form all the material Newton chose to retain. The notes at the beginning of the first booklet, however, are of considerable interest in showing the relationships Newton thought to hold between Egyptian, Greek, and Roman deities and

4. *Le triomphe hermetique* (2, n. 70); for Newton's correspondence with Nicolas Fatio de Duillier, see supra, Chapter 6.
5. Derek T. Whiteside in a personal communication kindly suggested to me a date of 1696 for the "Praxis," on the basis of the handwriting.
6. Newton was still buying alchemical books sometime between 1701 and 1705: Harrison, *Library* (1, n. 6), p. 9. Keynes MS 13 (6, n. 20), for example, contains Newton's final list of the best alchemical authors as well as other lists of alchemical writers; it also contains notes on Mint business.
7. Newton still had "a parcel of Chymical glasses" in his possession at his death, however: Richard de Villamil, *Newton: the Man*, foreword by Albert Einstein (London: Gordon D. Knox, [1931]), p. 50. He also told Conduitt sometime during his London years that, if he were younger, he would "have another touch at metals": quoted in Westfall, *Never at Rest* (1, n. 4), p. 531.

certain chemicals that were critical to the alchemical process. The verso of the front page (p. 2) and its spillover onto the unnumbered verso of the last page contain Newton's final version of these relationships; that is reproduced in Chapter 5 as Plate 10 and discussed there.

I have transcribed here the important formal treatise of pp. 3–20 of Babson MS 420, which one may assume to be Newton's climactic composition in alchemy. The general title of the treatise is "Praxis," and "Praxis" is also the title of the last chapter of the treatise. The word "praxis" is a postclassical derivation from the Greek that had entered into medieval Latin and was in common use in seventeenth-century English.[8] Its meaning was virtually identical to that of the word *practice*, the exercise of a technical subject or art as distinct from the theory of it, and was appropriately applied in the seventeenth century to the practice of medicine and to the practical art of chemistry. In the last chapter of his treatise Newton appropriately applied it to the practical art of alchemy, though not before leading the reader through the theory of alchemy (as he understood it in the 1690s) in the first four chapters on the spermatic materials, the first matter, the sulfur of the philosophers, and the first agent.

Editorial devices are designed to reproduce the the original manuscript pages as accurately as possible. Newton's many deletions are placed in angle brackets, as are editorial comments, but the editorial comments also are italicized to distinguish them from Newton's words. His interlineations are indicated by arrows up and down. In one place (p. 18) he entered the alternative number "twenty" above the number "thirty" in the text (for the number of days required for a particular fermentation and digestion); that alternative number is also indicated by arrows up and down. The spelling, abbreviations, and punctuation are Newton's own, except that I have expanded his abbreviations of "que" and of final consonants. I have also omitted all indications of ligatures except for that in the abbreviation for aqua fortis (nitric acid). The parentheses and square brackets are all Newton's; in a number of places he left some blank space in the square brackets for filling in exact citations, and I have retained some indication of these blanks. In places where he used his reserved marginal spaces for citations or other addenda, I have so indicated. I have kept Newton's Latin passages and his symbolic representations untranslated in the transcription, but the symbols are rendered into their verbal equivalents in Table E-1, in order of their first appearance. In this particular manuscript one frequently finds that a symbol may be translated either as the name of a metal or as the name of a planetary deity; because this ambiguity is sometimes significant, in the table I offer the reader both options where appropriate.

8. *Oxford English Dictionary* (2, n. 79), s.v. "praxis"; *Revised Medieval Latin Word-List from British and Irish Sources. Prepared by R. E. Latham, M. A., under the direction of a Committee appointed by The British Academy* (London: published for The British Academy by The Oxford University Press, Amen House, 1965), s.v. "praxis."

Table E-1. *Symbols used by Newton*

Symbol	Meaning
☿	quicksilver (argentum vivum), mercury (mercurius), Mercury
🜍	sulfur
☽	silver (argentum) or moon (luna), Diana
☉	gold (aurum) or sun (sol), Apollo
♂	iron (ferrum), Mars
♄	lead (plumbum), Saturn
♁	antimony: either the ore, which was probably stibnite (antimony trisulfide), or the metal
♀	copper (cuprum), Venus
⦶	niter, probably potassium nitrate
✳	(usually) sal ammoniac, ammonium chloride, or (sometimes) star regulus
♃	tin (stannum), Jupiter, Jove
♆	bismuth
Æ	aqua fortis, nitric acid
⊕	vitriol, a sulfate, usually that of iron or copper

[p. 3] Praxis
Cap. 1. De materiis ⟨et Sulphure, *deleted*⟩
↗ ⟨prima et ultima, *deleted*⟩ ↙ spermaticis.

Haec sunt fflamelli Dracones in agro nigro, foemina alata & mas sine alis, principia Philosophiae, Serpentes circa caduceum, ☿ et 🜍, ☽ et ☉, Dracones qui custodiunt poma Hesperidum, serpentes duo missi a Junone (natura metallica) quos Hercules (philosophus) in incunabulo (seu operis initio, efficiendo ut putrefiant) occidet; spermata duo metallorum masculinum et foemininum, quae ex una radice oriuntur et sola ad opus sufficiunt [Flamel H. Cap. 1 et 3]. Rex et Regina qui una cum Aquario ⟨Materiae, *deleted*⟩ (seu Caduceo) ↗ initio operis ↙ putrefiunt [Philaletha in Ripl. port. 1. p. 105, 106]. Claves Plutonis et Amphisbena quam Bacchus palmite vitis id est ⟨*illegible words, deleted*⟩ ☿ii Caduceo ⟨[, *deleted*⟩ occidit [Majer Septim. philosoph. Die 5. in serpente.] Lapides duo principales, ↗ (e septem a natura suppeditatis) ↙ quorum unus sulphur invisibile alter mercurium spiritualem infundit, ille calorem et siccitatem hic vero frigiditatem et humiditatem inducit. Prior ⟨qua, *deleted*⟩ (☿) in orientali plaga posterior in occidentali reperitur. Uterque tingendi ↗ (seu fermentandi) ↙ & multiplicandi vim habent: et nisi ⟨lapis, *deleted*⟩ ab illis primam tincturam lapis hauserit, neque tinget neque multiplicabit [Arcan. Hermet. sect 57]. These are ↗ Chalybs & Magnes ↙ ye fiery Dragon & vegetable Saturnia [Secr. Rev. cap. 2, 3, 4, 6 On Ripl. Gates p. 28, 29, 31, 35, 36.] Lapides solis et Lunae ex quibus una cum lapide ☿ii qui medium est conjungendi tincturas, magisterium componitur. [Clangor Bucc. p. 310 Arca Arcanorum p. 302. ⟨], *deleted*⟩ Com. on Ripl. Gates p. 28,

29 31, 35, 36] Lapides ex quibus fit Reg ♂^{tis} [Secr. Rev. [p. 4] cap. 2 & 7. Philal. on Ripl. Gates p. 31.
⟨*Here Newton left a space of about ½ inch before beginning the next chapter.*⟩

Chap. 2
De materia prima.

The matter (called Nature by Philaletha) is o^{r} terra virginaea foliata ↗ sal nitri sapientum, Chamaeleon et Proteus, ↙ spiritus corporalis et corpus spirituale, ↗ a ↙ fat viscous heavy juicy mineral, the first matter of all metals [Instructio de arbore solari p.] a metallic Gumm [Turba p] a stone ↗ & no water ↙ because friable & no stone but water because fluxible in ye fire [ ] It's metalline but void of metallic 🜍, fusible, fugitive, in no ways malleable, in colour sable wth intermixed argent glittering branches composed of a pure ☿ & feculent 🜍, the Green Lyon wch easily destroys iron & devours also ye companions of Cadmus. [On Praef. to Ripl. Gates p. 51, 52, 53. Marrow of Alk. part 2. pag. 5, 15] It must be prepared ↗ & ⟨purg, *deleted*⟩ separated from its feces & made totally volatile ↙ before it enter ye work ⟨[Mayer. Lib. 8 Symb.], *deleted*⟩ Our Vulcan Lunatique (or fire described by ⟨], *deleted*⟩ ⁺ ⟨*Newton wrote the following section "⁺ by Artephius . . . Maier. 1. 8. Symb." in the margins of pp. 4–5*⟩ [margin, p. 4] ⁺ by Artephius ⟨), *deleted*⟩ is of ye same nature wth o^{r} matter & both must be prepared by the Artist. Triomphe Hermetique p. 42 D. Thomas Antimonio Hispanico, non tamen absque singulari preparatione usus est, cum ex vulgaribus, ut Greverus loquitur, physica facienda sint. Quicenim opus philosophicum ingreditur, debet esse purum homogeneum ab omni sua heterogeneitate et terra superflua purgatum in substantiam clavam, operi physico convenientem quae vel tota volat vel tota in fundo maneat, pro operationis di [margin, p. 5]versitate. Maier. 1. 8 Symb. [p. 4] ⟨*illegible words, deleted*⟩ ↗ This preparation ↙ Philaletha ↗ hints by ↙ calling the Queen ye daughter of ye Waterbearer ↗ arising out of his loines ↙ & says that she is conteined invisibly ⟨*Two or three illegible letters deleted in midst of "invisibly"*⟩ in ye water of his ↗ silver coloured ↙ pitcher & arose out ⟨of ye water in wch saith he was seen a lamp burning, or a twinkling spark wch sent forth its beames from ye center & that to ye surfaces, *all deleted*⟩ of it. [On Ripl. Gates, p 115] ↗ & that ↙ ⟨ being thereby in this wise metamorphosed, *deleted*⟩ by a strange metamorphosis done by a magical vertue of nature [ib. p. 107] & that after this rise she was ⟨*illegible words, deleted*⟩ ↗ naked, that is divested of impurities [ib. p 107] & ↙ beautyfull, ⟨&, *deleted*⟩ [ib p 115] & thô a body yet she was all spirit [ib. p 109] & yet able to endure wthout [p. 5] hurt ye greatest fires that can be made [ib. p. 113] & in this state it is ↗ properly ↙ o^{r} matter in wch vulgar Chymists do not work, ⟨ma, *deleted*⟩ & wch is not to be found upon ye Earth of ye living, tis not that earth wch wee tread on but that w^{ch} ⟨[, *deleted*⟩ (by sublimation hangs over o^{r} heads, wch ye wise call their terra virginea foliata, supra quam sol [martialis] radios suos nunquam lancinavit, quamvis sit ejus pater et alba luna mater [i.e. 🜍 acidum & ☿ oleosus. Nam] Hermaphroditus est ex duabus naturis 🜍^{e} et ☿^{o} constans. [Instruct. de arb. solari. cap. 3.]

Now its preparation is thus described by ye author of Instructio de arbore solari cap. 4. Cognita tandem hac materia, saith he, necessum est in principio hanc materiam singulari et occulto artificio in aquam convertere & post quam naturaliter evaporaverit leni et naturali medio occulto in terram mutare: quo facto terrae virgineae sapientum possessor eris. Ex hac terra sapientes suum mercurium ⟨*illegible words, deleted*⟩ (simplicem) et suum mercurium duplicatum parant, et aquam siccam hauriunt quam aquosam ignem et igneam aquam vocant quia omnia corpora radicaliter solvit. Metalla enim non tingunt donec spiritus in illis inclusus extrahatur et ex centro solaris nostrae terrae Adamicae [i.e. ⟨Reg ♂ et metallorum ↗ ex minera ↙ , *all deleted*⟩ ↗ ex minera ♂^{tis} ↙] opera aquae nostrae albae depromatur. Est enim nostra [p. 6] aqua sicca de natura sulphuris nostri et Mercurij nostri ideoque ↗ iuncta ↙ se amant et uniuntur. Nam aqua illa et Mercurius sunt sorores unius originis et ex unica scaturigine oriundi ideoque se amant et postquam ex conformitate naturae uniunter duplatus mercurius noster nominantur ⟨[Instruc, *deleted*⟩ ⟨Hactanus Author Instructionis, *deleted* ⟩ Thus far that Author. Now the scaturigo of this ⟨water & this, *deleted*⟩ ☿ & this water is ye silver coloured water of ye pitcher in wch was seen a lamp burning, for no water saith sendivow ⟨was, *deleted*⟩ is profitable in this work, but y^t wch was drawn out of ye rays of ye Sun or Moon. But the terra virginea must be praepared out of ye rays of ye Moon because ⟨it, *deleted*⟩ ↗ tis a female & ↙ (as saith ye Author of instructio de arbore solari) ⟨it must be taken be, *deleted*⟩ the sun hath not shone upon it. ⟨But after tis prepared by turning ye water into earth it must be sublimed as Basil Valentine thus teaches. And tis, *all deleted*⟩ Its preparation ye same Sendivow hints by calling his menstruum aquam salis nitri de terra nostra in qua est rivulles et unda viva si ↗ [non ad centrum sed tantum] ↙ ad genua foveam foderis. + ⟨*Newton wrote the following "+ Our Antimony..." through the cancelled note on the second figure of Abraham the Jew (next paragraph) in the margins on pp. 6–7.*⟩ [margin, p. 6] +Our Antimony, sait Maier, is ye King wch cryes in the Sea Qui me liberabit ex aquis & in siccum [denuo] reducet, Ego hunc divitijs beabo. Maier.

Alij appellaverunt hanc Terram Draconem devorantem, congelantem vel mortificantem caudam suam i.e. suum argentum vivum; alij appellaverunt illum locum desertum quia depopulata est a suis spiritibus ⟨*the "ti" in "spiritibus" was interlineated*⟩ Lullius apud Maierum Symb. l. 9. Cadmus in Rhodum veniens Neptuno aedificavit templum & ollam aeream obtulit cum hâc inscriptione vatidica, Terram Rhodum a serpentibus vastatum iri: quae verba rite intellecta total artem continent. Maierus Hierogl. l. 1. p. 45. ⟨The same thing is signified in y^e *illegible symbol* second figure [margin, p. 7] figure of Abraham the Jew wh where by the serpents & Griffins *illegible word* that is liquors, & volatile salts making their neasts & aboud y^e, *all deleted*⟩ [p. 6] ⟨But after tis, *deleted*⟩ ↗ These Parables shew sufficiently how ye earth is to be ⟨"*earth is to be" spills over onto p. 7*⟩ ↙ prepared by turning it first into water & then into earth, But ↗ after tis thus prepared ↙ it must be sublimed as Basil Valentine thus teaches. Saturn will put into your hand a deep glittering mineral wch in his mine is grown of ye first matter of all metalls. If this minera after its preparation wch he will shew unto thee is set in a strong sublimation mixed wth three parts of [p. 7] bole or tyle meal, then riseth to ye highest mount a noble sublimate like feathers or alumen plumosum, wch in due time dissolveth into a strong & effectual water ↗ ⟨[or sp^t of Mercury], *deleted*⟩ ↙ ⟨wch bringeth thy seed ↗ [vizt ye extracted soul of common Gold] ↙ in a little putrefaction very suddenly into ye first volatility, if so there be added to it a due quantity of [that] water y^t it may be dissolved therein then they ascend are able to ascend above ye highest mountain & stay inseparably together a soul & spirit or a spirit & soul [y^t is ☿ duplatus] y^t is ye soul of Gold & spirit of ☿ , *all deleted*⟩ wch water he afterwards calls a spirit, ↗ & mercurial spirit ↙ that is ye spirit of ☿ & saith that it dissolves ye extracted soul of ⟨Common, *deleted*⟩ Gold vulgar & by a little putrefaction volatizes it & stays wth it inseparably, as it doth also wth ye body or salt of Gold, being ye volatile bird from whence all metals have their original. Thus far Basil pag 127, 128, 129. So then this sublimate is ye terra virginea wch is no where to be found upon ye Earth of ye living being not ye earth we tread on but that wch hangs over o^r heads, the terra alba foliata in wch saith Hermes we must sow o^r Gold, the true matter wch vulgar Chemysts never work in, Dame nature or ye naked metamorphosed Queen whose body is all spirit.

Cap. 3
De Sulphure ~~Phorum.~~

This sulphur is ↗ our Chalybs ↙ the true Key of o^r work [p. 8] wthout wch ye fire of ye Lamp cannot be kindled Tis the[a] minera of Gold even as ⟨y^e, *deleted*⟩ o^r Magnet is ye minera of this o^r Chalybs. ffor ↗ as tis hid in ye belly of our Magnet so ↙ being dissolved in o^r ☿ it passes into ⊙ by digestion. Tis a spirit very pure beyond others, o^r ffiery Dragon, our infernal secret fire in its kind most highly volatile [Secr. Rev. p. 7, 8, 15 ↗ Comment. on Ripl. Pref. p. 7 ⟨*illegible symbols, deleted*⟩ ↙ .] our Cadmus, the God of war, Mars [Marrow of Alk. part 2 pag. 4, 8, 9, 14, 17] ↗ stout ↙ Alcides [on Ripl. Gates. p. 52 53] the sulphur ↗ or Chalybs ↙ hid in ye belly ↗ or house ↙ of Aries [Marrow of Alk part. 2. pag. 6, 9. Secr. Rev. p. 28. ↗ On Ripl. Pref. p. 52. ↙ Sendivogij aenigma p 87] that Chalybs wch o^r Magnet chiefly attracts & swallows up in fusion to make ye starry Reg. of ♂ [Secr. Rev. p. 5, 7 16, 28. Comment on Ripl. Pref. p. 7, 31. Marrow of Alk. p. 17.] ⟨Now this Sulphur must be also be prepared as Maier told you, ffor 'tis Philaletha's Kin, *deleted*⟩ Tis the most digested metal next to Gold, for tis Philaletha's King whose Brethren in their passage to him were taken prisoners ↗ & are kept in bondage ↙ by impure 🜍 & must be redeemed by his flesh & blood [Phil. on Ripl. 1st Gate p. 111.

Now this Sulphur must be also prepared. For this ⟨sulphur, *deleted*⟩ King is son to both ye waterbearer & ye Queen, ↗ [Phil. on Ripl. Gate p. 133] ↙ being extracted ↗ not by or out of them both but ↙ by a substance common to them both. ⟨Our, *deleted*⟩ For o^r crude sperm flows from a trinity of ↗ immature ↙ substances in one essence of wch two ⟨ *illegible symbol, deleted*⟩ (♂ & ♄) are extracted ↗ out of ye earth of their nativity ↙ by ye third (♁) & then become

[a] ⟨*Marginal note*⟩ Vide Sendivogij aenigma p. 83.

a pure milky virgin-like Nature drawn from ye menstruum of o^r sordid whore. [Phil. on Ripl. Pref. p. 28, 29, 31. Manuduct. ad Rub. coelest.] Hic est Pisciculus Echinis rotundus ossibus et corticibus carens & habet in se pinguedinem mirificamque vir[p. 9]tutem qui coqui et aqua maris imbui & assari donec albescat et rubescat [AEnigm. 2 in calce Turbae] This fish lies hid in ye bottom of ye great sea of the world without blood & bones ↗ is very little ↙ & can stop the high ships of this sea, that is retain ye spirits of ye world. He is the treasure lying hid in ye caelestial aqua vitae of o^r sea & is extracted out of ye great sea by our Magnet: ⟨*illegible word, deleted*⟩ & if he be taken naturally, he is turned first into water & then into earth. To find this matter of o^r stone you must draw ye moon [sp^t of ♁] from ye firmament [As in distilling] & bring it from heaven up ye earth [of ♂] & turn it into water & then into earth [Instruct. de arb. solar ↗ c. 3. ↙ p. 172, 173. This 🜍 thus extracted is the King ↗ cloathed in a gay rayment as it were ↙ ⟨*illegible letter, deleted*⟩ a robe of ⟨gold, *deleted*⟩ ↗ beaten gold reaching ↙ down to ye grownd & crowned wth a crown of Gold [Philal. on Ripl. Gates p. 105, 106] the Carbuncle set in Mars's crown [Snyders Metamorph. of ye Plan. p. 23.] the martial ruby wch wth ye Venereal Emerauld were stuck in Diana's crown by the help of wch she was to bear solary children. [ib. p. 42.] This is the metallique fixt salt by the help of wch Mercury (after he had by his wings brought ye most high up to his throne & thereby become ye Caduceus) established an everlasting Kingdom ⟨(, *deleted*⟩ [Snyders ⟨Metam, *deleted*⟩ ib. ⟨p, *deleted*⟩ c. 6. p. 15.] This ⟨salt, *deleted*⟩ is the ⟨spe, *deleted*⟩ sharp spere of Mars & sith of Saturn. ffor it is ye salt of Mars by wch he [p. 10] gives ☿ work enough to do & ⟨by, *deleted*⟩ by w^{ch} ⟨Mars coa, *deleted*⟩ Saturn coagulates ☿ [Snyders Pharmac. cath. c. 11. p. 29] & c. 16. p. 48] For Mars must be reduced into his first matter by that wch he hath in him self [ib ⟨p, *deleted*⟩ c. ⟨7, *deleted*⟩ 11. p. 29.] This ↗ salt or red earth ↙ is therefore Flamels male Dragon wthout wings, for ↗ after it is thus extracted out of its ⟨minera, *deleted*⟩ native earth ↙ it is one of the thre substances of wch ye Sun & Moons bath is made [Philal on Ripl. ⟨Gate, *deleted*⟩ Pref. p. 28, 31.] And therefore it is also ye male serpent about ye Caduceus, called by Hermes. Rubor meridionalis & anima solis [capit. 1.]

Cap. 4

De agente primo.

The rod of Mercury reconciles the two serpents ↗ & makes them stick to it [Maier ↙ & therefore is ye medium of joyning their tinctures, whence it's called ye bond of Mercury [Secr. Rev. cap. 2] wch bond is Venus [Marrow of Alk] ⟨for this rod is, *deleted*⟩ It's volatile because anciently painted wth wings in this manner whence came this character ☿ for mercury Its a fluxible ⟨sal, *deleted*⟩ Menstruum because the means of joyning tinctures, ↗ & therefore its also ↙ ⟨*illegible words, deleted*⟩ a saline spirit. ffor its that ☿ wch they call ye salt of ye wise men ⟨*Newton seems to have inadvertently omitted "without" here*⟩ wch nothing is done [Secr. Rev. ch. 1.] & that salt w^{ch} ⟨gave

🜍 , *deleted*⟩ in the beginning of ye work gave 🜍 an incurable wound & wch is said to be ye Key & beginning of ye work, whereby ye prison is opened in w^{ch} 🜍 lies bound [Sendivog. de Sulph. p. 175, 213. Rosar. mag. p 146 [p. 11] 147.] wch Key is ye central salt of Venus [Marrow of Alk. part. 1. pag. 39] & of o^r Magnet [Secr. Rev. c. 4] the most purged salt of nature wherewith ye ofspring of Saturn abounds [ib. c. 11. p. 27] the first ens of mineral salts wch dissolves all metalls but stays not with them & wch alone ye Magi rejected not [ib. p. 25] & wth wch ye Reg. of Mars abounds [Philal on Ripl. Pref. p. 31.] the salt of nature found in saturns ofspring wth w^{ch} ⟨y^e, *deleted*⟩ & sulphur ye vulgar ☿ must be acuate & sol putrefied [Marr. Alk. part. 2. pag. 6] ye water wth w^{ch} ♂ & ♀ wrapt in ye net of Vulcan must be wet [ib. p. 17] ye waterbearer who in ye beginning of ye work must be digested wth ye King & Queen & who is all one wth is water & silver like pitcher in ye midst of wch was a ⟨burning lamp, *deleted*⟩ spark shining to ye surface like a burning lamp or star [Philal. on Ripl. Gates p. 106, 114] ye AEs Hermetis or Venus of ~~Phers~~ wch tingeth not unless it be tinged but if it be tinged then it tinges [in Turba p] the water in wch ye hard & dry substances of ye King & Queen must relent & putrefy [ ] the winged Dragon wch dies not unless slain by his brother & sister who are ye Sun & Moon [Maier. Embl. 25] the cup of love by wch ye brother & sister become fruitfull [ib. Embl. 4] Neptune ⟨w^{ch}, *deleted*⟩ wth his Trident [p. 12] who introduces us into y^e ~~Phers~~ garden [Sendiv. AEnig. p] the vine branch wch wch Bacchus killed the Amphisbena, that is, saith Maier, the liquor thereof [Maier.] the Golden ↗ bough or ↙ Rod upon wch ye two doves of Venus sit [Virg. AEn. 6 & Arcan. Herm. p. 17] the moist fire of ye magi in ye making of wch Sydera Veneris et corniculatae Dianae tibi propitia numina sunto [Triomphe Herm. p. 44] whose composition is thus described by Philaletha. Our fiery Dragon ⟨must, *deleted*⟩ concretes wth Saturnia into a wonderful body wch is volatile & yet in ye fire resembles a a molten metal. The Serpent must[a] then devour ye companions of Cadmus[a] & be[b] fixed by him to an hollow Oak & this oak by ([c]two) sublimations ⟨becoms Dianas Doves, *deleted*⟩ wth Venus (o^r green Lyon) becoms Dianas doves & (if ☿ be added) the winged Caduceus. For these Doves come flying & are infolded in ye arms of Venus & (as alcalies do acids) asswage ye Green Lion & like ye Eagles are numbred by sublimations. Now this ⟨*illegible letters, deleted*⟩ Caduceus is o^r Cupid wch strikes all in love & ye first fire of Sniders by wch saith he ye Metall must be brought into flux, that is by fermentation & digestion. Of its preparation see ye 3^d figure of Abraham ye Jew & Fons Chemicae Philosophiae p. 94, 96.

Chap. 5. Praxis

This rod & ye male & female serpents ⟨*illegible superscript, deleted*⟩ joyned in

[a] ⟨*No note appears.*⟩
[b] ⟨*No note appears.*⟩
[c] ⟨*No note appears.*⟩

y^{e} [a]proportion of 3, 1, 2 compose ye three headed Cerberus wch keeps ye gates of Hell. For being fermented & digested together they resolve & grow dayly more fluid for 15 or 20 days & in 25 or [p. 13] 30 days begin to lack breath ↗ & thicken ↙ & put on a green colour & ⟨turn, *deleted*⟩ in 40 days turn to a rotten black pouder. The green matter may be kept for ferment. Its spirit is ye blood of ye green Lion. The black pouder is our Pluto, ye God of wealth, o^{r} Saturn who beholds himself in ye looking glass of ♂, the calcination wch they call ye first gate, & ye sympathetick fire of Snyders, composed of two contrary fires 🜍 & ⦶ by ye mediation of his first fire. This pouder amalgams wth ☿ & purges out its feces if shaken together in a glass [Epist. N. Fatij]. It mixes also wth melted metalls & Regulus's & in a little quantity purifies y^{m} (as was ⟨said, *deleted*⟩ hinted) but in a greater, burns & calcines them & upon a certain sign, (vizt in ye beginning of ye calcination before ye resolved 🜍 of ye metal flys away & leaves ye Reg. dead like an electrum ↗ & relapsed into an hydrophoby ↙) if ⟨*illegible word, deleted*⟩ it be poured out into twice as much ☿ they amalgam & ye feces of both are purged out wch being well washed of & ye matter sublimed wth ✳ ye Reg will be found resolved into ☿, ⟨Excep, *deleted*⟩ that is its 🜍 & ☿, for the salt of ye metal will stay below, & may be eliviated. Thus may you make a ☿ of 7, 8, 9 or 10 Eagles ↗ wth y^{e} ✳ of ♂ ↙ for the work ⟨of ☉, *deleted*⟩ in common ☉ & by y^{e} ☿ of 1, 2, or 3 eagles resolve ♀ , ♃ & ♄ ↗ (or the ore of ♄ melted down wth ♁) ↙ into ☿ & of y^{t} ☿ sublimed wth ye salt of Venus make ye cold fire, & then wth ye black pouder calcine an amalgam of ☉ ↗ 1 part ↙ & ⟨*illegible symbol, deleted*⟩ y^{e} ☿ of 7 Eagles ↗ 2 parts ↙ & so soon as it beginns to calcine pour upon it 1 part of ye cold [p. 14] fire extracted out of Lead ore wth salt of Venus & not yet volatized, & so on till you have poured on eleven parts & all be calcined. ffor ye saturn will first resolve into water by fusion & then resolve y^{e} ☿^{ies} of the bodies ↗ into salt. ↙ & promote ye action of ye sympathetic fire ↗ that it may in ye calcination pierce them throughly. ↙ & then by sublimation & elixiviation of ye residue you will have ten parts of ye cold fire ↗ or Philosophers ☿ ↙ & one of y^{e} ↗ fixt ↙ salt of y^{e} ☉ & ↗ former ↙ ☿, ⟨wch are to be decocted. Thus have you y^{e} gate of calcination fully opened., *all deleted*⟩ ↗ ⟨wch are to be decocted together seven months to get o^{r} 🜍 ., *deleted*⟩ ↙ w^{ch} ↗ after purgation ↙ are to be decocted together^{+} [margin, p. 14] $^{+}$forty days & y^{n} y^{e} water drawn out till y^{e} matter become retentive ⟨*illegible letter, deleted*⟩ wch must then be decocted again [p. 14] first ↗ 5, 6 or ↙ seven months to get o^{r} 🜍 ⟨stone, *deleted*⟩ & then ten more ↗ ⟨in y^{e} blood of y^{e} green Lion or rather in y^{e} 3 principles, *deleted*⟩ in its sweat ↙ to get o^{r} ⟨stone, *deleted*⟩ tinging stone, & then to be multiplied ↗ by ye 3 principles. ↙ . ⟨This is ye via sicca & ye solutio in ☿ per ☿. The other, *all deleted*⟩ And thus you may understand what the first gate of calcination is & how in the calcination of perfect bodies wth ye first menstrue nothing unclean enters but ye green Lyon & how ye King after his resurrection is fed with the blood of this Lyon, & what is ye solutio violenta of Sendivoguis under wch all other solutions are comprehended, what his humidum radicale metallicum ⟨*illegible word, deleted*⟩ ye ashes of ye burnt

[a] ⟨*Marginal note.*⟩ Hermes capitul. 1. Dimidia quod agentis et illius adjice duplo Maier.

old man; in what sence his aurum is vivum y^{t} is by vertue of ye sympathetic fire; how he separates ye spirit from ye water & congeales ye water in heat & y^{n} adds ye spirit to it; ⟨what is his aqua salis nitri &, *deleted*⟩ how ye seed of o^{r} saturn purges ye matrix of his mother; what is ye aqua salis nitri & ye menstruum mundi; what his 10 parts of air or water & one of ⊙ , & how Diana first (that is ye Reg.) & then ye King falls into [p. 15] y^{e} ⟨fountain milky, *deleted*⟩ fountain of milk white water, ⟨*illegible words, deleted*⟩ & how Trevisan's Golden book falls in after ye King & then he draws out ye water with his bucket ⟨*illegible word, deleted*⟩ till they become retentive that is till two thirds of ye water be drawn out (or perhaps more) ↗ for the imbibitions ↙ & one third remain wth y^{e} ⊙ . This is ye via sicca & ye solution in ☿ per ☿ the other solution & ye via humida are as follows.

When ye Caduceus wth ye 2 serp^{ts} are set to putrefy & are resolved into ⟨liquor, *deleted*⟩ ↗ water ↙ & grown sufficiently liquid wch may be in 3 or 4 days or a week; put in the ☿ ^{ial} precip. of ye net ⟨*illegible word, deleted*⟩ ye scepter of ♃ & y^{e} ☿ ^{ial} precip of ♃ gradually. Let y^{e} ⟨rod, *deleted*⟩ scepter be equal to ye Caduceus & ye precipitates to each other. Or better, let a chaos be made of ye four Elem^{ts} ♂ , ♃ , ♀ , ♁ & quintessentia`|´, ⟨*illegible words, deleted*⟩ in equal proportion & put in first ye scepter wth as much of y^{e} ☿ ^{ial} ⟨sublimate, *deleted*⟩ ↗ precipitate ↙ of these. Or else after ye two former ⟨sublimat, *deleted*⟩ precipitates are fermented in some competent quantity, put in an amalgam of y^{e} ⟨*illegible word, deleted*⟩ Chaos. And note y^{t} this Chaos ⟨if made w^{t} ye Reg. of ♂ , *deleted*⟩ is ye hollow oak. But perhaps it must be made wth y^{e} ↗ two ↙ Reg^{s} of ♂ & ♃ . ⟨*illegible words, deleted*⟩ When ⟨*illegible word, deleted*⟩ this Chaos is fermented & sufficiently resolved into ☿ wch perhaps will be in ⟨*illegible letter or number, deleted*⟩ a few days, ⟨distill it, *deleted*⟩ wash away ye feces & distill it. & sublime it wth ye salt of saturn. But first impregnate ye salt of ♄ wth as much volatile salt of ⟨ ♄ , *deleted*⟩ ♂ as it will retain for this is ye sith wth wch he [p. 16] must cut of ye leggs of ☿ & coagulate him. This ⟨cold, *deleted*⟩ is ye cold fire, w^{ch} ⟨*illegible word, deleted*⟩ being fermented wth ye two dragons in a due proportion as ⟨yo, *deleted*⟩ was ye rod of ☿ ⟨will give, *deleted*⟩ & digested ↗ ten or 20 days or ↙ till ye green colour appear will by distilling give you ye blood of ye green Lyon, o^{r} Venus, ⟨ye 3^{d} fire, *deleted*⟩ o^{r} wine, our ⟨Mercury, *deleted*⟩ ↗ dry water our ↙ Mercurius duplatus. ⟨The third fire of, *deleted*⟩ Artephius his third fire, his ↗ Vinegar ↙ Antimonial saturnine mercurial ⟨argent vive, *deleted*⟩ ↗ & of salarmoniac ↙ in wch there is a double substance of argent vive ye one of Antimony the other of ☿ ⟨[& ♄], *deleted*⟩ sublimed [wth ♄].

In this ⟨*illegible word, deleted*⟩ water 2 digest ye Reg of ♂ 1 ye former ☿ 1, for a week till the ⟨*illegible letter or symbol, deleted*⟩ ♁ be resolved into ☿ (Or if ye Reg of ♂ will not ↗ amalgam & ↙ resolve wthout ye addition of ⟨*illegible letter or symbol, deleted*⟩ other metalls, ⟨digest mix it wth ♀ digest y^{e} *illegible letters, all deleted*⟩ mix it with (♀ , ♃ , & ♈ana.) ↗ or rather wth Luna ↙ & digest a week.) ⟨*illegible letters, deleted*⟩ distill away y^{e} ☿ , ⟨dissolve, *deleted*⟩ & rectify it, dissolve ye rest in Æ & you will have o^{r} Gold in a black pouder. The same may be got by ⟨*illegible word, deleted*⟩ digesting y^{e} ☿ of 7, 8, 9 or 10 Eagles but this way is about. ⟨Amalgam y^{e} *illegible words, all deleted*⟩ Wash & dry y^{e}

black pouder. Amalgam it wth ⟨the, *deleted*⟩ its rectified ☿ in a due proportion. ⟨Shut them, *deleted*⟩ Wash them well & Dry them ⟨*illegible words, deleted*⟩ & in a digestion of 7 months you will have o^{r} ⟨ 🜍 , *deleted*⟩ tinging 🜍 ⟨*illegible letters, deleted*⟩ In ye mean time if you putrefy again ye mercurial ⟨*illegible word, deleted*⟩ water till it be ⟨black as, *deleted*⟩ ↗ like melted ↙ pitch & then distill you will have a white & red spirit wch are Diana & Apollo, Aqua vitae & vinegre, the virgins milk & blood, & [p. 17] in ye bottom will remain a black earth wch is Latona our salt of tartar o^{r} Gold found in a dunghill, o^{r} Toad, o^{r} Bacchus. Rectify ye spirits 7 times & each time put the feces of ye white to ye red & those of ye red to ye black earth. Calcine this earth gently Extract its salt wth distilld water amalgam one part of this salt ↗ first ↙ wth 3 parts of ye white or red spirit & then wth 9 parts of ye stone of ye same colour & by a ↗ short ↙ digestion you will have ye stone multiplied. Thus you may multiply each stone 4 times & no more for they will then become oyles shining in ye dark & fit for magicall uses. You may ferment them wth ☉ & ☽ by keeping the stone & metall in fusion together for a day, & then project upon metalls. This is the multiplication of ye stone in vertue. To multiply it in weight ad to it of ye first Gold whether philosophic or vulgar. Thus ye Sulphur will every ⟨*illegible word, deleted*⟩ multiplication encrease ten times in vertue & if you multiply it wth y^{e} ☿ of ye 1st or 2^{d} rotation you may encrease it much more. If you want ye Philosophic Gold, you may ad to it Gold vulgar wth ye fixt salt & multiplying mercuries in ye hour of y^{e} ⟨stones, *deleted*⟩ ↗ mercuries ↙ nativity ↗ that is ↙ in ye beginning of ye Regimen of Luna.

Thus you must do for multiplication. But [p. 18] if you would whiten Latona then distill not ye red spirit but cohobate ye white spirit upon ye black matter wth interposed digestions till it bring over ↗ all ye red spirit ↙ with it wch you shall know by its the black matter a light dry pouder. Imbibe this pouder ⟨wth, *deleted*⟩ first wth an eighth part of its weight of ye animated spirit then wth a seventh then wth a sixt then wth a fift, ⟨*illegible words, deleted*⟩ & ever after wth a fourth, interposing a weeks digestion between every imbibition till ye matter be moderately dry & then distilling off ye flegm. And when by these imbibitions Latona grows white & fluxible as wax & will ascend, sublime her ↗ from ye feces ↙ , & you shall have ye plumbum album sapientum, the white Diana. ⟨*illegible letters, deleted*⟩ Imbibe one part of this sublimate wth three of ye spirit digest for 24 hours distill, imbibe ye remainder wth thrice its weight of new spirit, digest 24 hours & destill, ⟨*illegible words, deleted*⟩ Imbibe & digest a third & 4th time ⟨*illegible word, deleted*⟩ & all will ascend. ⟨Dig, *deleted*⟩ Circulate it for eight or nine weeks & you have the Alkahest.

Rectify & dilute wth rain water ye oyle of good & well purified Hungarian Vitriol. Therein dissolve a clean & unctuous ☿ . ↗ vizt the abovementioned living ☿ . ↙ fferment & digest it 30 ↗ 20 ↙ days. Draw of ye spirit ⟨& dig, *deleted*⟩ fferment & digest again & when ye matter is like molten pitch distill & cohobate ye spirit to extract ye soul & procee as ⟨before, *deleted*⟩ in ye work of ♄ & you shall have the true Alkahest. If you add ⟨*illegible word, deleted*⟩ the ↗ said ↙ solution of ☿ in oyle of ⊕ unto ye cold fire & putrefy them with ye two Dragons, you have Snyders his most general ☿ of both a solary & lunary nature, borrowing heat from Venus & coldnes from ♄ & conteining all ye vertues of ye Univers [margin, p. 18] ⟨*unkeyed note, evidently referring to Snyders*⟩ But to ↗ ☿ of ♄ & ↙ ye oyle of ⊕ & ☿ of ♄ he adds ye fixt salt of ye terra Adamica to obtain all in all., *all deleted*⟩ [p. 19] But to ye cold ☿ of ♄ & hot ☿ of Venus, he adds ye fixt salt of Terra Adamica to obtein all in all. ⟨*illegible word, deleted*⟩ So [a]Basil Valentine, Set Adam in a water bath, where Venus like her self one hath ↗ wch ye old Dragon hath prepared. ↙ . This earth he tells us in [b]another place is not likened to any thing y^t^ is grown, that is neither to stones nor minerals, ⟨& y^t^ it hath a scent of dead mens bones, *deleted*⟩ ↗ Manna saith tis not clay nor mud but a quintessentiall matter or Chaos out of wch man & all ye world was made & that tis called earth but is not so, ⟨*part of this interlineation spilled over into the margin of p. 19*⟩ ↙ . So Norton tells us that ⟨*unkeyed marginal note to Norton*⟩ Ordinal. p. 41 42, 43, 56. [p. 19] ⟨the, *deleted*⟩ many things help ye work but ⟨*illegible letters, deleted*⟩ yet there are only two materialls ↗ to the white stone ↙ , the mother & ye child, the female & ye male, besides salarmoniack & sulphur ⟨of kind a metall, *deleted*⟩ gotten out of metalls. The one is ⟨*illegible superscript, deleted*⟩ fixt ↗ in ye fire ↙ as stones are, but ⟨appears, *deleted*⟩ ↗ is ↙ not ⟨like, *deleted*⟩ a stone ↗ in handling nor ↙ in sight but a subtil earth broun ruddy & not bright, ↗ & yet ↙ ⟨&, *deleted*⟩ after some deale white. This ↗ is the chief ⟨stone &, *deleted*⟩ material & ↙ he calls it Markasite & after its ⟨preparation, *deleted*⟩ separation Litharge ↗ & saith its of no more value then a lump of clay ⟨*part of this interlineation spills into the margin of p. 19*⟩ ↙ . The other material is a stone in handling & in sight glorious fair & bright glittering wth perspicuity, being of wonderfull diaphanity, colore subalbido ↗ like pale urin ↙ or like in colour to Orrichine stone, yet glittering wth clearness, called Magnesia & being Res aeris in qua latet scientia divina. ⟨Whence y^t^ memorabl, *deleted*⟩ To all this agrees also that memorable saying, Visita Interiora Terrae Rectificando Invenies Occultum Lapidem Veram Medicinam. Let therefore the waters be compounded wth ye earth & ye compound be fermented by ye two Dragons in a due proportion, & in this water decoct ☉ & ☽ .

Artefius tells us that his fire dissolves & gives life to stones & ⟨that, *deleted*⟩ Pontanus that their ⟨*illegible words, deleted*⟩ fire is not transmuted wth their matter becaus it is not of their matter, but turns it wth all its feces [p. 20] into ye elixir. W^ch^ deserves well to be considered. For this is ye best explication of their saying that ye stone is made of one only thing.

⟨*Newton left the remainder of p. 20 blank.*⟩

[a] ⟨*Marginal note.*⟩ Key. 12.

[b] ⟨*Marginal note.*⟩ Microcosm p. 8.

Bibliography of works cited

Manuscripts

England
Cambridge
Fitzwilliam Museum: Fitzwilliam MS 276*
King's College, Cambridge: Keynes MSS 3, 12, 13, 14, 15, 16, 18, 20, 21, 23, 24, 25, 26, 27, 28, 30, 32, 35, 37, 38, 40, 41, 42, 43, 44, 46, 48, 49, 50, 53, 54, 59, 61, 63, 64
University Library, Cambridge: Portsmouth Collection Add. MSS 3965, 3970, 3973, 3975, 3990, 4003
London
British Library: Add. MS 44, 888
Royal Society: Journal; Gregory MSS.
Oxford
Bodleian Library: MS Don. b. 15; Ekins Papers, MS New College 361

Israel
Jerusalem
Jewish National and University Library: MS Var. 259; Yahuda MS Var. 1, Newton MSS 6, 15, 16, 17, 21, 30, 38, 41

Switzerland
Geneva
Bibliothèque publique et universitaire de Genève: MS. français 605; Papiers Fatio No. 3 (Inv. 526)
Zürich
Zentralbibliothek Zürich: Codex rhenovacensis 172

United States
California
Los Angeles
The William Andrews Clark Memorial Library: Isaac Newton, "Out of Cudworth"

Stanford

Stanford University Libraries, Department of Special Collections, Isaac Newton Collection (M132): Isaac Newton, "Philalethes"

Connecticut

New Haven

The Medical Historical Library, Cushing/Whitney Medical Library, Yale University: Isaac Newton, "Transcript of John de Monte Snyders, The Metamorphosis of the Planets"

District of Columbia

Washington

Smithsonian Institution Libraries: Dibner Collection MSS 1031 B, 1032 B, 1070 A

Illinois

Chicago

Department of Special Collections, The University of Chicago Library: The Joseph Halle Schaffner Collection of Scientific Manuscripts, Box 1, Folder 47

Massachusetts

Babson Park

Babson College Archives: Babson MSS 414, 420, 421, 434

Boston

The Francis A. Countway Library of Medicine, Boston Medical Library/Harvard Medical Library: Isaac Newton, "The regimens described wth ye times & signes"

Cambridge

Massachusetts Institute of Technology, Institute Archives and Special Collections Department: MS 246/N56/17–/

New York

New York

Columbia University Library: Isaac Newton, "The Three Mysterious Fires"

Pennsylvania

Bethlehem

Lehigh University Libraries: Isaac Newton, "MS on Miracles"

Philadelphia

University of Pennsylvania, The Van Pelt Library: Isaac Newton, "Annotations in Elias Ashmole, *Theatrum Chemicum Britannicum*"

Wisconsin

Madison

University of Wisconsin, Duveen Collection: Isaac Newton, "Annotations in Eirenaeus Philalethes, *Secrets Reveal'd*"

Published Works

Ahonen, Kathleen. "Review of B. J. T. Dobbs, *The Foundations of Newton's Alchemy*." *Annals of Science* 33 (1976), 615–17.

Aiton, E. J. "Newton's aether-stream hypothesis and the inverse square law of gravitation." *Annals of Science* 25 (1969), 255–60.

Alexander, H. G. "Introduction." In H. G. Alexander, Ed., *The Leibniz–Clarke Correspondence, Together with Extracts from Newton's Principia and Opticks*. With introduction and notes by H. G. Alexander. Philosophical Classics, General Ed.: Peter G. Lucas. Manchester: Manchester University Press, 1956, pp. ix–lv.

Ed. *The Leibniz–Clarke Correspondence, Together with Extracts from Newton's Principia and Opticks*. With introduction and notes by H. G. Alexander. Philosophical Classics, General Ed.: Peter G. Lucas. Manchester: Manchester University Press, 1956.

Ancient Christian Writers. The Works of the Fathers in Translation. Vols. 41 and 42. *St. Augustine. The Literal Meaning of Genesis*. Tr. and annotated by John Hammond Taylor, S.J. Two vols. New York: Newman Press, 1982.

Angus, S. *The Religious Quests of the Graeco-Roman World. A Study in the Historical Background of Early Christianity*. London: John Murray; New York: Charles Scribner's Sons, 1929.

Aratus. *Phaenomena*. In *Sky Signs: Aratus' Phaenomena*. Introduction and translation by Stanley Lombardo. With illustrations by Anita Volder Frederick. Berkeley, CA: North Atlantic Books, n.d.

Phaenomena. In Stanley Frank Lombardo. "Aratus' *Phaenomena*: An Introduction and Translation." The University of Texas at Austin: Ph. D. dissertation, 1976.

The Phaenomena of Aratus. Tr. by G. R. Mair. In *Callimachus, Hymns and Epigrams; Lycophron; Aratus*. The Loeb Classical Library. Ed. by E. H. Warmington. Rev. reprint of the 1921 ed. Cambridge, MA: Harvard University Press; London: William Heinemann, 1969.

Aris, Rutherford, H. Ted Davis, and Roger H. Stuewer, Eds. *Springs of Scientific Creativity. Essays on Founders of Modern Science*. Minneapolis: University of Minnesota Press, 1983.

Aristotle. *The Works of Aristotle Translated into English under the Editorship of J. A. Smith and W. D. (Sir David) Ross*. Twelve vols. Oxford: University

Press and Clarendon Press; London: Humphrey Milford, 1908–60. Vols. 3 and 5.

Artis avriferae, qvam chemiam vocant, Volumina duo, qvae continent Tvrbam Philosophorum, aliosqúe antiquissimos auctores, quae versa pagina indicat. Accessit nouiter volumen tertium, continens: 1. Lullij vltimum Testamentum. 2. Elucidationem Testam. totius ad R. Odoardum. 3. Potestatem diuitiarum, cum optima expositione Testamenti Hermetis. 4. Compendium Artis Magicae, quoad compositionem Lapidis. 5. De Lapide & oleo Philosophorum. 6. Modum accipiendi aurum potabile. 7. Compendium Alchimiae & naturalis Philosophiae. 8. Lapidarium. Item Alberti Magni secretorum Tractatus. Abbreuiationes quasdam de Secretis Secretorum Ioannis pauperum. Arnaldi Quaest. de Arte Transmut. Metall. eiusqúe Testamentum. Omnia hactenus nunquam visa nec edita. Cum Indicibus rerum & verborum locupletissimis. Three vols. in one. Basileae: Typis Conradi Waldkirchii, 1610.

Ashmole, Elias. *Theatrum Chemicum Britannicum. Containing severall poeticall pieces of our famous English philosophers, who have written the Hermetique mysteries in their owne ancient language....* London: 1652.

Aspelin, Gunnar. *Ralph Cudworth's Interpretation of Greek Philosophy. A Study in the History of English Philosophical Ideas.* Göteborgs Högskolas Årsskrift XLIX 1943:1. Göteborg: Wettergren & Kerbers Förlag, Elanders Boktryckeri Aktiebolag, 1943.

Atwood, Mary Anne. *Hermetic Philosophy and Alchemy: A Suggestive Inquiry into "The Hermetic Mystery" with a Dissertation on the More Celebrated of the Alchemical Philosophers.* Introduction by Walter Leslie Wilmhurst. Rev. ed. New York: The Julian Press, 1960.

Augustine. *The Literal Meaning of Genesis.* In *Ancient Christian Writers. The Works of the Fathers in Translation.* Vols. 41 and 42. Tr. and annotated by John Hammond Taylor, S.J. New York: Newman Press, 1982.

Avicenna. "Avicennae Tractatulus de Alchimia." In *Artis avriferae, qvam chemiam vocant, Volumina duo, qvae continent Tvrbam Philosophorum, aliosqúe antiquissimos auctores, quae versa pagina indicat. Accessit nouiter volumen tertium, continens: 1. Lullij vltimum Testamentum. 2. Elucidationem Testam. totius ad R. Odoardum. 3. Potestatem diuitiarum, cum optima expositione Testamenti Hermetis. 4. Compendium Artis Magicae, quoad compositionem Lapidis. 5. De Lapide & oleo Philosophorum. 6. Modum accipiendi aurum potabile. 7. Compendium Alchimiae & naturalis Philosophiae. 8. Lapidarium. Item Alberti Magni secretorum Tractatus. Abbreuiationes quasdam de Secretis Secretorum Ioannis pauperum. Arnaldi Quaest. de Arte Transmut. Metall. eiusqúe Testamentum. Omnia hactenus nunquam visa nec edita. Cum Indicibus rerum & verborum locupletissimis.* Three vols. in one. Basileae: Typis Conradi Waldkirchii, 1610. Vol. I, 260–79.

Bäumer, Änne, and Manfred Büttner, Eds. *Science and Religion / Wissenschaft und Religion. Proceedings of the Symposium of the XVIIIth International Congress of History of Science at Hamburg-Munich 1.–9. August 1989.* Bochum: Universitätsverlag Dr. N. Brockmeyer, 1989.

Bailey, Cyril, Ed. Lucretius, *Titi Lvcreti Cari De Rervm Natvra Libri Sex.* With

Prolegomena, Critical Apparatus, Translation, and Commentary by Cyril Bailey. Three vols. Oxford: Clarendon Press, 1947.

Bak, J. M. "Medieval symbology of the state: Percy E. Schramm's contribution." *Viator* 4 (1973), 33–63.

Baker, Herschel. *The Wars of Truth. Studies in the Decay of Christian Humanism in the Earlier Seventeenth Century*. Cambridge, MA: Harvard University Press, 1952.

Bartholomaeus Anglicus. *De proprietatibus rerum*. Tr. by John Trevisa. Westminster: Wynken de Worde, 1495.

Basilius Valentinus. *Azoth*.... Paris, 1659.

Les dovze clefs de philosophie.... Paris, 1660.

Bass, David. "'The Errors of our Conceits': Accommodation in the Wilkins–Ross Debate." Trinity Evangelical Divinity School: Th.M. thesis, 1989.

Bechler, Zev, Ed. *Contemporary Newtonian Research*. Studies in the History of Modern Science, 9. Dordrecht: D. Reidel, 1982.

Bernstein, Howard R. "Leibniz and the *Sensorium Dei*." *Journal of the History of Philosophy* 15 (1977), 171–82.

Berthelot, Marcellin Pierre Eugene. *Les origines de l'alchimie*. Reprint of the 1885 ed. Paris: Librairie des Sciences et des Arts, 1938.

Bibliotheque de philosophes [chymiques,] ou recueil des oeuvres des auteurs les plus approuvez qui ont ecrit de la pierre philosophale. Tome premier. Contenant sept Traitez qui sont énoncey dans la page suivante. Avec un Discourse, servant de Preface, sur la verité de la Science, & touchant les Auteurs qui sont dans ce Volume. Et une Liste des Termes de l'Art, & des Mots anciens qui se trouvent dans ces Traitez, avec leur explication. Par le Sieur S. D. E. M. Tome second. Qui contient cinq Traitez énoncey dans l'autre page, & nouvellement traduits. Avec des Remarques & les diverses Leçons. Une Lettre Latine sur le Livre intitulé Icon Philosophiae Occultae. Vne Preface sur l'obscurité des Philosophes, & sur les Traittez de ce Tome, & leurs Auteurs. Et une Table des Matieres. Par le Sieur S. Docteur en Medecine. Two vols. Paris: Chez Charles Angot, ruë Saint Jacques, au Lyon d'or, 1672–8.

Blunt, Anthony. "Blake's 'Ancient of Days.' The symbolism of the compasses." *Journal of the Warburg Institute* 2 (1938–9), 53–63.

Bonelli, M. L. Righini, and William R. Shea, Eds. *Reason, Experiment, and Mysticism in the Scientific Revolution*. New York: Science History Publications, 1975.

Brewster, David. *Memoirs of the Life, Writings, and Discoveries of Sir Isaac Newton*. Two vols. Edinburgh: Thomas Constable and Co.; Boston: Little, Brown, and Co., 1855.

Brooks, Richard Stoddard. "The Relationships between Natural Philosophy, Natural Theology, and Revealed Religion in the Thought of Newton and Their Historiographic Relevance." Northwestern University: Ph. D. dissertation, 1976.

Burnet, Thomas. *Telluris Theoria Sacra: Orbis Nostri Originem & Mutationes Generales, quas Aut jam subiit, aut olim subiturus est, complectens. Libri duo posteriores De Conflagratione Mundi, et De Futuro Rerum Statu*. Londini: Typis R. N. Impensis Gualt. Kettilby, ad Insigne Capitis Episcopi in Coemeterio Paulino, 1689.

Telluris Theoria Sacra: Orbis Nostri Originem & Mutationes Generales, quas Aut jam subiit, aut olim subiturus est, complectens. Libri duo priores De Diluvio & Paradiso. Londini: Typis R. N. Impensis Gualt. Kettilby, ad Insigne Capitis Episcopi in Coemetrio Paulino, 1681.

The Theory of the Earth: Containing an Account of the Original of the Earth, and of all the General Changes Which it hath undergone, or is to undergo, Till the Consummation of all Things. The Two First Books Concerning The Deluge, and Concerning Paradise [1684]. *The Two Last Books, Concerning the Burning of the World, and Concerning the New Heavens and New Earth* [1690]. Two vols. in one. London: printed by R. Norton, for Walter Kettilby, at the Bishop's Head in St. Paul's Church-Yard, 1684–90.

The Sacred Theory of the Earth. With an Introduction by Basil Willey. Centaur Classics. General Ed.: J. M. Cohen. Carbondale, IL: Southern Illinois University Press, 1965.

Burns, R. M. *The Great Debate on Miracles from Joseph Glanvill to David Hume.* Lewisburg: Bucknell University Press; London: Associated University Presses, 1981.

Burtt, Edwin Arthur. *The Metaphysical Foundations of Modern Science.* Reprint of 2nd rev. ed. Doubleday Anchor Books. Garden City, NY: Doubleday & Co., 1954.

Buttrick, George A., et al., Eds. *The Interpreter's Bible.* Twelve vols. Nashville, TN: Abington Cokesbury Press, 1951–7.

Campbell, Joseph, Ed. *Spirit and Nature. Papers from the Eranos Yearbooks.* Bollingen Series xxx.1. Princeton: Princeton University Press, 1982.

Casini, Paolo. "Newton: the Classical Scholia." *History of Science* 22 (1984), 1–58.

Caspar, Max. *Kepler.* Tr. and ed. by C. Doris Hellman. London: Abelard-Schuman, 1959.

Cassirer, Ernst. *The Platonic Renaissance in England.* Tr. by James P. Pettegrove. Austin: University of Texas Press, 1953.

Castillejo, David. *The Expanding Force in Newton's Cosmos As Shown in His Unpublished Papers.* Madrid: Ediciones de Arte y Bibliofilia, 1981.

Catalogue Général des Livres Imprimés de la Bibliothèque National. Auteurs. Ministère de l'Éducation Nationale. 231 vols. in 232. Paris: Imprimerie Nationale, 1897–1981.

Catalogue of the Newton Papers Sold by Order of the Viscount Lymington to Whom They Have Descended from Catherine Conduitt, Viscountess Lymington, Great-niece of Sir Isaac Newton. London: Sotheby and Co., 1936.

A Catalogue of the Portsmouth Collection of Books and Papers written by or belonging to Sir Isaac Newton, the scientific portion of which has been presented by the Earl of Portsmouth to the University of Cambridge. Drawn up by the Syndicate appointed the 6th November, 1872. Cambridge University Press, 1888.

Chenu, M.-D. *Nature, Man, and Society in the Twelfth Century. Essays on New*

Theological Perspectives in the Latin West. Preface by Etienne Gilson. Selected, ed., and tr. by Jerome Taylor and Lester K. Little. Chicago: The University of Chicago Press, 1968.

"The Old Testament in twelfth-century theology." In M.-D. Chenu. *Nature, Man, and Society in the Twelfth Century. Essays on New Theological Perspectives in the Latin West.* Preface by Etienne Gilson. Selected, ed., and tr. by Jerome Taylor and Lester K. Little. Chicago: The University of Chicago Press, 1968, pp. 146–61.

"Theology and the new awareness of history." In M.-D. Chenu. *Nature, Man, and Society in the Twelfth Century. Essays on New Theological Perspectives in the Latin West.* Preface by Etienne Gilson. Selected, ed., and tr. by Jerome Taylor and Lester K. Little. Chicago: The University of Chicago Press, 1968, pp. 162–201.

Chippendale, Christopher. *Stonehenge Complete.* Ithaca, NY: Cornell University Press, 1983.

Christianson, Gale E. *In the Presence of the Creator. Isaac Newton and His Times.* New York: The Free Press; London: Collier Macmillan, 1984.

Churchill, Mary S. "*The Seven Chapters*, with explanatory notes." *Chymia* 12 (1967), 29–57.

Cicero. *Cicero in Twenty-Eight Volumes. XIX. De natura deorum, Academica.* Tr. by H. Rackham. The Loeb Classical Library. Ed. by E. H. Warmington. Cambridge, MA: Harvard University Press; London: William Heinemann, 1972.

M. Tullii Ciceronis, De natura deorum libri tres. Erklaert von G. F. Schoemann. Leipzig: Weidmannsche Buchhandlung, 1850.

[Claveus, Gasto?] *Le filet d'Ariadne....* Paris, 1695.

Cleanthês. *Hymn to Zeus.* In *The Oxford Book of Greek Verse in Translation.* Ed. by T. F. Higham and C. M. Bowra. Oxford: At the Clarendon Press, 1938.

Cleidophorus Mystagogus. *Mercury's Caducean Rod: Or, The great and wonderful Office of the Universal Mercury, or God's Vicegerent, Displayed. Wherein is Shewn His Nativity, Life, Death, Renovation and Exaltation to an Immutable State; Being A true Description of the Mysterious Medicine of the Ancient Philosophers.* London: printed by W. Pearson, and sold by T. Northcott, in George-Ally in Lombard-street, 1702.

Clement of Alexandria. *Opera....* Lutetiae Parisiorum, 1641.

Clifford, Richard J. "The temple and the holy mountain." In *The Temple in Antiquity. Ancient Records and Modern Perspectives.* Ed., with introductory essay, by Truman G. Madsen. The Religious Studies Monograph Series, Vol. 9. Provo, UT: Brigham Young University Religious Studies Center, 1984, pp. 107–24.

Cochrane, Charles Norris. *Christianity and Classical Culture. A Study of Thought and Action from Augustus to Augustine.* Reprint of the 1944 ed. London: Oxford University Press, 1977.

Cohen, I. Bernard. "Hypotheses in Newton's philosophy." *Physis* 8 (1966), 163–84.

"Introduction." In Isaac Newton, *A Treatise of the System of the World Trans-*

lated into English. Introduction by I. Bernard Cohen. Facsimile reprint of the 2nd London ed. of 1731. London: Dawsons of Pall Mall, 1969.

Introduction to Newton's "Principia." Cambridge, MA: Harvard University Press; Cambridge University Press, 1971.

The Newtonian Revolution. With Illustrations of the Transformation of Scientific Ideas. Cambridge University Press, 1980.

"The *Principia*, universal gravitation, and the 'Newtonian style,' in relation to the Newtonian revolution in science: notes on the occasion of the 250th anniversary of Newton's death." In *Contemporary Newtonian Research.* Ed. by Zev Bechler. Studies in the History of Modern Science, 9. Dordrecht: D. Reidel, 1982, pp. 21–108.

and R. Taton, Eds. *Mélanges Alexandre Koyré, publiés à l'occasion de son soixante-dixième anniversaire.* Two vols. Paris: Hermann, 1964.

Colie, Rosalie L. *Light and Enlightenment. A Study of the Cambridge Platonists and the Dutch Arminians.* Cambridge University Press, 1957.

Collier, Katherine Brownell. *Cosmogonies of Our Fathers. Some Theories of the Seventeenth and Eighteenth Centuries.* Studies in History, Economics and Public Law, ed. by the Faculty of Political Science of Columbia University, No. 402. New York: Columbia University Press; London: P. S. King & Son, 1934; New York: Octagon Books, 1968.

The Compact Edition of the Oxford English Dictionary. Complete Text Reproduced Micrographically. Two vols. n.p.: Oxford University Press, 1971.

Conduitt, John. "Memoirs of Sir Isaac Newton, sent by Mr. Conduitt to Monsieur Fontenelle, in 1727." In Edmund Turnor, *Collections for the History of the Town and Soke of Grantham. Containing Authentic Memoirs of Sir Isaac Newton, Now First Published From the Original MSS. in the Possession of the Earl of Portsmouth.* London: printed for William Miller, Albemarle-Street, by W. Bulmer and Co. Cleveland-Row, St. James's, 1806, pp. 158–86.

Conger, George Perrigo. *Theories of Macrocosms and Microcosms in the History of Philosophy.* Reprint of the 1950 ed. New York: Russell & Russell, 1967.

Cook, Eleanor, Chaviva Hošek, Jay Macpherson, Patricia Parker, and Julian Patrick, Eds. *Centre and Labyrinth. Essays in Honour of Northrop Frye.* Toronto: University of Toronto Press in association with Victoria University, 1983.

Copenhaver, Brian P. "Jewish theologies of space in the scientific revolution: Henry More, Joseph Raphson, Isaac Newton and their predecessors." *Annals of Science* 37 (1980), 489–548.

Copernicus, Nicolaus. *De Revolutionibus Orbium Coelestium.* In *Copernicus: On the Revolutions of the Heavenly Spheres. A New Translation from the Latin with an Introduction and Notes by A. M. Duncan.* Newton Abbot: David & Charles; New York: Barnes & Noble, 1976.

Craven, J. B. *Count Michael Maier. Doctor of Philosophy and of Medicine, Alchemist, Rosicrucian, Mystic, 1568–1622. Life and Writings.* Kirkwall: William Peace & Son, Albert Street, 1910.

Cross, F. L., and E. A. Livingstone, Eds. *The Oxford Dictionary of the Christian Church.* 2nd ed. Oxford: Oxford University Press, 1983.

Cross, Frank Moore, Jr. "The priestly tabernacle in the light of recent research."

In *The Temple in Antiquity. Ancient Records and Modern Perspectives.* Ed., with introductory essay, by Truman G. Madsen. The Religious Studies Monograph Series, Vol. 9. Provo, UT: Brigham Young University Religious Studies Center, 1984, pp. 91–105.

Cudworth, Ralph. *The True Intellectual System of the Universe: The First Part; Wherein, All the Reason and Philosophy of Atheism is Confuted; and Its Impossibility Demonstrated.* London: printed for Richard Royston, bookseller to His most Sacred Majesty, 1678; Stuttgart-Bad Connstatt: Friedrick Frommann Verlag [Günther Holzboog], 1964.

Curry, Patrick, Ed. *Astrology, Science and Society. Historical Essays.* Woodbridge, Suffolk: The Boydell Press, 1987.

Cyril. *Opera....* Lutetiae Parisiorum, 1631.

Daniélou, Jean. *From Shadows to Reality. Studies in the Biblical Typology of the Fathers.* London: Burns & Oates, 1960.

d'Atremont, H. *Le tombeau de la pauvreté....* Paris, 1673.

Davis, Audrey B. *Circulation Physiology and Medical Chemistry in England 1650–1680.* Lawrence, KA: Coronado Press, 1973.

Debus, Allen G. *The Chemical Philosophy. Paracelsian Science and Medicine in the Sixteenth and Seventeenth Centuries.* Two vols. New York: Science History Publications, 1977.

"Edward Jorden and the fermentation of the metals: an iatrochemical study of terrestrial phenomena." In *Toward a History of Geology.* Ed. by C. E. Schneer. Cambridge, MA: Harvard University Press, 1969, pp. 100–21.

"Review of B. J. T. Dobbs, *The Foundations of Newton's Alchemy.*" *Centaurus* 21 (1977), 315–16.

Ed. *Science, Medicine, and Society in the Renaissance. A Festschrift in Honor of Walter Pagel.* Two vols. New York: Neale Watson Academic Publications, 1972.

"The sun in the universe of Robert Fludd." In *Le soleil à la Renaissance. Sciences et mythes. Colloque international tenu en avril 1963 sous les auspices de la Fédération Internationale des Instituts et Sociétés pour l'Étude de la Renaissance et du Ministère de l'Éducation nationale et de la Culture de Belgique. Publié avec le concours du Gouvernement belge.* Bruxelles: Presses universitaires de Bruxelles; Paris: Presses universitaires de France, 1965, pp. 259–77 and following plates.

de' Conti, Luigi. *Discours philosophiques....* Paris, 1678.

de la Chastre, René. *Le prototype ... de l'art Chimicq;....* Paris, 1620.

[de Respour, P. M.] *Rare experiences svr l'esprit mineral....* Paris, 1668.

Descartes, René. *Correspondance avec Arnauld et Morus. Texte latin et traduction. Introduction et notes par Geneviève Lewis.* Bibliothèque des textes philosophiques, Directeur: Henri Gouhier. Paris: Librairie Philosophique J. Vrin, 1953.

Oeuvres de Descartes publiées par Charles Adam & Paul Tannery. Eleven vols. Paris: Librairie Philosophique J. Vrin, 1964–74.

Principia philosophiae. In René Descartes, *Opera philosophica.* 3rd ed. Amsterdam, 1656.

Principles of Philosophy. Tr., with explanatory notes, by Valentine Rodger Miller and Reese P. Miller. Collection des Travaux de l'Académie Internationale d'Histoire des Sciences, No. 30. A Pallas Paperback. Dordrecht: D. Reidel, 1984.

A Descriptive Catalogue of the Grace K. Babson Collection of the Works of Sir Isaac Newton and the Material Relating to Him in the Babson Institute Library, Babson Park, Mass. With an Introduction by Roger Babson Webber. New York: Herbert Reichner, 1950.

[d'Espagnet, Jean.] *La philosophie natvrelle....* Paris: 1651.

Deux traitez nouveaux sur la philosophie naturelle.... Paris, 1689.

de Villamil, Richard. *Newton: the Man*. Foreword by Albert Einstein. London: Gordon D. Knox, [1931].

Devons, Samuel. "Newton the alchemist?" *Columbia Library Columns* 20 (1971), 16–26.

Digby, Kenelm. *Of Bodies, and of Mans Soul. To Discover the Immortality of Reasonable Sovls. With two Discourses Of the Powder of Sympathy, and Of the Vegetation of Plants*. London: printed by S. G. and B. G. for John Williams, and are to be sold in Little Britain over against St. Buttolphs-Church, 1669.

Two Treatises. London, 1658.

Dirks, D. R. *Early Greek Astronomy to Aristotle*. Aspects of Greek and Roman Life. General Ed. H. H. Schullard. Reprint of the 1970 ed. Cornell Paperbacks. Ithaca, NY: Cornell University Press, 1985.

Divers traitez de la philosophie naturelle. Sçavoir, la turbe des philosophes, ou le code de verité en l'art. La parole delaissée de Bernard Trevisan. Les deux traitez de Corneille Drebel Flaman. Avec le tres-ancien duel des Chevaliers. Nouvellement traduit en François, par un Docteur en Medicine. Paris: Chez Jean d'Houry a l'Image S. Jean, au bout du Pont-neuf, sur le Quay des Augustins, 1672.

Dixon, Laurinda S. *Alchemical Imagery in Bosch's Garden of Delights*. Studies in the Fine Arts: Iconography, 2; Linda Seidel, Series Ed. Ann Arbor, MI: UMI Research Press, 1981.

Dobbs, B. J. T. *Alchemical death & resurrection: the significance of alchemy in the age of Newton. A lecture sponsored by the Smithsonian Institution Libraries in conjunction with the Washington Collegium for the Humanities Lecture Series: Death and the Afterlife in Art and Literature. Presented at the Smithsonian Institution, February 16, 1988*. Washington, D.C.: Smithsonian Institution Libraries, 1990.

"Alchemische Kosmogonie und arianische Theologie bei Isaac Newton." Tr. by Christoph Meinel. *Wolfenbütteler Forschungen* 32 (1986), 137–50.

"Conceptual problems in Newton's early chemistry: a preliminary study." In *Religion, science, and worldview. Essays in honor of Richard S. Westfall*. Ed. by Margaret J. Osler and Paul Lawrence Farber. Cambridge University Press, 1985, pp. 3–32.

The Foundations of Newton's Alchemy, or "The Hunting of the Greene Lyon." Cambridge University Press, 1975.

"Newton and Stoicism." *The Southern Journal of Philosophy 23 Supplement* (1985), 109–23.
"Newton manuscripts at the Smithsonian Institution." *Isis 68* (1977), 105–7.
"Newton's alchemy and his 'active principle' of gravitation." In *Newton's Scientific and Philosophical Legacy.* Ed. by P. B. Scheuer and G. Debrock. International Archives of the History of Ideas, 123. Dordrecht: Kluwer Academic Publishers, 1988, pp. 55–80.
"Newton's alchemy and his theory of matter." *Isis 73* (1982), 511–28.
"Newton's 'Clavis': new evidence on its dating and significance." *Ambix 29* (1982), 190–202.
"Newton's *Commentary* on *The Emerald Tablet* of Hermes Trismegistus: its scientific and theological significance." In *Hermeticism and the Renaissance. Intellectual History and the Occult in Early Modern Europe.* Ed. by Ingrid Merkel and Allen G. Debus. Folger Books. Washington, D.C.: The Folger Shakespeare Library; London: Associated University Presses, 1988, pp. 182–91.
"Newton's copy of *Secrets Reveal'd* and the regimens of the work." *Ambix 26* (1979), 145–69.
"Newton's rejection of a mechanical aether for gravitation: empirical difficulties and guiding assumptions." In *Scrutinizing Science: Empirical Studies of Scientific Change.* Ed. by Arthur Donovan, Larry Laudan, and Rachel Laudan. Synthese Library Studies in Epistemology, Logic, Methodology, and Philosophy of Science, Vol. 193. Dordrecht: Kluwer Academic Publishers, 1988, pp. 69–83.
"Review of *Contemporary Newtonian Research*, edited by Zev Bechler." *Isis 74* (1983), 609–10.
Domson, Charles A. "Nicolas Fatio de Duillier and the Prophets of London: An Essay in the Historical Interaction of Natural Philosophy and Millennial Belief in the Age of Newton." Yale University: Ph. D. dissertation, 1972.
Donovan, Arthur, Larry Laudan, and Rachel Laudan, Eds. *Scrutinizing Science: Empirical Studies of Scientific Change.* Synthese Library Studies in Epistomology, Logic, Methodology, and Philosophy of Science, Vol. 193. Dordrecht: Kluwer Academic Publishers, 1988.
Doria, Charles, and Harris Lenowitz, Eds. *Origins. Creation Texts from the Ancient Mediterranean. A Chrestomathy.* Tr., with introduction and notes, by Charles Doria and Harris Lenowitz, and with a preface by Jerome Rothenberg. Anchor Books. Garden City, NY: Anchor Press/Doubleday, 1976.
Dorn, Gerhard. "Tractatus de naturae luce physica ex Genesi desumpta, Iuxta Sapientiam Theophrasti Paracelsi." In *Theatrum chemicum, praecipuos selectorum auctorum tractatus de chemiae et lapidis philosophici antiquitate, veritate, jure, praestantia, & operationibus, continens: In gratiam Verae Chemiae, & medicinae Chemicae studiosorum (ut qui uberrimam inde optimorum remediorum messem facere poterunt) congestum, & in Sex partes seu volumina digestum; singulis voluminibus, suo auctorum et librorum catalogo primis pagellis: rerum verò & verborum Indice postremis annexo.*

Six vols. Argentorati: Sumptibus Heredum Eberh. Zetzneri, 1659–61. Vol. I, 326–457.

Duby, Georges. *The Age of the Cathedrals. Art and Society, 980–1420.* Tr. by Eleanor Levieux and Barbara Thompson. Chicago: The University of Chicago Press, 1981.

Du Chesne, Joseph. *Recveil des plvs cvrievx et rares secrets....* Paris, 1648.

Traicté de la matiere.... Paris, 1626.

Dunbar, H. Flanders. *Symbolism in Medieval Thought and Its Consummation in the Divine Comedy.* New York: Russell & Russell, 1961.

Eirenaeus Philalethes. *Secrets Reveal'd: or, An Open Entrance to the Shut-Palace of the King: Containing The greatest Treasure in Chymistry. Never yet so plainly Discovered. Composed By a most famous English-man, Styling himself Anonymous, or Eyraeneus Philaletha Cosmopolita: Who, by Inspiration and Reading, attained to the Philosophers Stone at his Age of Twenty three Years, Anno Domini, 1645. Published for the Benefit of all English-men, by W. C. Esq.; a true Lover of Art and Nature.* London: printed by W. Godbid for William Cooper in Little St. Bartholomews, near Little-Britain, 1669.

Eliade, Mircea. *Images and Symbols. Studies in Religious Symbolism.* Tr. by Philip Mairet. London: Harvill Press, 1961.

The Myth of the Eternal Return or, Cosmos and History. Tr. from the French by Willard R. Trask. Bollingen Series XLVI. Reprint of the Princeton/Bollingen paperback ed. of 1971. Princeton: Princeton University Press, 1974.

Emerton, Norma E. *The Scientific Reinterpretation of Form.* Foreword by L. Pearce Williams. Cornell History of Science Series. Ithaca, NY: Cornell University Press, 1984.

English Renaissance Studies Presented to Dame Helen Gardner in Honour of Her Seventieth Birthday. Oxford: Clarendon Press, 1980.

Farley, John. *Gametes & Spores. Ideas about Sexual Reproduction 1750–1914.* Baltimore: The Johns Hopkins University Press, 1982.

The Spontaneous Generation Controversy from Descartes to Oparin. Baltimore: The Johns Hopkins University Press, n.d. [1974].

Fauvel, John, Raymond Flood, Michael Shortland, and Robin Wilson, Eds. *Let Newton Be!* Oxford: Oxford University Press, 1988.

Ferguson, George. *Signs & Symbols in Christian Art with Illustrations from Paintings of the Renaissance.* A Galaxy Book. New York: Oxford University Press, 1966.

Ferguson, John. *Bibliotheca Chemica: A Catalogue of the Alchemical, Chemical and Pharmaceutical Books in the Collection of the Late James Young of Kelly and Durris, Esq., Ll.D., F.R.S., F.R.S.E.* Two vols. Glasgow: James Maclehose and Sons, 1906; London: Derek Verschoyle Academic and Bibliographic Publications, 1954.

Ficino, Marsilio. *De sole.* In Paul Oskar Kristeller, *The Philosophy of Marsilius Ficino.* Tr. by Virginia Conant. Gloucester, MA: Peter Smith, 1964.

Ed. Plato. *Opera omnia quae exstant. M. Ficino interprete....* Francofurti, 1602.

Figala, Karin. "Newton as alchemist." *History of Science* 15 (1977), 102–37.
Fisch, Harold. *Jerusalem and Albion. The Hebraic Factor in Seventeenth-Century Literature.* London: Routledge & Kegan Paul, 1964.
"The scientist as priest: a note on Robert Boyle's natural theology." *Isis* 44 (1953), 252–65.
Five Treatesis of the Philosophers Stone. Two of Alphonso King of Portugall, as it was written with his own hand, and taken out of his Closset: Translated out of the Portuguez into English. One of John Sawtre a Monke, translated into English. Another written by Florianus Raudorff, a German Philosopher, and translated out of the same Language, into English. Also a Treatise of the names of the Philosophers Stone, by William Gratacolle, translated into English. To which is added the Smaragdine Table. By the Paines and Care of H. P. London: printed by Thomas Harper, and are to be sold by John Collins, in Little Brittain, near the Church door, 1652.
Flamel, Nicolas. *Les figvres hieroglyphiquves de Nicolas Flamel....* In *Trois Traictez de la Philosophie natvrelle, non encore imprimez. Scavoir, le secret livre dv tres-ancien Philosophe Artephivs, traictant de l'Art occulte & transmutation Metallique, Latin François. Plus les figvres hieroglyphiqves de Nicolas Flamel, ainsi qu'il les a mises en la quatriesme arche qu'il a bastie au Cimetiere des Innocens à Paris, entrant par la grande porte de la rue S. Denys, & prenant la main droite; auec l'explication d'icelles par iceluy Flamel. Ensemble Le vray Liure du docte Synesivs Abbé Grec, tiré de la Bibliotheque de l'Empereur sur le mesme sujet, le tout traduict par P. Arnavld, sieur de la Cheuallerie Poicteuin.* Paris: Chez la vefue M. Gvillemot & S. Thibovst, au Palais, en la galerie des prisonniers, 1612.
Le Livre des Figures Hiéroglyphiques attribué à Nicolas Flamel. In Claude Gagnon, *Description du Livre des Figures Hiéroglyphiques attribué à Nicolas Flamel, suivie d'une réimpression de l'édition originale et d'une reproduction des sept talismans du Livre d'Abraham, auxquels on a joint le Testament authentique dudit Flamel.* Montréal: Les Editions de l'Aurore, 1977.
Flammel, Nicholas, His Exposition of the Hieroglyphicall Figures which he caused to be painted upon an Arch in St. Innocents Church-yard, in Paris. Together with The secret Books of Artephivs, And the Epistle of Iohn Pontanus: Concerning both the Theoricke and the Practicke of the Philosophers Stone. Faithfully, and (as the Maiesty of the thing requireth) religiously done into English out of the French and Latine Copies. By Eirenaevs Orandvs, qui est, Vera veris enodans. London: imprinted at London by T. S. for Thomas Walkley, and are to bee solde at his Shop, at the Eagle and Childe in Britans Bursse, 1624.
Fludd, Robert. *Utriusque Cosmi Maioris scilicet et Minoris Metaphysica, Physica atqve Technica Historia In duo Volumina secundum Cosmi differentiam diuisa. Avthore Roberto Flud aliàs Fluctibus, Armigero, & in Medicina Doctore Oxoniensi.* Two vols. in three. Oppenhemii: AEre Johan-Theodori de Bry Typis Hieronymi Galleri, 1617–21.
Force, James E. "Newton and Deism." In *Science and Religion/Wissenschaft und Religion. Proceedings of the Symposium of the XVIIIth International Con-*

gress of History of Science at Hamburg-Munich, 1.–9. August 1989. Ed. by Anne Bäumer and Manfred Büttner. Bochum: Universitätsverlag Dr. N. Brockmeyer, 1989, pp. 120–32.

William Whiston, Honest Newtonian. Cambridge University Press, 1985.

and Richard H. Popkin. *Essays on the Context, Nature, and Influence of Isaac Newton's Theology*. International Archives of the History of Ideas, 129. Dordrecht: Kluwer Academic Publishers, 1990.

Foster, M. B. "The Christian doctrine of creation and the rise of modern science." In *Creation. The Impact of an Idea*. Ed. by Daniel O'Connor and Francis Oakley. Scribner Source Books in Religion. New York: Charles Scribner's Sons, 1969, pp. 29–53.

Fowler, Alastair, Ed. *Silent Poetry. Essays in Numerological Analysis*. London: Routledge and Kegan Paul, 1970.

Frank, Robert G., Jr. *Harvey and the Oxford Physiologists. Scientific Ideas and Social Interaction*. Berkeley: University of California Press, 1980.

Froehlich, Karlfried. "Pseudo-Dionysius and the reformation of the sixteenth century." In Pseudo-Dionysius, *The Complete Works*. Tr. by Colm Luibheid; foreword, notes, and translation collaboration by Paul Rorem; preface by Rene Roques; introductions by Jaroslav Pelikan, Jean Leclercq, and Karlfried Froehlich. The Classics of Western Spirituality, A Library of the Great Spiritual Masters. New York: Paulist Press, 1987, pp. 33–46.

Froom, LeRoy Edwin. *The Prophetic Faith of our Fathers. The Historical Development of Prophetic Interpretation*. Four vols. Washington, D.C.: Review and Herald, 1946–54.

Fruton, Joseph S. "From ferments to enzymes." In Joseph S. Fruton, *Molecules and Life. Historical Essays on the Interplay of Chemistry and Biology*. New York: Wiley-Interscience, 1972, pp. 22–86.

Molecules and Life. Historical Essays on the Interplay of Chemistry and Biology. New York: Wiley-Interscience, 1972.

Frye, Northrop. *The Great Code: The Bible and Literature*. New York: Harcourt Brace Jovanovich, 1982.

Funkenstein, Amos. "Descartes, eternal truths, and the divine omnipotence." *Studies in History and Philosophy of Science* 6 (1975), 185–99.

Gagnabin, Bernard. "De la cause de la pesanteur. Mémoire de Nicolas Fatio de Duillier, Présenté à la Royal Society le 26 février 1690. Reconstitué et publié avec une introduction par Bernard Gagnebin, Conservateur des manuscrits à la Bibliothèque publique et universitaire de Genève." *Notes and Records of the Royal Society of London* 6 (1949), 105–60.

Gagnon, Claude. *Description du Livre des Figures Hiéroglyphiques attribué à Nicolas Flamel, suivie d'une réimpression de l'édition originale et d'une reproduction des sept talismans du Livre d'Abraham, auxquels on a joint le Testament authentique dudit Flamel*. Montréal: Les Editions de l'Aurore, 1977.

Galilei, Galileo. *Discoveries and Opinions of Galileo, Including The Starry Messenger (1610), Letter to the Grand Duchess Christina (1615), And Excerpts from Letters on Sunspots (1613), The Assayer (1623)*. Tr., with introduction and notes by Stillman Drake. Doubleday Anchor Books. Garden City, NY: Doubleday & Co., 1957.

"Letter to Madame Christina of Lorraine, Grand Duchess of Tuscany, Concerning the Use of Biblical Quotations in Matters of Science [1615]." In *Discoveries and Opinions of Galileo, Including The Starry Messenger (1610), Letter to the Grand Duchess Christina (1615), And Excerpts from Letters on Sunspots (1613), The Assayer (1623).* Tr., with introduction and notes, by Stillman Drake. Doubleday Anchor Books. Garden City, NY: Doubleday & Co., 1957.

Gasking, Elizabeth B. *Investigations into Generation 1651–1828.* Baltimore: The Johns Hopkins University Press, 1967.

Genuth, Sara Schechner. "Comets, teleology, and the relationship of chemistry to cosmology in Newton's thought." *Annali dell' Instituto e Museo di Storia della Scienza di Firenze 10,* (1985), 31–65.

Georghegan, D. "Some indications of Newton's attitude towards alchemy." *Ambix 6* (1957–58), 102–06.

Gilbert, Creighton, Ed. *Renaissance Art.* Reprint of the 1970 ed. New York: Harper & Row, 1974.

Gilson, Etienne. *History of Christian Philosophy in the Middle Ages.* New York: Random House, 1955.

Gleick, James. *Chaos. Making a New Science.* New York: Viking Penguin, 1987.

Godwin, Joscelyn. *Robert Fludd. Hermetic Philosopher and Surveyor of Two Worlds.* Boulder, CO: Shambhala, 1979.

Gombrich, E. H. *Symbolic Images. Studies in the Art of the Renaissance, with 170 Illustrations. II.* Reprint of the 1972 ed. Chicago: University of Chicago Press, 1985.

Goodenough, Erwin R. *An Introduction to Philo Judaeus.* 2nd ed. Oxford: Basil Blackwell, 1962.

Gouk, Penelope. "The harmonic roots of Newtonian science." In *Let Newton Be!* Ed. by John Fauvel, Raymond Flood, Michael Shortland, and Robin Wilson. Oxford: Oxford University Press, 1988, pp. 100–25.

"Newton and music: from the microcosm to the macrocosm." *International Studies in the Philosophy of Science 1* (1986), 36–59.

Gould, Stephen Jay. *Time's Arrow, Time's Cycle. Myth and Metaphor in the Discovery of Geological Time.* The Jerusalem–Harvard Lectures sponsored by the Hebrew University of Jerusalem and Harvard University Press. Cambridge, MA: Harvard University Press, 1987.

Grant, Edward. "The condemnation of 1277, God's absolute power, and physical thought in the Late Middle Ages." *Viator 10* (1979), 211–44.

Much Ado about Nothing. Theories of Space and Vacuum from the Middle Ages to the Scientific Revolution. Cambridge University Press, 1981.

Grant, Robert M. *Miracle and Natural Law in Graeco-Roman and Early Christian Thought.* Amsterdam: North-Holland Publishing Co., 1952.

Gregg, Robert C., and Dennis E. Groh. *Early Arianism – A View of Salvation.* Philadelphia: Fortress Press, 1981.

Gregory, J. C. "Chemistry and alchemy in the natural philosophy of Sir Francis Bacon, 1561–1626." *Ambix* 2 (1938), 93–111.

Grosseteste, Robert. *Hexaëmeron*. Ed. by Richard C. Dales and Servus Gieben. Auctores Britannici Medii Aevi, VI. London: published for The British Academy by The Oxford University Press, 1982.

Guerlac, Henry. *Essays and Papers in the History of Modern Science*. Baltimore: The Johns Hopkins University Press, 1977.

"Francis Hauksbee: Expérimentateur au Profit de Newton." *Archives Internationales d'Histoire des Sciences 16* (1963), 1113–28. Also in: Henry Guerlac, *Essays and Papers in the History of Modern Science*. Baltimore: The Johns Hopkins University Press, 1977, pp. 107–19.

Newton et Epicure. Conférence donnée au Palais de la Découverte, Université de Paris, le 2 Mars 1963, Histoire des Science. Paris: Sur les presses de l'imprimerie Alençonnaise, 1963. Also in: Henry Guerlac. *Essays and Papers in the History of Modern Science*. Baltimore: The Johns Hopkins University Press, 1977, pp. 82–106.

"Newton's optical aether: his draft of a proposed addition to his *Opticks*." *Notes and Records of the Royal Society of London* 22 (1967), 45–57. Also in: Henry Guerlac, *Essays and Papers in the History of Modern Science*. Baltimore: The Johns Hopkins University Press, 1977, pp. 120–30.

"Review of B. J. T. Dobbs, *The Foundations of Newton's Alchemy*." *Journal of Modern History* 49 (1977), 130–3.

"Sir Isaac and the ingenious Mr. Hauksbee." In *Mélanges Alexandre Koyré, publiés à l'occasion de son soixante-dixième anniversaire*. Ed. by I. B. Cohen and R. Taton. Two vols. Paris: Hermann, 1964, vol. I, 228–54.

"Theological voluntarism and biological analogies in Newton's physical thought." *Journal of the History of Ideas* 44 (1983), 219–29.

and M. C. Jacob. "Bentley, Newton, and providence (the Boyle lectures once more)." *Journal of the History of Ideas* 30 (1969), 307–18.

Guthrie, W. K. C. *In the Beginning. Some Greek Views on the Origins of Life and the Early State of Man*. Ithaca, NY: Cornell University Press, 1957.

Gwatkin, Henry Melvill. *Studies of Arianism, Chiefly Referring to the Character and Chronology of the Reaction which Followed the Council of Nicaea*. 2nd ed. Cambridge: Deighton Bell and Co.; London: George Bell and Sons, 1900.

Hager, Paul. "Chrysippus' theory of the pneuma." *Prudentia 14* (1982), 97–108.

Hahm, David E. *The Origins of Stoic Cosmology*. n.p.: Ohio State University Press, 1977.

Hall, A. Rupert. "Newton as alchemist." *Nature* 266 (28 April 1977), 78.

Philosophers at War. The Quarrel between Newton and Leibniz. Cambridge University Press, 1980.

"Sir Isaac Newton's note-book, 1661–65." *Cambridge Historical Journal* 9 (1948), 239–50.

and Marie Boas Hall. "Introduction to MS Add. 4003." In Isaac Newton, *Unpublished Scientific Papers of Isaac Newton: A Selection from the Portsmouth Collection in the University Library, Cambridge. Chosen, edited, and*

Unpublished Scientific Papers of Isaac Newton: A Selection from the Portsmouth Collection in the University Library, Cambridge. Chosen, edited, and translated by A. Rupert Hall and Marie Boas Hall. Cambridge University Press, 1962, pp. 89–90.

and Marie Boas Hall. "Introduction to part II." In Isaac Newton, *Unpublished Scientific Papers of Isaac Newton: A Selection from the Portsmouth Collection in the University Library, Cambridge. Chosen, edited, and translated by A. Rupert Hall and Marie Boas Hall.* Cambridge University Press, 1962, pp. 75–85.

and Marie Boas Hall. "Introduction to part III, theory of matter." In Isaac Newton, *Unpublished Scientific Papers of Isaac Newton: A Selection from the Portsmouth Collection in the University Library, Cambridge. Chosen, edited, and translated by A. Rupert Hall and Marie Boas Hall.* Cambridge University Press, 1962, pp. 187–9.

and Marie Boas Hall, Eds. Isaac Newton, *Unpubished Scientific Papers of Isaac Newton: A Selection from the Portsmouth Collection in the University Library, Cambridge. Chosen, edited, and translated by A. Rupert Hall and Marie Boas Hall.* Cambridge University Press, 1962.

Hall, James. *Dictionary of Subjects and Symbols in Art.* Intro. by Kenneth Clark. Rev. ed. Icon Editions. New York: Harper & Row, 1979.

Hall, Marie Boas. "Review of B. J. T. Dobbs, *The Foundations of Newton's Alchemy.*" *British Journal for the History of Science 10* (1977), 262–64.

and A. Rupert Hall. "Newton's electric spirit: four oddities." *Isis 50* (1959), 473–6.

Hall, Thomas S. "Descartes' physiological method: position, principles, examples." *Journal of the History of Biology 3* (1970), 53–79.

Ideas of Life and Matter. Studies in the History of General Physiology 600 B.C. – *1900* A.D. Two vols. Chicago: The University of Chicago Press, 1969.

Hanen, Marsha P., Margaret J. Osler, and Robert G. Weyant, Eds. *Science, Pseudo-Science and Society.* Waterloo, Ontario: Wilfrid Laurier University Press, 1980.

Harris, John. *Lexicon Technicum or an Universal English Dictionary of Arts and Sciences.* Facsimile of the 1704–10 ed. Two vols. New York: Johnson Reprint Corp., 1966.

Harrison, John. *The Library of Isaac Newton.* Cambridge University Press, 1978.

Harvey, E. Newton. *A History of Luminescence from the Earliest Times until 1900.* Memoirs of the American Philosophical Society Held at Philadelphia for Promoting Useful Knowledge, Vol. 44. Philadelphia: The American Philosophical Society, 1957.

Harvey, William. *Anatomical Exercises on the Generation of Animals.* In *Great Books of the Western World.* Robert Maynard Hutchins, Editor-in-Chief. 54 vols. Chicago: William Benton for Encyclopedia Britannica, 1952. Vol. XXVIII.

Hawes, Joan L. "Newton and the 'Electrical Attraction Unexcited.'" *Annals of Science 24* (1968), 121–30.

"Newton's revival of the aether hypothesis and the explanation of gravitational attraction." *Notes and Records of the Royal Society of London* 23 (1968), 200–12.

"Newton's two electricities." *Annals of Science* 27 (1971), 95–103.

Heilbron, J. L. *Electricity in the 17th and 18th Centuries. A Study of Early Modern Physics*. Berkeley: University of California Press, 1979.

Physics at the Royal Society during Newton's Presidency. Los Angeles: The William Andrews Clark Memorial Library, University of California, Los Angeles, 1983.

Heninger, S. K., Jr. *The Cosmographical Glass. Renaissance Diagrams of the Universe*. San Marino, CA: The Huntington Library, 1977.

Touches of Sweet Harmony. Pythagorean Cosmology and Renaissance Poetics. San Marino, CA: The Huntington Library, 1974.

Herberg, Will. "Biblical faith as 'Heilsgeschichte.' The meaning of redemptive history in human existence." In Will Herberg, *Faith Enacted as History. Essays in Biblical Theology*. Ed. with introduction by Bernhard W. Anderson. Philadelphia: The Westminster Press, 1976, pp. 32–42.

Faith Enacted as History. Essays in Biblical Theology. Ed. with introduction by Bernhard W. Anderson. Philadelphia: The Westminster Press, 1976.

Herivel, John. *The Background to Newton's Principia. A Study of Newton's Dynamical Researches in the Years 1664–84*. Oxford: Clarendon Press, 1965.

Hermes Trismegistus. *Corpvs Hermeticvm. Texte établi par A. D. Nock et traduit par A.-J. Festugière*. Rev. ed. of 1946–54. Four vols. Collection des Universités de France publiée sous le patronage de l'Association Guillaume Budé. Paris: Société d'Édition «Les Belles Lettres», 1980–3.

The Emerald Tablet. In: (1) *Bibliotheque de philosophes [chymiques]....* Two vols. Paris, 1672–8. Vol. I. (2) *Five Treatesis of the Philosophers Stone....* London, 1652. (3) *Theatrum chemicum....* Six vols. Argentorati, 1659–61. Vols. I and VI.

Hermetica. The Ancient Greek and Latin Writings which contain Religious or Philosophic Teachings ascribed to Hermes Trismegistus. Edited with English Translation and Notes by Walter Scott. 1st ed., 1924. Reprint of the 1982 ed. Four vols. Boston: Shambhala, 1985.

The Seven Chapters. In: *Bibliotheque de philosophes [chymiques]....* Two vols. Paris, 1672–8. Vol. II.

Hesse, Mary B. *Forces and Fields: The Concept of Action at a Distance in the History of Physics*. Reprint of the 1962 ed. Westport, CT: Greenwood Press, 1970.

Hierocles. *Commentarius in aurea Pythagoreorum carmina....* Parisiis, 1583.

Higham, T. F., and C. M. Bowra, Eds. *The Oxford Book of Greek Verse in Translation*. Oxford: Clarendon Press, 1938.

Himrod, David Kirk. "Cosmic Order and Divine Activity: A Study in the Relation between Science and Religion, 1850–1950." University of California, Los Angeles: Ph.D. dissertation, 1977.

Hiscock, W. G., Ed. *David Gregory, Isaac Newton and Their Circle: Extracts from David Gregory's Memoranda 1677–1708*. Oxford: printed for the editor, 1937.

Hoheisel, Karl. "Christus und der philosophische Stein. Alchemie als über- und nichtchristlicher Heilsweg." *Wolfenbütteler Forschungen* 32 (1986), 61–84.

The Holy Bible containing the Old and New Testaments Translated out of the Original Tongues: and with the former translations diligently compared and revised by His Majesty's special command. Appointed to be read in churches. The Oxford Self-Pronouncing Bible. Authorized King James Version. Oxford: printed at the University Press; London: Humphrey Milford, Oxford University Press; New York/Toronto, n.d.

Home, R. W. "Force, electricity, and the powers of living matter in Newton's mature philosophy of nature." In *Religion, Science, and Worldview. Essays in Honor of Richard S. Westfall*. Ed. by Margaret J. Osler and Paul Lawrence Farber. Cambridge University Press, 1985, pp. 95–117.

"Newton on electricity and the aether." In *Contemporary Newtonian Research*. Ed. by Zev Bechler. Studies in the History of Modern Science, 9. Dordrecht: D. Reidel, 1982, pp. 191–213.

Hooykaas, Reijer. *Natural Law and Divine Miracle. The Principle of Uniformity in Geology, Biology and Theology*. 2nd impression. Leiden: E. J. Brill, 1963.

Religion and the Rise of Modern Science. Edinburgh: Scottish Academic Press, 1973.

Hunt, H. A. K. *A Physical Intrepretation of the Universe. The Doctrines of Zeno the Stoic*. Melbourne: Melbourne University Press, 1976.

Hunter, William B., Jr., General Ed. *A Milton Encyclopedia*. With John T. Shawcross and John M. Steadman, Co-editors; and Purviss E. Boyette and Leonard Nathanson, Associate Editors. Eight vols. Lewisburg, PA: Bucknell University Press; London: Associated University Presses, 1978–80.

Hurlbutt, Robert H., III. *Hume, Newton, and the Design Argument*. Rev. ed. Lincoln: University of Nebraska Press, 1985.

Hutchins, Robert Maynard, Ed. *Great Books of the Western World*. 54 vols. Chicago: William Benton for Encyclopedia Britannica, 1952.

Hutchison, Keith. "Supernaturalism and the mechanical philosophy." *History of Science* 21 (1983), 297–333.

Hutin, Serge. *Henry More. Essaie sur les doctrines théosophiques chez les Platoniciens de Cambridge*. Studien und Materialien zur Geschichte der Philosophie, Herausgegeben von Heinz Heimsoeth, Dieter Henrich, und Giorgio Tonelli, Band 2. Hildesheim: Georg Olms Verlagsbuchhandlung, 1966.

Hutton, James. "Spenser's 'Adamantine Chains': a cosmological metaphor." In *The Classical Tradition. Literary and Historical Studies in Honor of Harry Caplan*. Ed. by Luitpold Wallach. Ithaca, NY: Cornell University Press, 1966, pp. 572–94.

Huxley, H. H., Ed. *Virgil: Georgics I & IV*. London: Methuen & Co., 1963.

Iamblichus. *De mysterius liber....* Oxonii, 1678.

De vita Pythagorica liber.... Amstelodami, 1707.

The International Style. The Arts in Europe around 1400. October 23–December

2, 1962. *The Walters Art Gallery, Baltimore.* Baltimore: The Walters Art Gallery, 1962.

Isler, Hansruedi. *Thomas Willis 1621–1675. Doctor and Scientist.* New York: Hafner, 1968.

Jackson, Samuel Macauley, Editor-in-Chief. *The New Schaff–Herzog Encyclopedia of Religious Knowledge, Embracing Biblical, Doctrinal, and Practical Theology, and Biblical, Theological, and Ecclesiastical Biography from the Earliest Times to the Present Day. Based on the Third Edition of the Realencyklopa¨die Founded by J. J. Herzog, and Edited by Albert Hauck.* Twelve vols. New York: Funk and Wagnalls Co., 1908–12.

Jacob, Margaret C. "Newton and the French prophets: new evidence." *History of Science* 16 (1978), 134–42.

The Newtonians and the English Revolution 1689–1720. Hassocks, Sussex: The Harvester Press, 1976.

"Review of B. J. T. Dobbs, *The Foundations of Newton's Alchemy.*" *The Eighteenth Century: A Current Bibliography*, n.s. 1 (1975, published 1978), 345–7.

Johnson, Samuel. *A Dictionary of the English Language in which The Words are deduced from their Originals, and Illustrated in their Different Significations by Examples from the best Writers. To which are prefixed, A History of the Language, and An English Grammar.* 2nd ed. Two vols. London: printed by W. Strahan, for J. and J. Knapton; T. and T. Longman; C. Hitch and L. Hawes; A. Millar; and R. and J. Dodsley, 1755–6.

Jonas, Hans. "Jewish and Christian elements in the western philosophical tradition." In *Creation. The Impact of an Idea.* Ed. by Daniel O'Connor and Francis Oakley. Scribner Source Books in Religion. New York: Charles Scribner's Sons, 1969, pp. 241–58.

Jones, Inigo. *The Most Notable Antiquity of Great Britain Vulgarly Called Stonehenge (1655).* Introd. note by Graham Parry. A Scolar Press Facsimile. Reprint of the 1972 ed. Menston, Yorkshire: The Scolar Press, 1973.

Jung, C. G. *Psychology and Alchemy.* Tr. by R. F. C. Hull. Bollingen Series XX. In *The Collected Works of C. G. Jung.* Vol. 12. Ed. by Herbert Read, Michael Fordham, Gerhard Adler, and William McGuire. Reprint of the 1968 ed. Princeton/Bollingen Paperback. Princeton: Princeton University Press, 1980.

Justin Martyr. *Exhortation to the Greeks.* In S. K. Heninger, Jr., *Touches of Sweet Harmony. Pythagorean Cosmology and Renaissance Poetics.* San Marino, CA: The Huntington Library, 1974.

Opera. . . . Coloniae, 1686.

Kahn, Charles H. *Anaximander and the Origins of Greek Cosmology.* New York: Columbia University Press, 1960.

Kargon, Robert Hugh. *Atomism in England from Hariot to Newton.* Oxford: Clarendon Press, 1966.

Keynes, John Maynard, Lord Keynes. "Newton the man." In *The Royal Society Tercentenary Celebrations 15–19 July 1946.* Cambridge University Press for the Royal Society, 1947, pp. 27–34.

Klaaren, Eugene M. *Religious Origins of Modern Science. Belief in Creation in Seventeenth-Century Thought.* Grand Rapids, MI: William B. Eerdmans Publishing Co., 1977.

Kocher, Paul H. *Science and Religion in Elizabethan England.* San Marino, CA: The Huntington Library, 1953.

Korshin, Paul J. *Typologies in England 1650–1820.* Princeton: Princeton University Press, 1982.

Koyré, Alexandre. "Newton and Descartes." In Alexandre Koyré, *Newtonian Studies.* Cambridge, MA: Harvard University Press, 1965, pp. 53–114.

Newtonian Studies. Cambridge, MA: Harvard University Press, 1965.

"Newton's 'Regulae Philosophandi.'" In Alexandre Koyré, *Newtonian Studies.* Cambridge, MA: Harvard University Press, 1965, pp. 261–72.

and I. Bernard Cohen. "The case of the missing *Tanquam.* Leibniz, Newton, and Clarke." *Isis 52* (1961), 555–66.

and I. Bernard Cohen, Eds. Isaac Newton, *Isaac Newton's Philosophiae naturalis principia mathematica. The Third Edition (1726) with Variant Readings. Assembled and Edited by Alexandre Koyré and I. Bernard Cohen with the Assistance of Anne Whitman.* Two vols. Cambridge, MA: Harvard University Press, 1972.

and I. Bernard Cohen. "Newton & the Leibniz–Clarke correspondence with notes on Newton, Conti, & Des Maizeaux." *Archives internationales d'histoire des sciences 15* (1962), 63–126.

and I. Bernard Cohen. "Newton's 'Electric and Elastic Spirit.'" *Isis 51* (1960), 337.

Kristeller, Paul Oskar. *The Philosophy of Marsilius Ficino.* Tr. by Virginia Conant. Gloucester, MA: Peter Smith, 1964.

Renaissance Thought and Its Sources. Ed. by Michael Mooney. New York: Columbia University Press, 1979.

Kubrin, David Charles. "Newton and the cyclical cosmos: providence and the mechanical philosophy." *Journal of the History of Ideas 28* (1967), 325–46.

"Providence and the Mechanical Philosophy: The Creation and Dissolution of the World in Newtonian Thought. A Study of the Relations of Science and Religion in Seventeenth Century England." Cornell University: Ph.D. dissertation, 1968.

La Brecque, Mort. "Fractal applications." *Mosaic 17*, No. 4 (Winter 1986/7), 34–48.

l'Agneau, D. *Harmonie mystiqve....* Paris, 1636.

Lamy, Guilliaume. *Dissertation sur l'antimoine.* Paris, 1682.

Lapidge, Michael. "*Archai* and *Stoicheia*: a problem in Stoic cosmology." *Phronesis 18* (1973), 240–78.

"Stoic cosmology." In *The Stoics.* Ed. by John M. Rist. Berkeley: University of California Press, 1978, pp. 161–85.

"A Stoic metaphor in late Latin poetry: the binding of the cosmos." *Latomus 39* (1980), 817–37.

Leclercq, Jean. "Influence and noninfluence of Dionysius in the Western Middle

Ages." In Pseudo-Dionysius, *The Complete Works*. Tr. by Colm Luibheid; foreword, notes, and translation collaboration by Paul Rorem; preface by Rene Roques; introductions by Jaroslav Pelikan, Jean Leclercq, and Karlfried Froehlich. The Classics of Western Spirituality, A Library of the Great Spiritual Masters. New York: Paulist Press, 1987, pp. 25–32.

[Leibniz, G. W. von.] *Hypothesis physica nova... Auctore G. G. L. L.* Londini, 1671.

The Leibniz–Clarke Correspondence, Together with Extracts from Newton's Principia and Opticks. Ed. with introduction and notes by H. G. Alexander. Philosophical Classics, General Ed.: Peter G. Lucas. Manchester: Manchester University Press, 1956.

Lemery, Nicolas. *Traité de l'antimoine....* Paris, 1707.

Lemni, Charles W. *The Classic Deities in Bacon: A Study in Mythological Symbolism*. Reprint of the 1933 ed. New York: Octagon Books, 1971.

Lewis, C. S. *The Discarded Image. An Introduction to Medieval and Renaissance Literature*. Reprint of the 1964 ed. Cambridge University Press, 1979.

[Limojon, A. T. (Sieur de St. Didier).] *Lettre d'un philosophe sur le secret du grand oeuvre, ecrite Au sujet de ce qu'Aristée a laissé par écrit à son Fils, touchant le Magistere philosophique. Le Nom d l'Autheur est en Latin dans cett' Anagramme. Dives Sicut Ardens, S.* La Haye: Chez Adrian Moetjens, 1686.

Le tres-ancien duel des Chevaliers. In: *Divers traitez de la philosophie naturelle....* Paris: Chez Jean d'Houry a l'Image S. Jean, au bout du Pont-neuf, sur le Quay des Augustins, 1672.

Le triomphe hermetique, Ou La Pierre Philosophale victorieuse. Traité Plus complet & plus intelligible, qu'il y en ait eu jusques ici, touchant le Magistère hermetique. Amsterdam: Chez Henry Wetstein, 1689.

Le triomphe hermetique, Ou La Pierre Philosophale victorieuse. Traité Plus complet & plus intelligible, qu'il y en ait eu jusques ici, touchant le Magistère hermetique. Amsterdam: Chez Henry Wetstein, 1699.

Lindberg, David C. "The genesis of Kepler's theory of light: light metaphysics from Plotinus to Kepler." *Osiris*, Second Series, 2 (1986), 5–42.

and Ronald L. Numbers, Eds. *God and Nature. Historical Essays on the Encounter between Christianity and Science*. Berkeley: University of California Press, 1986.

Linden, Stanton J. "Alchemy and eschatology in seventeenth-century poetry." *Ambix* 31 (1984), 102–24.

Lindsay, Jack. *The Origins of Alchemy in Graeco-Roman Egypt*. New York: Barnes & Noble, 1970.

Lipsius, Justus. *Epistolarum selectarum chilias....* Genevae, 1611.

Justi Lipsi V. C. Opera omnia, postremum ab ipso aucta et recensita: nunc primum copioso rerum indice illustrata. Four vols. Versaliae: Apud Andraeam ab Hoogenhuysen et Societatem, 1675.

Physiologiae stoicorum libri tres: L. Annaeo Senecae alliisq. scriptoribus illustrandis. In Justus Lipsius, *Justi Lipsi V. C. Opera omnia, postremum ab ipso aucta et recensita: nunc primum copioso rerum indice illustrata*. Four vols. Vesaliae: Apud Andraeam ab Hoogenhuysen et Societatem, 1675, vol. IV, 823–1006.

Roma illustrata, sive Antiquitatum Romanarum breviarium ... Ex nova recensione A. Thysii ... Postrema ed. Amstelodami, 1657.

Loemker, Leroy E. *Struggle for Synthesis. The Seventeenth Century Background of Leibniz's Synthesis of Order and Freedom.* Cambridge, MA: Harvard University Press, 1972.

Lombardo, Stanley Frank. "Aratus' *Phaenomena*: An Introduction and Translation." The University of Texas at Austin: Ph.D. dissertation, 1976.

Ed. *Sky Signs: Aratus' Phaenomena.* Introduction and translation by Stanley Lombardo, with illustrations by Anita Volder Frederick. Berkeley, CA: North Atlantic Books, n.d.

Long, A. A. *Hellenistic Philosophy: Stoics, Epicureans, Sceptics.* London: Duckworth, 1974.

"Heraclitus and Stoicism." *Philosophia (Athens) 5/6* (1975–6), 134–56.

Love, Rosaleen. "Herman Boerhaave and the element-instrument concept of fire." *Annals of Science 31* (1974), 547–59.

"Some sources of Herman Boerhaave's concept of fire." *Ambix 19* (1972), 157–74.

Lucretius. *De rerum natura libri VI....* Cantabrigiae, 1686.

Titi Lvcreti Cari De Rervm Natvra Libri Sex. Ed., with Prolegomena, Critical Apparatus, Translation, and Commentary, by Cyril Bailey. Three vols. Oxford: Clarendon Press, 1947.

La lumière sortant par soy méme des tenebres ou veritable theorie de la Pierre des Philosophes écrite en vers Italiens, & amplifiée en Latin par un Auteur Anonyme, en forme de Commentaire; le tout traduit en François par B. D. L. Paris: Chez Laurent D'Houry, ruë S. Jacques, devant la Fonteine S. Severin, au S. Esprit, 1687.

Lundquist, John M. "The common temple ideology of the ancient Near East." In *The Temple in Antiquity. Ancient Records and Modern Perspectives.* Ed. with introductory essay by Truman G. Madsen. The Religious Studies Monograph Series, Vol. 9. Provo, UT: Brigham Young University Religious Studies Center, 1984, pp. 53–76.

Luttmer, Frank. "Enemies of God: Atheists and Anxiety about Atheists in England, 1570–1640." Northwestern University: Ph.D. dissertation, 1987.

MacRae, George. "The temple as a house of revelation in the Nag Hammadi texts." In *The Temple in Antiquity. Ancient Records and Modern Perspectives.* Ed. with introductory essay by Truman G. Madsen. The Religious Studies Monograph Series, Vol. 9. Provo, UT: Brigham Young University Religious Studies Center, 1984, pp. 175–90.

McEvoy, James. *The Philosophy of Robert Grosseteste.* Oxford: Clarendon Press, 1982.

McGuire, J. E. "Existence, actuality and necessity: Newton on space and time." *Annals of Science 35* (1978), 463–508.

"Force, active principles, and Newton's invisible realm." *Ambix* 15 (1968), 154–208.

"Neoplatonism and active principles: Newton and the *Corpus Hermeticum*." In Robert S. Westman and J. E. McGuire, *Hermeticism and the Scientific Revolution: Papers Read at a Clark Library Seminar, March 9, 1974*. Los Angeles: Clark Memorial Library, University of California, 1977, pp. 93–142.

"The origin of Newton's doctrine of essential qualities." *Centaurus* 12 (1968), 233–60.

"Transmutation and immutability: Newton's doctrine of physical qualities." *Ambix* 14 (1967), 69–95.

and P. M. Rattansi. "Newton and the 'Pipes of Pan.'" *Notes and Records of the Royal Society of London* 21 (1966), 108–43.

and Martin Tamny. *Certain Philosophical Questions: Newton's Trinity Notebook*. Cambridge University Press, 1983.

and Martin Tamny. "Gravitation, attraction, and cohesion." In J. E. McGuire and Martin Tamny, *Certain Philosophical Questions: Newton's Trinity Notebook*. Cambridge University Press, 1983, pp. 275–95.

and Martin Tamny. "Infinity, indivisibilism, and the void. In J. E. McGuire and Martin Tamny, *Certain Philosophical Questions: Newton's Trinity Notebook*. Cambridge University Press, 1983, pp. 26–126.

and Martin Tamny. "Introduction." In J. E. McGuire and Martin Tamny, *Certain Philosophical Questions: Newton's Trinity Notebook*. Cambridge University Press, 1983, pp. 3–25.

and Martin Tamny. "The origin of Newton's optical thought and its connection with physiology." In J. E. McGuire and Martin Tamny, *Certain Philosophical Questions: Newton's Trinity Notebook*. Cambridge University Press, 1983, pp. 241–74.

and Martin Tamny. "Physiology and Hobbesian epistemology." In J. E. McGuire and Martin Tamny, *Certain Philosophical Questions: Newton's Trinity Notebook*. Cambridge University Press, 1983, pp. 216–40.

McMullin, Ernan. *Newton on Matter and Activity*. Notre Dame: University of Notre Dame Press, 1978.

Mackenzie, Ann Wilbur. "A word about Descartes' mechanistic conception of life." *Journal of the History of Biology* 8 (1975), 1–13.

Mackenzie, Elizabeth. "The growth of plants. A seventeenth-century metaphor." In *English Renaissance Studies Presented to Dame Helen Gardner in Honour of her Seventieth Birthday*. Oxford: Clarendon Press, 1980, pp. 194–211.

Madsen, Truman G., Ed. *The Temple in Antiquity. Ancient Records and Modern Perspectives*. With introd. essay by Truman G. Madsen. The Religious Studies Monograph Series, Vol. 9. Provo, UT: Brigham Young University Religious Studies Center, 1984.

Mahdihassan, S. "Imitation of creation by alchemy and its corresponding symbolism." *Abr-Nahrain (Melbourne)* 12 (1972), 95–117.

Malville, J. McKim. *The Fermenting Universe. Myths of Eternal Change*. New York: The Seabury Press, 1981.

Manuel, Frank E. *Isaac Newton, Historian*. Cambridge, MA: The Belknap Press of Harvard University Press, 1963.

A Portrait of Isaac Newton. Cambridge, MA: The Belknap Press of Harvard University Press, 1968.

The Religion of Isaac Newton. The Fremantle Lectures 1973. Oxford: Clarendon Press, 1974.

Manuscripts of the Dibner Collection in the Dibner Library of the History of Science and Technology of the Smithsonian Institution Libraries. Smithsonian Institution Libraries, Research Guide No. 5. Washington, D.C.: Smithsonian Institution Libraries, 1985.

Martin, Jean. *Histoire du texte des Phénomènes d'Aratos*. Études et Commentaires XXII. Paris: Librairie C. Klincksieck, 1956.

Meagher, Paul Kevin, OP, S.T.M.; Thomas C. O'Brien; and Sister Consuelo Maria Akerne, SSJ; Eds. *Encyclopedic Dictionary of Religion*. 3 vols. Washington, D.C.: Corpus Publications, 1979.

Meibomius, Marcus. *Antiquae musicae auctores septem....* Two vols. Amstelodami, 1652.

Meiss, Millard. "Light as form and symbol in some fifteenth-century paintings." In *Renaissance Art*. Ed. by Creighton Gilbert. Reprint of the 1970 ed. New York: Harper and & Row, 1974, pp. 43–68.

Mellor, Anne Kostelanetz. *Blake's Human Form Divine*. Berkeley: University of California Press, 1974.

Merkel, Ingrid, and Allen G. Debus, Eds. *Hermeticism and the Renaissance. Intellectual History and the Occult in Early Modern Europe*. Folger Books. Washington, D.C.: The Folger Shakespeare Library; London: Associated University Presses, 1988.

Miller, Stephen G. *The Prytaneion. Its Function and Architectural Form*. Berkeley: University of California Press, 1978.

Millington, E. C. "Theories of cohesion in the seventeenth century." *Annals of Science* 5 (1941–47), 253–69.

Montgomery, John Warwick. *Cross and Crucible. Johann Valentin Andreae (1586–1654), Phoenix of the Theologians. Volume I. Andreae's Life, World-view, and Relations with Rosicrucianism and Alchemy. Volume II. The Chymische Hochzeit with Notes and Commentary*. International Archives of the History of Ideas, no. 55. The Hague: Martinus Nijhoff, 1973.

More, Henry. *The Immortality of the Soul, So farre forth as it is demonstrable from the Knowledge of Nature and the Light of Reason*. London: printed by J. Flesher, for William Morden Bookseller in Cambridge, 1659.

A Plain and Continued Exposition Of the several Prophecies or Divine Visions of the Prophet Daniel, Which have or may concern the People of God, whether Jew or Christian; Whereunto is annexed a Threefold Appendage, Touching Three main Points, the First, Relating to Daniel, the other Two to the Apocalypse. London: printed by M. F. for Walter Kettilby, at the Bishop's-Head in Saint Paul's Church-Yard, 1681.

Morrison, Philip. "Review of B. J. T. Dobbs, *The Foundations of Newton's Alchemy*." *Scientific American* 235 (August 1976), 113–15.

Munby, A. N. L. "The Keynes collection of the works of Sir Isaac Newton at King's College, Cambridge." *Notes and Records of the Royal Society of London* 10 (1952), 40–50.

Musaeum hermeticum reformatum et amplificatum. Facsimile reprint of the Frankfort ed. of 1678. Introduction by Karl R. H. Frick. Introduction tr. by C. A. Burland. Graz: Akademische Druck-u. Verlagsanstalt, 1970.

Musaeum hermeticum reformatum et amplificatum, omnes sopho-spagyricae artis discipulos fidelissimè erudiens, quo pacto Summa illa veraque lapidis philosophici Medicina, qua res omnes qualemcunque defectum patienties, instaurantur, inveniri & haberi queat, Continens tractatus chimicos XXI. Praestantissimos, quorum Nomina & Seriem versa pagella indicabit. Francofurti: Apud Hermannum à Sande, 1678.

Nauert, Charles G., Jr. *Agrippa and the Crisis of Renaissance Thought.* Illinois Studies in the Social Sciences, No. 55. Urbana: University of Illinois Press, 1965.

The New English Bible With the Apocrypha. N.P.: Oxford University Press and Cambridge University Press, 1970.

Newman, William. "Newton's 'Clavis' as Starkey's 'Key.'" *Isis* 78 (1987), 564–74.

Newton, Isaac. *The Chronology of Ancient Kingdoms Amended. To which is Prefix'd, A Short Chronicle from the First Memory of Things in Europe, to the Conquest of Persia by Alexander the Great.* London: printed for J. Tonson in the Strand, and J. Osborn and T. Longman in Paternoster Row, 1728.

The Correspondence of Isaac Newton. Ed. by H. W. Turnbull, J. P. Scott, A. R. Hall, and Laura Tilling. Seven vols. Cambridge: published for the Royal Society at the University Press, 1959–77.

De la gravitation ou les fondements de la mécanique classique. Introduction, Traduction et Notes de Marie-Françoise Biarnais. Ouvrage publié avec le concours du CNRS. Science et Humanisme. Paris: Les Belles Lettres, 1985.

Isaaci Newtoni Opera quae exstant omnia. Commentariis illustrabat Samuel Horsley, LL. D. R. S. S. Reverendo admodum in Christo Patri Roberto Episcopo Londinensi a Sacris. Five vols. Londini: Excudebat Joannes Nichols, 1779–85.

Isaac Newton's Philosophiae naturalis principia mathematica. The Third Edition (1726) with Variant Readings. Assembled and Edited by Alexandre Koyré and I. Bernard Cohen with the Assistance of Anne Whitman. Two vols. Cambridge, MA: Harvard University Press, 1972.

The Mathematical Papers of Isaac Newton. Ed. by Derek T. Whiteside with the assistance in publication of M. A. Hoskin. Eight vols. Cambridge University Press, 1967–80.

Observations upon the Prophecies of Daniel, and the Apocalypse of St. John. In Two Parts. London: printed by J. Darby and T. Browne in Bartholomew-Close. And sold by J. Roberts in Warwick-lane, J. Tonson in the Strand, W. Innys and R. Manby at the West End of St. Paul's Church-Yard, J. Osborn and T. Longman in Pater-Noster-Row, J. Noon near Mercers Chapel in

Cheapside, T. Hatchett at the Royal Exchange, S. Harding in St. Martin's lane, J. Stagg in Westminster-Hall, J. Parker in Pall-mall, and J. Brindley in New Bond-street, 1733.

Optice: sive de Reflexionibus, Refractionibus, Inflexionibus & Coloribus Lucis libri tres. Authore Isaaco Newton, Equite Aurato. Latine reddidit Samuel Clarke, A. M. Reverendo admodum Patri ac D^{no} Joanni Moore Episcopo Nowicensi a Sacris Domesticis. Accedunt Tractatus duo ejusdem Authoris de Speciebus & Magnitudine Figurarum Curvilinearum, Latine scripti. Londini: Impensis Sam. Smith & Benj. Walford, Regiae Societatis Typograph. ad Insignia Principis in Coemeterio D. Pauli, 1706.

Opticks, or A Treatise of the Reflections, Refractions, Inflections & Colours of Light. Foreword by Albert Einstein, intro. by Sir Edmund Whittaker, preface by I. Bernard Cohen, analytical table of contents by Duane H. D. Roller. Based on the 4th London ed. of 1730. New York: Dover, 1952.

Opticks: or, A Treatise of the Reflections, Refractions, Inflections & Colours of Light. Also Two Treatises of the Species and Magnitude of Curvilinear Figures. London: printed for Sam. Smith, and Benj. Walford, printers to the Royal Society, at the Prince's Arms in St. Paul's Church-yard, 1704.

Opticks: or, A Treatise of the Reflections, Refractions, Inflections & Colours of Light. The Second Edition, with Additions. By Sir Isaac Newton, Knt. London: printed by W. Bowyer for W. Innys at the Prince's Arms in St. Paul's Church-Yard, 1717.

Opticks: or, A Treatise of the Reflections, Refractions, Inflections & Colours of Light. The Second Edition, with Additions. By Sir Isaac Newton, Knt. London: printed for W. and J. Innys, printers to the Royal Society, at the Prince's Arms in St. Paul's Church-Yard, 1718.

Philosophiae naturalis principia mathematica. Londini: Jussu Societatis Regiae ac Typis Josephi Streater. Prostat apud plures Bibliopoles, 1687.

Sir Isaac Newton Theological Manuscripts. Selected and ed. with an introduction by H. McLachlan. Liverpool: At the University Press, 1950.

Sir Isaac Newton's Mathematical Principles of Natural Philosophy and His System of the World. Tr. 1729 by Andrew Motte. Ed. by Florian Cajori. Reprint of the 1934 ed. Two vols. Berkeley: University of California Press, 1962.

A Treatise of the System of the World Translated into English. Intro. by I. Bernard Cohen. Facsimile reprint of the 2nd London ed. of 1731. London: Dawsons of Pall Mall, 1969.

Über die Gravitation ... Texte zu den philosophischen Grundlagen der klassischen Mechanik. Text lateinisch-deutsch Übersetzt und erläutert von Gernot Böhme. Frankfurt/M.: Vittorio Klostermann, 1988.

Unpublished Scientific Papers of Isaac Newton: A Selection from the Portsmouth Collection in the University Library, Cambridge. Chosen, edited, and translated by A. Rupert Hall and Marie Boas Hall. Cambridge University Press, 1962.

Nibley, Hugh W. "What is a temple?" In *The Temple in Antiquity. Ancient Records and Modern Perspectives.* Ed. with introductory essay by Truman

G. Madsen. The Religious Studies Monograph Series, Vol. 9. Provo, UT: Brigham Young University Religious Studies Center, 1984, pp. 19–37.

Nock, A. O., Ed. *Corpvs Hermeticvm. Texte e'tabli par A. D. Nock et traduit par A.-J. Festugiere.* Rev. ed. of 1946–54. Four vols. Collection des Universités de France publiée sous le patronage de l'Association Guillaume Budé. Paris: Société d'Édition «Les Belles Lettres», 1980–3.

O'Connor, Daniel. "Introduction: two philosophies of nature." In *Creation. The Impact of an Idea.* Ed. by Daniel O'Connor and Francis Oakley. Scribner Source Books in Religion. New York: Charles Scribner's Sons, 1969, pp. 15–28.

and Francis Oakley. "General introduction." In *Creation. The Impact of an Idea.* Ed. by Daniel O'Connor and Francis Oakley. Scribner Source Books in Religion. New York: Charles Scribner's Sons, 1969, pp. 1–12.

and Francis Oakley, Eds. *Creation. The Impact of an Idea.* Scribner Source Books in Religion, New York: Charles Scribner's Sons, 1969.

Oakley, Francis. "Christian theology and the Newtonian science: the rise of the concepts of the laws of nature." In *Creation. The Impact of an Idea.* Ed. by Daniel O'Connor and Francis Oakley. Scribner Source Books in Religion. New York: Charles Scribner's Sons, 1969, pp. 54–83.

Omnipotence, Covenant, & Order. An Excursion in the History of Ideas from Abelard to Leibniz. Ithaca, NY: Cornell University Press, 1984.

Osler, Margaret J. "Descartes and Charleton on nature and God." *Journal of the History of Ideas* 40 (1979), 445–56.

"Eternal truths and the laws of nature: the theological foundations of Descartes' philosophy of nature." *Journal of the History of Ideas* 46 (1985), 349–62.

"Providence and divine will in Gassendi's views on scientific knowledge." *Journal of the History of Ideas* 44 (1983), 549–60.

and Paul Lawrence Farber, Eds. *Religion, Science, and Worldview. Essays in Honor of Richard S. Westfall.* Cambridge University Press, 1985.

Pagel, Walter. *Joan Baptista van Helmont. Reformer of Science and Medicine.* Cambridge University Press, 1982.

Paracelsus: An Introduction to Philosophical Medicine in the Era of the Renaissance. Basel: Karger, 1958.

Palter, Robert. "Saving Newton's text: documents, readers, and the ways of the world." *Studies in History and Philosophy of Science* 18 (1987), 385–439.

Paracelsus. *Opera omnia medico chemico-chirurgica....* Genevae, 1658.

Le Parnasse assiegé.... Lyon, 1697.

Pelikan, Jaroslav. "The odyssey of Dionysian spirituality." In Pseudo-Dionysius, *The Complete Works.* Tr. by Colm Luibheid; foreword, notes, and translation collaboration by Paul Rorem; preface by Rene Roques, introductions by Jaroslav Pelikan, Jean Leclercq, and Karlfried Froehlich. The Classics of Western Spirituality, A Library of the Great Spiritual Masters. New York: Paulist Press, 1987, pp. 11–24.

Perkins, William. *A Resolution to the Countrey-man.* In William Perkins, *The Workes of That Famous and Worthy Minister of Christ... M. W. Perkins.* Three vols. Cambridge: 1608–31, vol. III.

Peters, Edward, Ed. *Heresy and Authority in Medieval Europe. Documents in Translation.* With introduction by Edward Peters. Philadelphia: University of Pennsylvania Press, 1980.

"'The Heretics of Old': the definition of orthodoxy and heresy in Late Antiquity and the Early Middle Ages." In *Heresy and Authority in Medieval Europe. Documents in Translation.* Ed., with introduction, by Edward Peters. Philadelphia: University of Pennsylvania Press, 1980, pp. 13–56.

Philo. *Allegorical Interpretation of Genesis II., III (Legum Allegoria).* In *Philo in Ten Volumes with an English Translation by F. H. Colson and G. H. Whitaker.* The Loeb Classical Library. London: William Heinemann; New York: G. P. Putnam's Sons, 1929. Vol. I.

Omnia quae extant opera.... Lutetiae Parisiorum, 1640.

Philosophie naturelle de trois anciens philosophes renommez Artephius, Flamel, & Synesius. Traitant de l'Art occulte, & de la Transmutation metallique. Augmentée d'un petit Traité du Mercure, & de la Pierre de Philosophes de G. Ripleus, nouvellement traduit en François. Derniere edition. Paris: Chez Laurent d'Houry, sur le Quay des Augustins, à l'Image Saint Joan, 1682.

La physique des anciens. [By D. R.] Paris, 1701.

Pico della Mirandola, Giovanni. *Heptaplus or Discourse on the Seven Days of Creation.* Tr. with introduction and glossary by Jessie Brewer McGaw. New York: Philosophical Library, 1977.

Plato. *Opera omnia quae exstant. M. Ficino interprete....* Francofurti, 1602.

Plato's Cosmology: The Timaeus of Plato. Ed. and tr. by Francis MacDonald Cornford. Reprint of the 1937 ed. Indianapolis: Bobbs Merrill, n.d.

Popkin, Richard H. *The History of Scepticism from Erasmus to Spinoza.* Berkeley: University of California Press, 1979.

Isaac La Peyrère (1596–1676). His Life, Work and Influence. Brill's Studies in Intellectual History, Vol. 1, General Ed.: A. J. Vanderjagt. Leiden: E. J. Brill, 1987.

Ed. *Millenarianism and Messianism in English Literature and Thought (1650–1800). Clark Library Lectures 1981–82.* Publications from the Clark Library Professorship, UCLA, No. 10. Leiden: E. J. Brill, 1988.

"Newton's biblical theology and his theological physics." In *Newton's Scientific and Philosophical Legacy.* Ed. by P. B. Scheuer and G. Debrock. International Archives of the History of Ideas, 123. Dordrecht: Kluwer Academic Publishers, 1988, pp. 81–97.

"The third force in seventeenth-century philosophy: scepticism, science, and biblical prophecy." *Nouvelle République des Lettres 1* (1983), 35–63.

Pseudo-Dionysius. *The Complete Works.* Tr. by Colm Luibheid, foreword, notes, and translation collaboration by Paul Rorem, preface by Rene Roques, introductions by Jaroslav Pelikan, Jean Leclercq, and Karlfried Froehlich. The Classics of Western Spirituality, A Library of the Great Spiritual Masters. New York: Paulist Press, 1987.

Dionysius the Areopagite, The Divine Names and The Mystical Theology. Tr. by C. E. Rolt, preface and conclusion by W. J. Sparrow-Simpson, intro. by

C. E. Rolt. Reprint of the 1920 ed. Translations of Christian Literature. London: SPCK, 1979.

Ptolemy, Claudius. *Harmonicorum libri III*.... Oxonii, 1682.

Pulver, Max. "The experience of the pneuma in Philo." Tr. by Ralph Manheim. In *Spirit and Nature. Papers from the Eranos Yearbooks*. Ed. by Joseph Campbell. Bollingen Series XXX.1. Princeton: Princeton University Press, 1982, pp. 107–21. Originally published in German in *Eranos-Jahrbuch 13* (1945).

Quinn, Arthur. *The Confidence of British Philosophers. An Essay in Historical Narrative*. Studies in the History of Christian Thought, Vol. 17. Ed. by Heiko A. Oberman, in cooperation with Henry Chadwick, Edward A. Dowey, Jaroslav Pelikan, Brian Tierney, and E. David Willis. Leiden: E. J. Brill, 1977.

"On reading Newton apocalyptically." In *Millenarianism and Messianism in English Literature and Thought 1650–1800. Clark Library Lectures 1981–82*. Ed. by Richard H. Popkin. Publications from the Clark Library Professorship, UCLA, No. 10. Leiden: E. J. Brill, 1988, pp. 176–92.

Ramnoux, Clémence. "Héliocentrisme et Christocentrisme (sur un texte du Cardinal de Bérulle)." In *Le soleil à la Renaissance. Sciences et mythes. Colloque international tenu en avril 1963 sous les auspices de la Fédération Internationale des Instituts et Sociétés pour l'Étude de la Renaissance et du Ministère de l'Éducation nationale et de la Culture de Belgique. Publié avec le concours du Gouvernement belge*. Bruxelles: Presses universitaires de Bruxelles; Paris: Presses universitaires de France, 1965, pp. 447–61.

Rattansi, P. M. "Last of the magicians." *Times Higher Education Supplement*, June 1976.

"Newton as chymist." *Science* 192 (No. 4240, 14 May 1976), 689–90.

Redwood, John. *Reason, Ridicule and Religion. The Age of Enlightenment in England 1660–1750*. London: Thames and Hudson, 1976.

Rees, Graham. "The fate of Bacon's cosmology in the seventeenth century." *Ambix* 24 (1977), 27–38.

"Francis Bacon on verticity and the bowels of the earth." *Ambix* 26 (1979), 202–11.

"Francis Bacon's semi-Paracelsian cosmology." *Ambix* 22 (1975), 81–101.

"Francis Bacon's semi-Paracelsian cosmology and the Great Instauration." *Ambix* 22 (1975), 161–73.

"Matter theory: a unifying factor in Bacon's natural philosophy?" *Ambix* 24 (1977), 110–25.

Regnéll, Hans. *Ancient Views on the Nature of Life. Three Studies in the Philosophies of the Atomists, Plato and Aristotle*. Library of Theoria, No. 10. Lund: CWK Gleerup, 1967.

Reiss, Edmund. *The Art of the Middle English Lyric. Essays in Criticism*. Athens, GA: University of Georgia Press, 1972.

Revised Medieval Latin Word-List from British and Irish Sources. Prepared by R. E. Latham, M. A., under the direction of a Committee appointed by the

British Academy. London: Published for the British Academy by The Oxford University Press, Amen House, 1965.

Rhenanus, Joannus, Ed. *Harmoniae chymico-philosophicae, sive philosophorvm antiquorum consentientium, hactenus quidem plurrimùm desideratorum, sed nondum in lucam publicam emissorum, Decas II. Collecta studio & industria Ioannis Rhenani M. D. Cum elencho singulorum huius Decadis* Francofurti: Apud Conradvm Eifridvm, 1625.

Opera chymiatrica, quae hactenus in lucem prodierunt omnia à plurimus, quae in prioribus editionibus irrepserant, mendis vindicata & selectissimis Medicamentis aucta, inq; unum fasciculum collecta, Quorum Catalogum versa indicabit pagina. Francofurti: Apud Jacobum Gothofredum Seylerum, 1668.

Rist, John M., Ed. *The Stoics*. Berkeley: University of California Press, 1978.

Robbins, Frank Egleston. *The Hexaemeral Literature. A Study of the Greek and Latin Commentaries on Genesis. A Dissertation Submitted to the Faculty of the Graduate School of Arts and Literature in Candidacy for the Degree of Doctor of Philosophy (Department of Greek)*. Chicago: The University of Chicago Press, 1912.

Røstvig, Maren-Sofie. "Structure as prophecy: the influence of biblical exegesis upon theories of literary structure." In *Silent Poetry. Essays in Numerological Analysis*. Ed. by Alastair Fowler. London: Routledge & Kegan Paul, 1970, pp. 32–72.

Rosenau, Helen. *Vision of the Temple. The Image of the Temple of Jerusalem in Judaism and Christianity*. London: Oresko Books, 1979.

Rosenfeld, L. "Newton's views on aether and gravitation." *Archive for History of Exact Sciences* 6 (1969), 29–37.

Rossi, Paoli. "Hermeticism, rationality and the scientific revolution." In *Reason, Experiement, and Mysticism in the Scientific Revolution*. Ed. by M. L. Righini Bonelli and William R. Shea. New York: Science History Publications, 1975, pp. 247–73.

Rykwert, Joseph. *On Adam's House in Paradise. The Idea of the Primitive Hut in Architectural History*. 2nd ed. Cambridge, MA: The MIT Press, 1981.

Sailor, Danton B. "Newton's debt to Cudworth." *Journal of the History of Ideas* 49 (1988), 511–18.

Sambursky, Shmuel. *Physics of the Stoics*. Reprint of the 1959 ed. London: Hutchinson, 1971.

Sander, Leonard M. "Fractal growth." *Scientific American* 256, No. 1 (January 1987), 94–100.

Sandmel, Samuel. *Philo of Alexandria. An Introduction*. New York: Oxford University Press, 1979.

Saunders, Jason Lewis. *Justus Lipsius, The Philosophy of Renaissance Stoicism*. New York: The Liberal Arts Press, 1955.

Schaffer, Simon. "Newton's comets and the transformation of astrology." In *Astrology, Science and Society. Historical Essays*. Ed. by Patrick Curry. Woodbridge, Suffolk: The Boydell Press, 1987, pp. 219–43.

Scheuer, P. B., and G. Debrock, Eds. *Newton's Scientific and Philosophical Legacy*. International Archives of the History of Ideas, 123. Dordrecht: Kluwer Academic Publishers, 1988.

Schneer, C. E., Ed. *Toward a History of Geology*. Cambridge, MA: Harvard University Press, 1969.

Schuler, Robert M. "Some spiritual alchemies of seventeenth-century England." *Journal of the History of Ideas* 41 (1980), 293–318.

Schwartz, Hillel. *The French Prophets. The History of a Millenarian Group in Eighteenth-Century England*. Berkeley: University of California Press, 1980.

Knaves, Fools, Madmen, and That Subtile Effluvium. A Study of the Opposition to the French Prophets in England, 1706–1710. University of Florida Monographs, Social Science Number 62. A University of Florida Book. Gainesville: The University Presses of Florida, 1978.

Scott, Walter, Ed. *Hermetica. The Ancient Greek and Latin Writings Which Contain Religious or Philosophic Teachings Ascribed to Hermes Trismegistus*. English tr. and notes by Walter Scott. 1st ed. 1924. Reprint of 1982 ed. Four vols. Boston: Shambhala, 1985.

Seneca. *Natural Questions*. In *Seneca in Ten Volumes, Volumes VII and X*. Tr. by Thomas H. Corcoran. The Loeb Classical Library. Ed. by E. H. Warmington. Cambridge, MA: Harvard University Press; London: William Heinemann, 1971–72.

Senior. "Senioris antiquissimi philosophi libellvs, ut Brevis ita artem discentibus & exercentibus utilissimus, & verè aureus, Dixit Senior Zadith filius Hamuel." In *Theatrum chemicum, praecipuos selectorum auctorum tractatus de chemiae et lapidis philosophici antiquitate, veritate, jure, praestantia, & operationibus, continens: In gratiam Verae Chemiae, & medicinae Chemicae studiosorum (ut qui uberrimam inde optimorum remediorum messem facere poterunt) congestum, & in Sex partes seu volumina digestum; singulis voluminibus, suo auctorum et librorum catalogo primis pagellis: rerum verò & verborum Indice postremis annexo*. Six vols. Argentorati: Sumptibus Heredum Eberh. Zetzneri, 1659–61. Vol. V, 193–239.

Shapiro, Alan E. "Beyond the Dating Game: Watermark Clusters and the Composition of Newton's *Opticks*." In press.

"Light, pressure, and rectilinear propagation: Descartes' celestial optics and Newton's hydrostatics." *Studies in History and Philosophy of Science* 5 (1974), 239–96.

Sheppard, H. J. "Gnosticism and alchemy." *Ambix* 6 (1957), 86–101.

Smith, C. U. M. *The Problem of Life. An Essay in the Origins of Biological Thought*. A Halsted Press Book. New York: John Wiley & Sons, 1976.

Le soleil à la Renaissance. Sciences et mythes. Colloque international tenu en avril 1963 sous les auspices de la Fédération Internationale des Instituts et Sociétés pour l'Étude de la Renaissance et du Ministère de l'Éducation nationale et de la Culture de Belgique. Publié avec le concours du Gouvernement belge. Bruxelles: Presses universitaires de Bruxelles; Paris: Presses universitaires de France, 1965.

Spanneut, M. *Le Stoïcisme des pères de l'Eglise: de Clément de Rome à Clément d'Alexandrie*. Paris: Seuil, 1957.

Spargo, P. E. "Review of B. J. T. Dobbs, *The Foundations of Newton's Alchemy*." *Ambix* 24 (1977), 175–6.

Stanley, Thomas. *Historia philosophiae Orientalis*. Amstelodami, 1690.

History of Philosophy. Two vols. London, 1656–60.

History of Philosophy. 3rd ed. London, 1701.

Steneck, Nicholas H. *Science and Creation in the Middle Ages. Henry of Langenstein (d. 1397) on Genesis*. Notre Dame: University of Notre Dame Press, 1976.

Stock, R. D. *The Holy and the Daemonic from Sir Thomas Browne to William Blake*. Princeton: Princeton University Press, 1982.

Stow, John, and Edmund Howe. *The Annales or Generall Chronicle of England*. London 1615.

Tachenius, Otto. *Hippocrates chimicus*. Venice, 1666.

[?] *La lumière par soy méme des tenebres ou veritable theorie de la Pierre des Philosophes écrite en vers Italiens, & amplifiée en Latin par un Auteur Anonyme, en forme de Commentaire; le tout traduit en François par B. D. L.* Paris: Chez Laurent D'Houry, ruë S. Jacques, devant la Fonteine S. Severin, au S. Esprit, 1687.

Tailor, Thomas. *Christ Revealed: Or The Old Testament Explained. A Treatise of the Types and Shadowes of our Saviovr contained throughout the whole Scriptvre: All opened and made usefull for the benefit of Gods Church.* London: printed by M. F. for R. Dawlman and L. Fawne at the signe of the Brazen serpent in Pauls Churchyard, 1635.

Taylor, Frank Sherwood. "An alchemical work of Sir Isaac Newton." *Ambix* 5 (1956), 59–84.

"The idea of the quintessence." In *Science, Medicine, and History. Essays on the Evolution of Scientific Thought and Medical Practice Written in Honour of Charles Singer*. Collected and ed. by E. Ashworth Underwood. Two vols. London: Oxford University Press, Geoffrey Cumberlege, 1953, vol. I, 247–65.

Taylor, Thomas. *Christ Revealed. A Facsimile Reproduction with an Introduction by Raymond A. Anselment*. Delmar, NY: Scholars' Facsimiles & Reprints, 1979.

Le texte d'alchymie et le songe-verd. Paris, 1695.

Thackray, Arnold. *Atoms and Powers: An Essay on Newtonian Matter-Theory and the Development of Chemistry*. Cambridge, MA: Harvard University Press, 1970.

Theatrum chemicum, praecipuos selectorum auctorum tractatus de chemiae et lapidis philosophici antiquitate, veritate, jure, praestantia, & operationibus, continens: In gratiam Verae Chemiae, & medicinae Chemicae studiosorum (ut qui uberrimam inde optimorum remediorum messem facere poterunt) congestum, & in Sex partes seu volumina digestum; singulis voluminibus, suo auctorum et librorum catalogo primis pagellis: rerum verò & verborum

Indice postremis annexo. Six vols. Argentorati: Sumptibus Heredum Eberh. Zetzneri, 1659–61.

Todd, Robert M. *Alexander of Aphrodisias on Stoic Physics. A Study of the De mixtione with Preliminary Essays, Texts, Translation and Commentary*. Leiden: E. J. Brill, 1976.

"Monism and immanence: the foundations of Stoic physics." In *The Stoics*. Ed. by John M. Rist. Berkeley: University of California Press, 1978, pp. 137–60.

Toulmin, Stephen, and June Goodfield. *The Architecture of Matter*. Reprint of the 1962 ed. Harper Torchbooks. New York: Harper & Row, 1966.

Trevisan, Bernard. *Taicté... de l'oevf des Philosophes....* Paris, 1659.

Trois Traictez de la Philosophe natvrelle, non encore imprimez. Scavoir, le secret livre dv tres-ancien Philosophe Artephivs, traictant de l'Art occulte & transmutation Metallique, Latin François. Plus les figvres hieroglyphiqves de Nicolas Flamel, ainsi qu'il les a mises en la quatriesme arche qu'il a bastie au Cimetiere des Innocens à Paris, entrant par la grande porte de la rue S. Denys, & prenant la main droite; auec l'explication d'icelles par iceluy Flamel. Ensemble Le vray Liure du docte Synesivs Abbé Grec, tiré de la Bibliotheque de l'Empereur sur le mesme sujet, le tout traduict par P. Arnavld, sieur de la Cheuallerie Poicteuin. Paris: Chez la vefue M. Gvillemot & S. Thibovst, au Palais, en la galerie des prisonniers, 1612.

Turnor, Edmund. *Collections for the History of the Town and Soke of Grantham. Containing Authentic Memoirs of Sir Isaac Newton, Now First Published From the Original MSS. in the Possession of the Earl of Portsmouth*. London: printed for William Miller, Albemarle-Street, by W. Bulmer and Co. Cleveland-Row, St. James's, 1806.

Underwood, E. Ashworth, Ed. *Science, Medicine, and History. Essays on the Evolution of Scientific Thought and Medical Practice Written in Honour of Charles Singer*. Collected by the editor. Two vols. London: Oxford University Press, Geoffrey Cumberlege, 1953.

van Helmont, Jan Baptista. *Ortus medicinae, id est initia physicae inaudita.... edente F. M. van Helmont....* 4th ed. Lugduni, 1667.

Causae et initia naturalium. In *Ortus medicinae....* Lugduni, 1667.

van Lennep, J. *Art & Alchimie. Étude de l'iconographie hermétique et de ses influences*. Préface de Serge Hutin. Collection Art et Savoir. Bruxelles: Éditions Meddens avec le concours de la Fondation Universitaire de Belgique, 1966.

van Pelt, R. J. *Tempel van de Wereld, de Kosmische symboliek van de tempel van Salomo*. HES Uitgevers/Utrecht: H & S, 1984.

Varenius, Bernhardus. *Geographia generalis. In qua affectiones generales Telluris explicantur Autore Bernhi: Varenio Med: D*. Amstelodami: Apud Ludovicum Elzevirium, 1650.

Geographia generalis. In qua affectiones generales Telluris explicantur Autore Bernhi: Varenio Med: D. Amstelodami: Ex Officina Elzeviriana, 1671.

Bernhardi Vareni Med. D. Geographia generalis, In qua affectiones generales

Telluris explicantur, Summâ curâ quam plurimis in locis emendata, & XXXIII Schematibus novis, aere incisis, unà cum Tabb. aliquot quae desiderabantur aucta & illustra. Ab Isaaco Newton Math. Prof. Lucasiano Apud Cantabrigienses. Cantabrigiae: Ex Officina Joann. Hayes, Celeberrimae Academiae Typographi, Sumptibus Henrici Dickinson Bibliopolae, 1672.

Bernhardi Vareni Med. D. Geographia generalis, In qua Affectiones Generales Telluris Explicantur, Summâ curâ quam plurimis in locis Emendata, & XXXIII Schematibus Novis, AEre incisis, unà cum Tabb. aliquot quae desiderabantur Aucta & Illustrata, Ab Isaaco Newton Math. Prof. Lucasiano Cantabrigienses. Editio Secunda Auctior & Emendatior. Cantabrigiae: Ex Officina Joann. Hayes, Celeberrimae Academiae Typographi, Sumptibus Henrici Dickinson Bibliopolae, 1681.

Venette, Nicolas. *Traité des pierres....* Amsterdam, 1701.

Verbeke, G. *L'evolution de la doctrine du pneuma du stoicisme à S. Augustin. Étude philosophique.* Bibliothèque de l'Institut Supérieur de Philosophie Université de Louvain. Paris: Desclée De Brouwer; Louvain: Éditions de l'Institut Supérieur de Philosophie, 1945.

Viner, Jacob. *The Role of Providence in the Social Order: An Essay in Intellectual History. Jayne Lectures for 1966.* Foreword by Joseph R. Strayer. Memoirs of the American Philosophical Society Held at Philadelphia For Promoting Useful Knowledge, Vol. 90. Philadelphia: American Philosophical Society, Independence Square, 1972.

Virgil. *The Aeneid of Virgil. A Verse Translation by Allen Mandelbaum. With Thirteen Drawings by Barry Moser.* Berkeley: University of California Press, 1981.

The Georgics. Tr. by Robert Wells. Manchester: Carcanet New Press, 1982.

Virgil: Georgics I & IV. Ed. by H. H. Huxley. London: Methuen & Co., 1963.

Virgil with an English Translation by H. Rushton Fairclough in Two Volumes. I. Eclogues, Georgics, Aeneid I–VI. II. Aeneid VII–XII, The Minor Poems. Rev. ed. The Loeb Classical Library. Ed. by T. E. Page, E. Capps, L. A. Post, W. H. D. Rouse, and E. H. Warmington. London: William Heinemann; Cambridge, MA: Harvard University Press, 1938–42.

von Franz, Marie-Louise. "The idea of the macro- and microcosmus in the light of Jungian psychology." *Ambix 13* (1965), 22–34.

Patterns of Creativity Mirrored in Creation Myths. Zürich: Spring Publications, 1972.

von Simson, Otto. *The Gothic Cathedral. Origins of Gothic Architecture and the Medieval Concept of Order.* 2nd ed. rev. Bollingen Series XLVIII. New York: Pantheon Books, 1962.

Wagar, W. Warren, Ed. *The Secular Mind. Transformations of Faith in Modern Europe. Essays Presented to Franklin L. Baumer, Randolph W. Townsend Professor of History, Yale University.* New York: Holmes & Meier, 1982.

Waite, Arthur Edward. *The Hermetic Museum, Restored and Enlarged: Most faithfully instructing all Disciples of the Sopho-Spagyric Art how that*

Greatest and Truest Medicine of the Philosopher's Stone may be found and held. Now first done into English from the Latin Original published at Frankfort in the Year 1678. Containing Twenty-two celebrated Chemical Tracts. Reprint of the 1893 ed. Two vols. New York: Samuel Weiser, 1973.

Walker, D. P. *The Ancient Theology: Studies in Christian Platonism from the Fifteenth to the Eighteenth Century.* London: Duckworth, 1972.

"Medical spirits: four lectures." Boston Colloquium for the Philosophy of Science, 27 October–19 November 1981.

Spiritual and Demonic Magic from Ficino to Campanella. Reprint of the 1969 ed. Notre Dame: University of Notre Dame Press, 1975.

Wallach, Luitpold, Ed. *The Classical Tradition. Literary and Historical Studies in Honor of Harry Caplan.* Ithaca, New York: Cornell University Press, 1966.

Ward, Benedicta. *Miracles and the Medieval Mind: Theory, Record, and Event, 1000–1215.* Philadelphia: University of Pennsylvania Press, 1982.

Webster, Charles. *From Paracelsus to Newton. Magic and the Making of Modern Science. The Eddington Memorial Lectures Delivered at Cambridge November 1980.* Cambridge University Press, 1982.

Webster, John. *Metallographia: or, An History of Metals, Wherein is declared the signs of Ores and Minerals both before and after digging, their kinds, sorts, and differences; with the description of sundry new Metals, or Semi Metals, and many other things pertaining to Mineral knowledge. As also, The Handling and shewing of their Vegetability, and the discussion of the most difficult Questions belonging to Mystical Chymistry, as of the Philosophers Gold, their Mercury, the Liquor Alkahest, Aurum potabile, and such like. Gathered forth of the most approved Authors that have written in Greek, Latine, or High-Dutch; With some Observations and Discoveries of the Author himself.* London: Printed by A. C. for Walter Kettilby at the Bishopshead in St. Pauls Churchyard, 1671.

Westfall, Richard S. "Alchemy in Newton's library." *Ambix 31* (1984), 97–101.

"The changing world of the Newtonian industry." *Journal of the History of Ideas 37* (1976), 175–84.

The Construction of Modern Science. Mechanisms and Mechanics. Reprint of the 1971 ed. Cambridge University Press, 1977.

Force in Newton's Physics. The Science of Dynamics in the Seventeenth Century. London: Macdonald; New York: American Elsevier, 1971.

"The foundations of Newton's philosophy of nature." *British Journal for the History of Science 1* (1962/63), 171–82.

"The influence of alchemy on Newton." In *Science, Pseudo-Science and Society.* Ed. by Marsha P. Hanen, Margaret J. Osler, and Robert G. Weyant. Waterloo, Ontario: Wilfrid Laurier University Press, 1980, pp. 145–69.

"Isaac Newton's *Index Chemicus.*" *Ambix 22* (1975), 174–85.

"Isaac Newton's *Theologiae Gentilis Origines Philosophicae.*" In *The Secular Mind. Transformations of Faith in Modern Europe. Essays presented to Franklin L. Baumer, Randolph W. Townsend Professor of History, Yale*

University. Ed. by W. Warren Wagar. New York: Holmes & Meier, 1982, pp. 15–34.
Never at Rest. A Biography of Isaac Newton. Cambridge University Press, 1980.
"Newton and the Hermetic tradition." In *Science, Medicine, and Society in the Renaissance. A Festschrift in Honor of Walter Pagel*. Ed. by Allen G. Debus. Two vols. New York: Neale Watson Academic Publications, 1972, vol. II, 183–98.
"Newton's development of the *Principia*." In *Springs of Scientific Creativity. Essays on Founders of Modern Science*. Ed. by Rutherford Aris, H. Ted Davis, and Roger H. Stuewer. Minneapolis: University of Minnesota Press, 1983, pp. 21–43.
"Newton's theological manuscripts." In *Contemporary Newtonian Research*. Ed. by Zev Bechler. Studies in the History of Modern Science, 9. Dordrecht: D. Reidel, 1982, pp. 129–43.
"Review of B. J. T. Dobbs, *The Foundations of Newton's Alchemy*." *Journal of the History of Medicine and Allied Sciences 31* (1976), 473–74.
"The role of alchemy in Newton's career." In *Reason, Experiment and Mysticism in the Scientific Revolution*. Ed. by M. L. Righini Bonelli and William R. Shea. New York: Science History Publications, 1975, pp. 189–232.
Westman, Robert S. "The Copernicans and the Churches." In *God and Nature. Historical Essays on the Encounter between Christianity and Science*. Ed. by David C. Lindberg and Ronald L. Numbers. Berkeley: University of California Press, 1986, pp. 76–113.
and J. E. McGuire. *Hermeticism and the Scientific Revolution: Papers Read at a Clark Library Seminar, March 9, 1974*. Los Angeles: Clark Memorial Library, University of California, 1977.
Whiteside, Derek T. "Before the *Principia*: the maturing of Newton's thoughts on dynamical astronomy, 1664–1684." *Journal for the History of Astronomy 1* (1970), 5–19.
"From his claw the Greene Lyon." *Isis 68* (1977), 116–21.
"Isaac Newton: birth of a mathematician." *Notes and Records of the Royal Society of London 19* (1964), 53–62.
Ed. *The Mathematical Papers of Isaac Newton*. With the assistance in publication of M. A. Hoskin. Eight vols. Cambridge University Press, 1967–80.
"Newton's early thoughts on planetary motion: a fresh look." *British Journal for the History of Science 2* (1964), 117–37.
Whyte, Lancelot Law. *Essay on Atomism: From Democritus to 1960*. Reprint of the 1961 ed. Harper Torchbooks. New York: Harper & Row, 1963.
Willard, Thomas. "Alchemy and the Bible." In *Centre and Labyrinth. Essays in Honour of Northrop Frye*. Ed. by Eleanor Cook, Chaviva Hošek, Jay Macpherson, Patricia Parker, and Julian Patrick. Toronto: University of Toronto Press in association with Victoria University, 1983, pp. 115–27.
Williams, Arnold. *The Common Expositor. An Account of the Commentaries on Genesis 1527–1633*. Chapel Hill: The University of North Carolina Press, 1948.

Willis, Thomas. *Diatribae Duae Medico-Philosophicae*. London: 1659.

Dr Willis's Practice of Physick, Being all the Medical Works of that Renowned and Famous Physician: Containing These Ten several Treatises, viz. I. Of Fermentation. II. Of Feavours. III. Of Urines. IV. Of the Accension of the Bloud. V. Of Musculary Motion. VI. Of the Anatomy of the Brain. VII. Of the Description and Use of the Nerves. VIII. Of Convulsive Diseases. IX. Pharmaceutice Rationalis the 1st and 2d Part. X. Of the Scurvy. Wherein most of the Diseases belonging to the Body of Man are treated of, with excellent methods and Receipts for the Cure of the same. Fitted to the meanest Capacity by an Index for the Explaining of all the hard and unusual Words and Terms of Art, derived from the Greek, Latine, or other Languages, for the benefit of the English Reader, with a large Alphabetical Table to the whole. With Thirty two Copper Plates. Done into English by S. P. Esq;. London: printed for T. Dring, C. Harper, and J. Leigh, and are to be sold at the Corner of Chancery-lane, and the Flower-de-Luce over against St Dunstans Church in Fleet-street, 1681.

Pathologiae cerebri. Amstelodami, 1668.

Wilson, Curtis. "How did Kepler discover his first two laws?" *Scientific American* 226 (1972), No. 3, 93–106.

"Kepler's derivation of the elliptical path." *Isis* 59 (1968), 5–25.

Winston, David. *Logos and Mystical Theology in Philo of Alexandria*. Cincinnati: Hebrew Union College Press, 1985.

Wolfson, Harry Austryn. *Philo. Foundations of Religious Philosophy in Judaism, Christianity, and Islam*. Structure and Growth of Philosophic Systems from Plato to Spinoza II. Two vols. Cambridge, MA: Harvard University Press; London: Geoffrey Cumberlege, Oxford University Press, 1947.

Woodbridge, John D. "German responses to the biblical critic Richard Simon: from Leibniz to J. S. Semler." *Wolfenbütteler Forschungen* 41 (1988), 65–87.

Yates, Frances A. *Giordano Bruno and the Hermetic Tradition*. Chicago: The University of Chicago Press; London: Routledge & Kegan Paul; Toronto: The University of Toronto Press, 1964.

Zafiropulo, Jean, and Catherine Monod. *Sensorium dei dans l'hermetisme et la science*. Collection d'études anciennes publiée sous le patronage de l'Association Guillaume Budé. Paris: Société d'Édition «Les Belles Lettres», 1976.

Zanta, Léontine. *La renaissance du stoïcisme au XVIe* siècle. Bibliothèque littéraire de la renaissance, nouvelle série, tome V. Paris: Librairie ancienne Honoré Champion, Édouard Champion, 1914.

Index